# Peanuts: Processing Technology and Product Development

# Peanuts: Processing Technology and Product Development

Edited by

## Qiang Wang

Institute of Food Science and Technology
Chinese Academy of Agricultural Sciences
Beijing, China

Science Press
Beijing

AMSTERDAM • BOSTON • HEIDELBERG • LONDON
NEW YORK • OXFORD • PARIS • SAN DIEGO
SAN FRANCISCO • SINGAPORE • SYDNEY • TOKYO

Academic Press is an imprint of Elsevier

ELSEVIER

Academic Press is an imprint of Elsevier
125 London Wall, London EC2Y 5AS, United Kingdom
525 B Street, Suite 1800, San Diego, CA 92101-4495, United States
50 Hampshire Street, 5th Floor, Cambridge, MA 02139, United States
The Boulevard, Langford Lane, Kidlington, Oxford OX5 1GB, UK

**Notices**
Knowledge and best practice in this field are constantly changing. As new research and experience broaden our understanding, changes in research methods, professional practices, or medical treatment may become necessary.

Practitioners and researchers must always rely on their own experience and knowledge in evaluating and using any information, methods, compounds, or experiments described herein. In using such information or methods they should be mindful of their own safety and the safety of others, including parties for whom they have a professional responsibility.

To the fullest extent of the law, neither the Publisher nor the authors, contributors, or editors, assume any liability for any injury and/or damage to persons or property as a matter of products liability, negligence or otherwise, or from any use or operation of any methods, products, instructions, or ideas contained in the material herein.

**British Library Cataloguing-in-Publication Data**
A catalogue record for this book is available from the British Library

**Library of Congress Cataloging-in-Publication Data**
A catalog record for this book is available from the Library of Congress

ISBN: 978-0-12-809595-9

For information on all Academic Press publications
visit our website at https://www.elsevier.com/

Working together
to grow libraries in
developing countries

www.elsevier.com • www.bookaid.org

*Publisher:* Nikki Levy
*Acquisition Editor:* Simon Tian
*Editorial Project Manager:* Naomi Robertson
*Production Project Manager:* Julie-Ann Stansfield
*Designer:* Victoria Pearson Esser

Typeset by Thomson Digital

**To my group
without whom this book would
not have been completed so early**

# Contents

## 3. Peanut Oil Processing Technology

*Q. Wang, H. Liu, H. Hu, R. Mzimbiri, Y. Yang, Y. Chen*

## 4. Peanut Protein Processing Technology

*H. Liu, A. Shi, L. Liu, H. Wu, T. Ma, X. He, W. Lin, X. Feng, Yuanyuan Liu*

## 5.  Peanut By-Products Utilization Technology

*Q. Wang, A. Shi, H. Liu, L. Liu, Y. Zhang, N. Li, K. Gong,
M. Yu, L. Zheng*

# List of Contributors

**X. Chen**, Institute of Food Science and Technology, Chinese Academy of Agricultural Sciences, Beijing, China

**Y. Chen**, Institute of Food Science and Technology, Chinese Academy of Agricultural Sciences, Beijing, China

**Y. Cong**, College of Food Science and Nutrition Engineering, China Agricultural University, Beijing, China

**L. Deng**, Institute of Food Science and Technology, Chinese Academy of Agricultural Sciences, Beijing, China

**Y. Du**, Institute of Food Science and Technology, Chinese Academy of Agricultural Sciences, Beijing, China

**X. Feng**, Institute of Food Science and Technology, Chinese Academy of Agricultural Sciences, Beijing, China

**K. Gong**, Institute of Food Science and Technology, Chinese Academy of Agricultural Sciences, Beijing, China

**Y. Guo**, Institute of Food Science and Technology, Chinese Academy of Agricultural Sciences, Beijing, China

**X. He**, Institute of Food Science and Technology, Chinese Academy of Agricultural Sciences, Beijing, China

**H. Hu**, Institute of Food Science and Technology, Chinese Academy of Agricultural Sciences, Beijing, China

**N. Li**, Institute of Food Science and Technology, Chinese Academy of Agricultural Sciences, Beijing, China

**W. Lin**, Institute of Food Science and Technology, Chinese Academy of Agricultural Sciences, Beijing, China

**H. Liu**, Institute of Food Science and Technology, Chinese Academy of Agricultural Sciences, Beijing, China

**L. Liu**, Institute of Food Science and Technology, Chinese Academy of Agricultural Sciences, Beijing, China

**Yuanyuan Liu**, Institute of Food Science and Technology, Chinese Academy of Agricultural Sciences, Beijing, China

**Yunhua Liu**, Institute of Food Science and Technology, Chinese Academy of Agricultural Sciences, Beijing, China

**T. Ma**, Institute of Food Science and Technology, Chinese Academy of Agricultural Sciences, Beijing, China

**R. Mzimbiri**, Institute of Food Science and Technology, Chinese Academy of Agricultural Sciences, Beijing, China

**X. Sheng**, Institute of Food Science and Technology, Chinese Academy of Agricultural Sciences, Beijing, China

**A. Shi**, Institute of Food Science and Technology, Chinese Academy of Agricultural Sciences, Beijing, China

**J. Wang**, Institute of Food Science and Technology, Chinese Academy of Agricultural Sciences, Beijing, China

**L. Wang**, Institute of Food Science and Technology, Chinese Academy of Agricultural Sciences, Beijing, China

**Q. Wang**, Institute of Food Science and Technology, Chinese Academy of Agricultural Sciences, Beijing, China

**H. Wu**, Institute of Food Science and Technology, Chinese Academy of Agricultural Sciences, Beijing, China

**W. Xue**, College of Food Science and Nutrition Engineering, China Agricultural University, Beijing, China

**Y. Yang**, Institute of Food Science and Technology, Chinese Academy of Agricultural Sciences, Beijing, China

**M. Yu**, Institute of Food Science and Technology, Chinese Academy of Agricultural Sciences, Beijing, China

**J. Zhang**, Institute of Food Science and Technology, Chinese Academy of Agricultural Sciences, Beijing, China

**Y. Zhang**, Institute of Food Science and Technology, Chinese Academy of Agricultural Sciences, Beijing, China

**L. Zheng**, Institute of Food Science and Technology, Chinese Academy of Agricultural Sciences, Beijing, China

# Preface

Currently, China is the world's largest peanut producer, with peanut acreage reaching about 4,651,599 ha and yield about 17,018,964t (FAO, 2014). Peanuts contain a great variety of nutrients and bioactive components, including functional protein, amino acids, monounsaturated fatty acids, resveratrol, phytosterols, folic acid, and vitamin E. All of these components play an important role in regulating body functions, suppressing abnormal platelet aggregation, and preventing cardiovascular and cerebrovascular diseases. Fat makes up 50% to 60% of peanuts, and the defatted peanut meal, which is the by-product of oil production, contains approximately 50% protein and is a great source of protein mass in China. These factors reinforce that studies on the processing technology of peanuts and the by-products are of great significance in improving people's nutrition and health.

Recently, both the editor and the research team have presided over a number of major national projects or subjects, including special scientific research for the public service sector (industry) including "Study on Processing Characteristics of Bulk Agricultural Products and Quality Evaluation Technology (200903043)—Study on Peanut Processing Characteristics and Quality Evaluation Technology" and "10th Five-Year Plan," research for the Major National Science And Technology Subject "Study on Peanut Processing Characteristics and Construction of Quality Evaluation Index System (2001BA501A32)," research on a special subject of social welfare research "Study on Processing Quality Evaluation Technology for Main Agricultural and Livestock Products (2005DIA4J035)," research on a subject of National Science and Technology Support Program "12th Five-Year Plan" and "Edible Agricultural Products Processing Suitability Assessment and Risk Monitoring Technology Demonstration (2012BAD29B03)," key international cooperation project of Ministry of Technology "Key Technological Cooperation in Preparing Functional Components from Agricultural and Sideline Products (S2012ZR0302)," and a project for transforming agricultural technological achievements funded by the Ministry of Agriculture "Pilot Test of Functional Peptide Processing Technology for Peanuts and Industrialized Production (2008GB23260385)." The editor has conducted an in-depth study in the field of peanut processing over 10 years and was awarded the second prize of the 2014 State Technological Invention Award, the highest academic award of the 2012 International Association for Cereal Science and Technology (ICC)—the Harald Perten Prize—first prize of the 2013 China Agricultural Science and Technology Award, the first prize of the 2011 Science and Technology Achievement Award of the Chinese Academy of Agricultural Sciences, the second prize of the 2009 Science and Technology Achievement Award of the Chinese Academy of Agricultural Sciences, and the first prize of the 2012 Science and Technology Award of the Chinese Cereals and Oils Association. The editor's research has formulated six standards for agricultural industry, such as Grades and Specifications for Processed Peanuts and Terminologies for Plant Protein and its Products, etc. More than 50 papers have been published in academic journals, such as *Food Chemistry, Journal of the Science of Food and Agriculture, International Journal of Food Properties, Advanced Materials Research, Food Science and Technology*, and *Food Research International*, etc. Nine national invention patents have been obtained and three monographs and books have been published, such as *Introduction of Bioactive Substances of Peanuts, Peanut Aflatoxin Control Technology*, and *Establishment of Whole Process Quality Control System*, etc. On this basis, the book *Peanut Deep Processing Technology* is compiled through systematic reorganization.

The book contains six chapters. The first chapter, provides general information on the distribution and utilization of peanut varieties both at home and abroad, peanut components, and progress in peanut processing and research. The second chapter, Peanut Processing Quality Evaluation Technology, introduces the evaluation model and standard for peanuts suitable to be processed into gel-type protein, soluble protein, and peanut oil, as well as peanut varieties for export. The third chapter, Peanut Oil Processing Technology, introduces the oil production method, pressing technique, and key equipment. The fourth chapter, Peanut Protein Processing Technology, introduces the processing and modification technologies for peanut protein powder, tissue peanut, protein concentrate, and protein isolate, etc., as well as the key production equipment. The fifth chapter, Peanut By-Products Utilization Technology, introduces the processing technology of functional factors in such by-products as peanut peptides, polyphenols, and polycarbohydrates, etc. Finally, the sixth Chapter describes food allergy in which it include peanut allergen species and components of other peanut allergens. It also provides details on allergic mechanisms and clinical manifestations of peanut allergy, allergen detection method and allergen fingerprints on rapid detection using heat treatment, enzymatic technique, irradiation treatment and genetic engineering have also been discussed.

This book, compiled on the basis of the graduation thesis of four doctoral students and eight postgraduate students, is systematic, novel, creative, forward-looking, and directional. The book intends to provide favorable references and guidance for the study and utilization of peanut processing so as to provide technological support for the sound development of the peanut processing industry in China.

The first chapter is compiled by Qiang Wang, Li Liu, Li Wang, Yalong Guo, and Jun Wang; the second chapter is compiled by Aimin Shi, Qiang Wang, Hongzhi Liu, Li Wang, Jianshu Zhang, Yin Du, and Xue Chen; the third chapter is compiled by Qiang Wang, Hongzhi Liu, Hui Hu, Rehema Mzimbiri, Ying Yang, and Yan Chen; the fourth chapter is compiled by Hongzhi Liu, Aimin Shi, Li Liu, Haiwen Wu, Tiezheng Ma, Xuanhui He, Weijing Lin, Xiaolong Feng, and Yuanyuan Liu; the fifth chapter is compiled by Qiang Wang, Aimin Shi, Hongzhi Liu, Li Liu, Yuhao Zhang, Ning Li, Kuijie Gong, Miao Yu, and Liyou Zheng; the sixth chapter is compiled by Wentong Xue, Yanjun Cong, Aimin Shi, Lei Deng, Xiaojing Sheng, and Yunhua Liu. Ying Yang, Liyou Zheng, Lei Deng, Xiaojing Sheng, Yunhua Liu, Yalong Guo, Jun Wang, and Bo Jiao also participated in the compilation of the book. We would also like to take advantage of this opportunity to express our heartfelt thanks to the relevant experts and scholars both at home and abroad whose books and papers we consulted during the compilation.

Due to the restrictions of materials, means, study methods, and authors' capacity, the book is inevitably subject to some inadequacies in the aspects of viewpoints and conclusions, and we sincerely hope that the readers can give us criticism and help correct any of the problems detected during the reading process.

Qiang Wang and the Research Team

# List of Abbreviations

| | |
|---|---|
| ABS | Avidin biobin system |
| ACE | Angiotensin converting enzyme |
| AOAC | Association of Official Analytical Chemists |
| BV | Biological value |
| CCD/CMOS | Charge coupled device complementary metal oxide semiconductor |
| CNDF | Cold neutral detergent fiber extraction |
| CPP | Casein phospho-peptide |
| DBPCFC | Double-blind placebo-controlled food challenges |
| DH | Degree of hydrolysis |
| DIA | Dot-immunobinding assay |
| DMSO | Dimethyl sulfoxide |
| DNA | Deoxyribonucleic acid |
| EAACI | Academy of Allergy and Clinical Immunology |
| EAST | Enzyme-linked allergo sorbent assays |
| EB | Elongation at break |
| EITB | Immune electro transfer blot |
| ELISA | Enzyme-linked immunosorbent assay |
| ESI | Electro spray ionization |
| FAB | Fast atom bombardment |
| FAO | Food and Agricultural Organization |
| FDA | Food and Drug Authority |
| FID | Flame ionization detector |
| GC-MS | Gas chromatography mass spectrography |
| GLC | Gas liquid chromatography |
| HEP | Histamine equivalent prick |
| HRT | Histamine releasing test |
| IBT | Immuno-blotting test |
| ICRISAT | International Crops Research Institute for the Semiarid Tropics |
| IDF | Insoluble dietary fiber |
| IEF | Isoelectric focusing |
| MALDI | Matrix-assisted laser desorption interpretation |
| MES | 2-($N$-morpholino) ethanesulfonic acid |
| MS | Mass spectrum |
| NDF | Neutral dietary fiber |
| NIR | Near infra-red |
| NMF | Natural moisturizing factor |
| NMR | Nuclear magnetic resonance |
| NPU | Net protein utilization |
| NSI | Nitrogen solution index |

| | |
|---|---|
| OFC | Open food challenge |
| PBS | Phosphate buffered saline |
| PCR | Polymerase chain reaction |
| PNA | Peanut agglutinin |
| POD | Peroxidase |
| PPC | Peanut protein concentrate |
| PPI | Peanut protein isolate |
| RAST | Radio Allergo Sorbent Test |
| RP-HPLC | Reverse phase high-performance liquid chromatography |
| RSD | Relative standard deviation |
| RSM | Response surface methodology |
| SBPCFC | Single-blind placebo-controlled food challenge |
| SDF | Soluble dietary fiber |
| SEM | Scanning electron microscope |
| SDS PAGE | Sodium dodecyl sulfate polyacrylamide gel electrophoresis |
| S/EDD | Seed extracted deodorized distillates |
| SFE | Supercritical fluid extraction |
| SPT | Skin Prick Test |
| TCA-NSI | Trichloroacetic acid-nitrogen solution index |
| TDF | Total dietary fiber |
| TLC | Thin-layer chromatography |
| Tris-HCl | Trisaminomethane hydrochloride |
| TP | Tensile properties |
| TPA | Texture profile analysis |
| TS | Tensile strength |
| UHF | Ultrahigh frequency |
| USD | United State dollars |
| UV | Ultraviolet |
| WHO | World Health Organization |
| WVP | Water vapor permeability |

# Chapter 1

# Introduction

Q. Wang, L. Liu, L. Wang, Y. Guo, J. Wang

*Institute of Food Science and Technology, Chinese Academy of Agricultural Sciences, Beijing, China*

## Chapter Outline

## 1.1 WORLD PEANUT PRODUCTION

### 1.1.1 World Peanut Production, Processing, and Utilization

1. World Peanut Germplasm Resources

   According to incomplete statistics, the total amount of peanut germplasm collection in the world has exceeded 40,000 portions. Currently, the institutions and countries that are in possession of large quantities of peanut germplasm resources are, in order of size, the International Crops Research Institute for the Semiarid Tropics (15,342 portions), the United States (8719 portions), China (7490 portions, Taiwan Province excluded temporarily), Argentina (2200 portions), Indonesia (1730 portions), Brazil (1300 portions), Senegal (900 portions), Uganda (900 portions), and the Philippines (753 portions), etc. (Yu, 2008).

2. World Peanut Acreage and Yield

   The peanut is widely planted in the world, throughout Asia, Europe, Africa, Americas, and Oceania, etc. According to the statistics of United Nations Food and Agricultural Organization (FAO), the peanut acreage was

24,709,500 ha and the yield was 41,185,900 t in 2012. The top five countries in terms of peanut acreage are India, China, Nigeria, Sudan, and Myanmar, and the top five countries in terms of peanut yield are China, India, Nigeria, the United States, and Myanmar.

3.  International Trade of Peanuts and Peanut Products Worldwide
    According to the FAO statistics, the world's trade of peanut and its processed products is on the rise every year. In 2011, the world exports of peanut cake and oil decreased by 50.09% and 26.77%, respectively, compared with 2001, while the exports of shelled peanuts and peanut butter increased by 52.88% and 20.64%, respectively; in 2011, the import volume of peanut cake and oil decreased by 45.54% and 9.71%, respectively, compared with 2001, while the import volume of shelled peanuts and peanut butter increased by 33.99% and 80.35%, respectively (Table 1.1).

4.  Peanut Utilization in Foreign Countries
    The peanut is rich in fat and protein, as well as functional active substances, such as vitamins and resveratrol, and it contains fewer antinutritional factors than the soybean. Therefore, it has a high value of comprehensive utilization, extensive diversified beneficial uses, and high value-added potential. The major peanut utilization and processing means include edible oil

**TABLE 1.1 World Peanuts and Peanut Products Export**

| Type | Export/Import | 2001 | 2011 | Variation % |
|---|---|---|---|---|
| Peanut cake | | 93.01 | 46.42 | −50.09 |
| Shelled peanut | | 325.20 | 497.15 | 52.88 |
| Peanut oil | Export/10,000 t | 79.54 | 58.25 | −26.77 |
| Peanut butter | | 14.1 | 17.01 | 20.64 |
| Peanut cake | | $1.26 \times 10^8$ | $1.44 \times 10^8$ | 14.29 |
| Shelled peanut | | $2.00 \times 10^9$ | $6.64 \times 10^9$ | 232 |
| Peanut oil | Export value/USD | $5.77 \times 10^8$ | $9.60 \times 10^8$ | 66.38 |
| Peanut butter | | $2.36 \times 10^8$ | $4.79 \times 10^8$ | 102.97 |
| Peanut cake | | 104.89 | 57.12 | −45.54 |
| Shelled peanut | | 372.91 | 499.66 | 33.99 |
| Peanut oil | Import/10,000 t | 75.01 | 67.73 | −9.71 |
| Peanut butter | | 13.13 | 23.68 | 80.35 |
| Peanut cake | | $1.69 \times 10^8$ | $2.09 \times 10^8$ | 23.67 |
| Shelled peanut | | $2.40 \times 10^9$ | $6.77 \times 10^9$ | 182.08 |
| Peanut oil | | $5.79 \times 10^8$ | $1.17 \times 10^9$ | −79.79 |
| Peanut butter | | $2.12 \times 10^8$ | $6.28 \times 10^8$ | 196.23 |

production, eating (eaten uncooked or boiled, deep-fried, fried, and roasted, etc.), deep-processing (peanut butter, peanut drink, candies, and cakes, etc.), peanut protein processing (protein powder, protein concentrate, protein isolate, and peanut peptides, etc.), and comprehensive utilization of peanut by-products (peanut shells used for mushroom cultivation, production of feed and fuel, as well as extraction of active substances, such as resveratrol and proanthocyanidins, etc.). There are different requirements for peanuts of different uses. For example, the peanuts for food are required to have a high content of protein and carbohydrate, a low content of fat, and a low oleic/linoleic (O/L) ratio; the peanuts for oil production are required to have a high content of fat and unsaturated fatty acids; the peanuts for export are required to have a high O/L ratio and a high carbohydrate content; and so forth.

The main peanut consuming countries have different peanut utilization patterns and great variations exist in the peanut utilization. Food and oil production are two main uses of peanuts, with the proportion of the former gradually decreasing while that of the latter is increasing significantly. In the developed countries, peanuts are mainly used for food. In the United States, 65% of peanuts are used for food and the country has great varieties of peanut products, including peanut biscuits of different flavors, peanut ice cream, and peanut candies, etc. Peanut oil in the United States includes refined peanut oil (commonly used by fast food chain stores), gourmet peanut oil (unrefined, with a high content of vitamins and sterol), and 100% pure peanut oil, etc. Since peanut oil has a fragrant flavor and high nutrition, and can make fried products "crisp outside and tender inside," it is often used for processing traditional food, such as turkey (http://www.peanut-institute.org/). In Japan and Western Europe, almost all peanuts are used for food. In the 1970s and the 1980s, the proportion of peanuts used for oil production in Europe dropped from 53.4 to 19.2%, while the proportion of peanuts used for food increased from 46.3 to 80% (Chen et al., 2009). In the developing countries, peanuts are mainly used for oil production. In recent years, 49% of peanuts are used for oil production in India, and most of the remaining peanuts are used for the processing of salted peanuts, peanut candies, peanut butter, roasted peanuts, full-fat peanut cake, and selected peanuts (Govindaraj and Jain, 2011); in Indonesia, the proportion of peanuts used for oil production has decreased from 9.5 to 5.1% while the proportion of peanuts used for food has increased from 79.8 to 83.8%. Indonesia has become the peanut-producing country with the highest peanut consumption rate in the world (Chen et al., 2009). Worldwide, in most cases peanuts can be eaten raw, used in recipes, made into oils, textile materials, and peanut butter, as well as many other uses.

**1.** Boiled peanuts

It is a popular snack in places where peanut is cultivated, for example, in the southern United States, India, and West Africa. Boiled peanuts are often prepared in briney water, and sold in street side stands; fully matured but undried peanuts are commonly used.

**2.** Dry roasted peanuts

It is usually eaten as snacks or as sauces with many foods such as meat. The favorable roasting condition for peanuts in the shell or shelled is 350°F or 177°C for 15–20 min (shelled) and 20–25 min (in shell).

**3.** Peanut flour

It is made from crushed, fully, or partly defatted peanuts. It is highly protein dense, providing up to 31.32 g per cup (60 g), depending on the quantity of fat removed. It is used as a thickener for soups, a flavor and aromatic enhancer in breads, pastries, and main dishes.

**4.** Peanut oil

It is often used in cooking, because it has a mild flavor and relatively high smoke point. Due to its high content of monounsaturated, does not oxidized or get rancid easily and it is considered as healthier than saturated fat and is resistant to rancidity. There are several types of peanut oil, including aromatic roasted peanut oil, refined peanut oil, extra virgin or cold-pressed peanut oil, and peanut extract.

**5.** Peanut butter

It is a food paste made primarily from ground dry roasted peanuts. Some varieties contain added salt, seed oils, emulsifiers, and sugar, while the natural type of peanut butter consists exclusively of ground peanuts. It is mainly used as a sandwich spread, sometimes in combination with other spreads, such as jam, honey, chocolate (in various forms), vegetables, or cheese.

**6.** Other uses

Peanuts are eaten as they are, either salted or sweetened, sprinkled over desserts, such as sundaes and ice cream, and can be employed in confectionery as an addition to breads, biscuits, sweets, muffins, and cakes.

**7.** Unusual uses of peanuts

Peanut soap—the monounsaturated fat in peanut oil, given its thick and sticky base, can be used in the production of soap to remove dirt and germs from our bodies.

Biodiesel fuel—large quantities of concentrated peanut oil can be used to power biodiesel motors.

Peanut laxative—by eating peanuts, the bulk of the peanut, the oil of the nuts, will eventually act in the stomach and intestines and improve bowel movements. It is not as potent as some of the other known medicines but certainly delicious.

Peanut dye—peanut oil can act as the base of dyes. Its slightly viscous and sticky fluid ensures that the color of the dye is able to stick to many surfaces. Other ingredients can be added to increase its viscosity.

Peanut shampoo—peanut oil is combined with truffle oil to provide a remedy for hair shafts.

Peanut shell charcoal—peanut shells can be used as a substitute for charcoal.

Peanut glue—peanut oil can be mixed with flour and corn oil before being transferred to a frying pan and boiling water is added while stirring the solution thoroughly. Once it stiffens, the glue is ready to use. The glue may not be the most efficient adhesive but it is easy to make.

## 1.1.2    Peanut Production, Processing, and Utilization

1. Peanut Germplasm Resources in China

   Among the over 7490 portions of peanut germplasm resources collected by China, 4638 portions are domestic varieties from 22 provinces, and 2852 are introduced from 31 countries (units), such as International Crops Research Institute for the Semiarid Tropics, the United States, and Thailand, etc. (Yu, 2008). Most of the domestic resources are Virginia type and Spanish type peanuts, followed by Peruvian type and irregular type peanuts, and the Valencia type peanuts account for the smallest proportion. Most of the Virginia type peanuts are planted in Hebei, Shandong, Jiangsu, Henan, and Anhui; most of the Spanish type peanuts are distributed in Guangdong, Guangxi, Jiangxi, Hubei, Hunan, Sichuan, Yunnan, and Guizhou; most of the Peruvian type peanuts are concentrated in Guangxi, Sichuan, and Jiangxi; most of the irregular type peanuts are mainly distributed in the peanut production area of the lower reach of Yangtze River and the large peanut plantation area in North China; and the Valencia type peanuts are mainly distributed in the special precocious peanut plantation area of Northwest China and inland peanut plantation area of Northwest China.

2. Peanut Acreage and Yield in China

   According to the Statistical Year book of China, the peanut production area was 4,855,307 ha in 2000, and the yield was 14,436,600 t; the peanut production area was 4,639,000 ha in 2012, and the yield was 16,692,000 t, with the production area decreasing by 4.46% and the yield increasing by 15.6%. The peanut plantation area increased most significantly in Liaoning, Jilin, and Heilongjiang and the peanut yield increased most significantly in Liaoning, Jilin, and Heilongjiang.

3. International Trade of Peanuts and Peanut Products of China

   Over the last 10 years, the peanut exports of China have been generally on the increase despite small fluctuations, having assumed a competitive edge in the international market and gained broad market development prospects. China is the largest peanut exporter in the world, and the main products for export include shelled peanuts, raw peanuts, processed peanuts, peanut oil, and peanut cake, which are sold to more than 120 countries and regions, including Europe, Southeast Asia, Japan, and the Middle East, accounting for about 30% of the world's peanut market share. Peanuts and their products are one of the few internationally competitive grain and oil products of China. In 2011, the annual exports of peanuts and peanut products were about 519,600 t, earning foreign exchange of USD 863 million. According to the statistics of FAO, from 2000 to 2010, the exports of shelled peanuts decreased year by year from 33,090 to 116,000 t; the exports of peanut oil increased from 13,600 to 68,500 t, and the export amount increased from USD13,080,000 to USD 86,785,000. The peanut varieties for export were increased year by year, and the export structure changed from the previously raw-material-dominated export to exports equally composed of raw peanuts and finished peanut products. Exports of food made of peanuts showed a yearly increase, and the scope of exports was also widening every year. The peanut products for export developed from the previously unitary graded peanut kernels to several varieties, such as graded peanut kernels, processed peanuts, and peanut oil, etc.

**4.** Peanut Utilization in China

The constant improvement of peanut productivity promotes the total volume of peanut processing and utilization, and the approaches and scope of peanut utilization are also expanding. In the 1990s, the annual average processing volume of peanuts in China had increased by nearly 40% compared with those of the 1980s, and most peanuts were used for oil production. Due to the continuous improvement of processing technology and peanut quality, both the oil yield and oil quality are on the rise constantly. With the increase of various peanut processing means, various peanut foods come into being, and the proportion of peanuts used for oil production is declining every year, but the proportion of peanuts used for peanut food processing and direct consumption is increasing every year. In the 1990s, an average of 58% of peanuts were used for oil production, decreasing by 6% compared with the 1980s; however, an average of 42% of peanuts were used for peanut food processing and direct consumption, increasing by 6% compared with the 1980s (Zhou, 2005a). In recent years, the domestic peanut consumption has included mainly peanut oil, peanut protein powder, baked peanuts, peanut kernel, and peanut drink, where 55% peanuts are used for oil production, which is the main approach of peanut utilization. Please refer to Fig. 1.1 (FAO Statistics) for specific information.

Among the peanuts used for eating, 37% are used for peanut butter, 32% are used for roasted peanuts, 6% are used for peanut protein, 7% are used for peanut milk, and 18% are used for raw eating. As the largest peanut exporter, the development of the peanut processing industry in China lags far behind that of other countries.

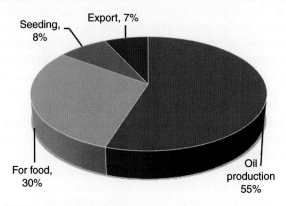

**FIGURE 1.1**    China peanut consumption chart.

## 1.2    PEANUT COMPONENTS

Peanuts have always enjoyed the reputation in China of "longevity nuts," "plant meat," "vegetarian meat," and "green milk," etc. and it is a world-recognized health food. The peanut is the fourth largest oil-bearing crop in the world and the third major source of protein (Jamdar et al., 2010). The peanut kernel is very nutritious, containing 38–60% fat, 24–36% protein, 10–23% carbohydrate, about 3% minerals, as well as bioactive components, such as vitamins, polyphenols, phytosterols, active polycarbohydrates, phospholipids, and dietary fiber (Wan, 2007; Zhang and Zhu, 2007). The peanut is categorized as an A+ grade crop by nutritionists.

For a whole peanut, the shell accounts for approximately 28–32% and the kernel accounts for approximately 68–72% of the peanut. In the peanut kernel, the kernel skin accounts for 3.0–3.6%, the cotyledon accounts for 62.1–64.5%, and the germ accounts for 2.9–3.9% (Zhou, 2003). The main nutrients in the peanut kernel include protein, fat, carbohydrate, vitamins, and minerals.

### 1.2.1    Protein

The peanut contains 24–36% protein, the content of which is only secondary to soybean compared with the other major oil-bearing crops, and higher than sesame and oilseed rape (Table 1.2). The peanut protein contains about 10% water-soluble protein, which is called albumin, and the remaining 90% is salt-soluble protein (Gao et al., 1995). The salt-soluble protein is composed of arachin and conarachin, with the former accounting for 73% and the latter accounting for 27% (Du et al., 2013). The study on the nutritional value of peanut protein proves that the biological value of peanut protein is 59, the net protein utilization is 51, and pure digestibility reaches up to 90% (Wan et al., 2004). The nutritional value of peanut protein powder resembles animal protein, and the protein content is 2.7 times that of beef, 3.3 times that of lean meat, 3.8 times that of egg, and 16.7 times that of milk (refer to Table 1.3 for the nutrition facts of peanut protein powder, meat, egg, milk, and other foods). The nutritional function of peanut protein resembles that of soybean protein, but it is more easily absorbed; the content of indigestible carbohydrate, such as raffinose and stachyose, in peanut protein is only one-seventh of that found in soybean protein, and there are no discomforts such as bloating and belching after eating peanut protein. Therefore, peanut protein is considered highly as a potential alternative for protein-based material, cow's milk, and other animal milk for lactose intolerant consumers (Zhang et al., 2008b). In addition, with a high nitrogen solubility index, peanut protein can be added into animal or plant food to improve food quality and strengthen food nutrition. In addition, the peanut has inherent fragrance, which makes it have wide market prospect. Peanut protein contains eight kinds of essential amino acids, all of which except methionine have reached the FAO-specified criteria. The peanut is rich in arginine, leucine, phenylalanine, glutamic acid, and aspartic acid. The content of lysine is higher

**TABLE 1.2 Chemical Components of Peanuts and Seeds of Other Oil-Bearing Crops (%)**

| Crop | Fat | Protein | Carbohydrate | Crude Fiber | Ash | Water |
|---|---|---|---|---|---|---|
| Peanut | 44.27–53.86 | 23.94–36.35 | 9.89–23.62 | 2.67–6.40 | 1.75–2.58 | 5.33–9.16 |
| Soybean | 14.95–22.14 | 41.18–53.61 | 17.81–30.47 | 4.22–6.40 | 3.89–5.72 | 5.71–12.50 |
| Oilseed rape | 28.15–48.08 | 19.13–27.17 | 16.61–38.86 | 4.58–11.22 | 3.34–7.84 | 6.53–10.53 |
| Sesame | 45.17–57.16 | 19.87–24.25 | 9.59–19.91 | 4.00–7.52 | 4.49–6.87 | 4.35–8.50 |
| Cottonseed | 17.46–23.07 | 24.27–37.66 | 19.14–33.33 | 1.12–3.56 | 5.12–6.12 | 9.42–12.09 |

**TABLE 1.3** Nutrition Facts of Peanut Protein Powder, Meat, Egg, and Other Foods

| Name | Protein (g) | Fat (g) | Carbohydrate (g) | Calcium (mg) | Magnesium (mg) | Iron (mg) |
|---|---|---|---|---|---|---|
| Peanut protein powder/100 g | 55.0 | 4.5 | 23.0 | 93 | 510 | 2.6 |
| Carp/100 g | 13.0 | 1.1 | 0.1 | 54 | 203 | 2.5 |
| Dried shrimp/100 g | 58.1 | 2.1 | 4.6 | 577 | 614 | 13.1 |
| Lean meat/100 g | 16.7 | 28.8 | 1.0 | 11 | 177 | 2.4 |
| Beef/100 g | 20.1 | 10.2 | 0.1 | 7 | 170 | 0.9 |
| Milk/100 g | 3.3 | 4.0 | 5.0 | 600 | 465 | 1.0 |
| Chicken/100 g | 21.5 | 2.5 | 0.7 | 11 | 190 | 1.5 |
| Egg/100 g | 14.6 | 11.6 | 1.6 | 55 | 210 | 2.7 |

than that of rice, wheat flour, and corn, and the effective utilization rate reaches up to 98.8%, whereas the effective utilization rate of lysine in soybean protein is only 78% (Chen et al., 2007b).

According to the research, the protein content of different peanuts is vastly different. Jiang et al. (2007) collected 6390 portions of planted peanuts. The analysis results suggest that the peanuts from Fujian and Jiangxi have high protein content, while the degree of genetic diversity of peanuts from Hubei, Henan, and Guangxi is higher than the peanuts from other regions. The protein content of the peanuts collected by Zang et al. (2003) ranges from 13.60 to 34.82%, and great difference exists among peanut kernels of different varieties. The Virginia type peanuts have a great variation in protein content, followed by Spanish type peanuts, the irregular type peanuts have the smallest variation in protein content, and the remaining peanut varieties are within the middle range. The protein content of Virginia type peanuts reaches up to 34.75%, the average protein content of Spanish type peanuts is 29.28%, and the protein content of Peruvian type peanuts is 21.06%.

Andersen et al. (1998) analyzed the amino acid content of six peanut varieties of high oleic acid content and 10 peanut varieties of normal oleic acid content. The results indicate that the disparity of threonine and methionine in each peanut group is 1.65 times and 2.1 times, respectively. Luan and Han (1986) analyzed the amino acid content of 379 portions of peanuts from Shandong, and it was found that the arginine content of different peanut varieties ranges from 0.96% to 6.35% and the variation coefficient is 15.07%; the methionine content variation is 0.26–1.44% and the variation coefficient is 23.27%. It can be seen that the amino acid content of various peanut varieties is vastly different.

Li et al. (1998) carried out SDS-PAGE and two-dimensional electrophoresis analysis on the protein of 46 peanut varieties and found four major types of protein composition patterns. The main difference in the protein composition of the peanut kernel comes from the different compositions of globulin subunits in the peanuts. The type I peanut globulin mainly contains 41, 38.5, and 2 18 subunits; type II mainly contains 41, 38.5, 37.5, and 3 18 kDa subunits; type III mainly contains 41, 38.5, 36.5, and 3 18 kDa subunits; and type IV mainly contains 41, 38.5, 37.5, 36.5, and 318 kDa subunits. Shokarii et al. (1991), Krishna et al. (1986), and Du et al. (2013) also reported similar results.

Protein is a spatially structured high-molecular polymer composed of various amino acids, and its physiochemical properties (molecular size and shape, amino acid composition and sequence, charge distribution, intermolecular and intramolecular interaction, and effective hydrophobic interactions) are closely related to its functional properties (Yuan et al., 2005). Monteiro and Prakash (1994) carried out analysis for the amino acids of different peanut protein components, and the results indicate that each component contains more than 17 kinds of amino acids. The total protein has a relatively high content of aspartic acid, glutamic acid, and arginine and low content of cysteine, methionine, tyrosine, and lysine. On the whole, conarachin I has a relatively higher content of serine, glycine, and lysine, with a glycine content of six times that found in the total protein. However, the content of aspartic acid, proline, alanine, valine, isoleucine, and arginine is lower than that found in other protein components. The amino acid content varies greatly among different peanut protein components, and since the peanut protein components of different varieties have different subunits, the functional properties of the peanut protein of different varieties are vastly different. However, there is currently limited research concerning the relationship among the three and further analysis is required. Lu et al. (2000) analyzed the amino acid composition of three main components of the peanut protein from two different sources. The results indicate that both of them contain 17 amino acids, with a high content of aspartic acid, glutamic acid, and arginine but an extremely low content of methionine and cysteine, just as Monteiro and Prakash (1994) had concluded.

Peanut protein, due to its functional properties, nutritional values, and good flavor, has been widely used in many foods. Currently, the peanut protein products are mainly in the form of peanut protein powder, which is mainly used as the basic raw material in food processing. It is used in meat product processing, grain bakery products processing, and the production of plant protein drinks, cold drinks, and foods, as well as nonstaple foods, condiments, and snack foods. The functional properties of the protein can be reinforced through modification methods so as to improve the product texture and structure and to improve the products' nutritional values. Due to the difference in consumption habits, the North American, European, and other developed countries have undertaken relatively more studies on peanut butter, peanut candy, and peanut-related diet foods, and they have selected special varieties for peanut candy processing, but they have relatively fewer studies on peanut protein products (Wu, 2009). Therefore, the studies on the protein components, subunit composition, protein structure, modifications, and functional properties of different peanut varieties, as well as the establishment of evaluation methods for the functional properties of peanut protein, have gradually become the focus of studies in this field.

## 1.2.2  Fat

The peanut kernel generally contains 44–54% fat, and compared with other oil-bearing crops, the fat content in peanut is only secondary to that in sesame, and higher than oilseed rape, soybean, and cottonseed. The studies of Zang et al. (2003) indicate that the fat content of peanuts from different areas is vastly different. The peanuts from Henan and Zhejiang have high oil content, while those from Sichuan and Guangxi have low oil content. The studies of Moore and Knauft (1989) and Ókeefe et al. (1993) indicate that the fat content of different peanut varieties is vastly different. The Valencia type peanuts have a high fat content, with an average of 51.09% but the irregular type peanuts have a low fat content, with an average of 46.28%.

The fatty acid in peanut includes saturated fatty acids (palmitic acid 6–18%, stearic acid 1.3–6.5%, arachidic acid 1.0–3.0%, behenic acid, wood tar, and myristic) and unsaturated fatty acids (oleic acid 35–72%, linoleic acid 20–45%, and arachidonic acid), and the total amount of unsaturated fatty acid reaches over 85% (Liu et al., 2008; Bockisch, 1998; Jiang et al., 1998). The fatty acid composition resembles that of olive oil (the content of palmitic acid, stearic acid, oleic acid, linoleic acid, and arachidic acid is respectively 6.9, 2.3, 84.4, 4.6, and 0.1%). Olive oil can reduce the risk of cardiovascular diseases, and therefore, peanut oil is also reputed as the "economic olive oil for the Chinese" (Yao, 2005). Oleic acid is a monounsaturated fatty acid and it has good resistance to rancidity, for example, due to the high content of fatty acid and low content of polyunsaturated fatty acids (only containing 2–4% polyunsaturated fatty acids) in sunflower seed oil with rich oleic acid, the sunflower seed oil has excellent oxidation stability (Shi, 1999). The total content of oleic acid and linoleic acid in peanut is about 80%, which is basically stable (Bovi, 1983; Basha, 1992). The Virginia type peanuts, among others, have the highest content of oleic acid, with an average of 49.32%, and the Valencia type and Spanish type peanuts have the lowest content of oleic acids, with an average of about 38.0%. The content of linoleic acid is the opposite of that of oleic acid among the peanut varieties (Luan and Han, 1990; Han and Luan, 1998). Table 1.4 describes the fatty acid composition and content of several common plant oils. It can be seen that, compared with other plant oils, the peanut oil contains long-chain fatty acids (C20:0–C24:0), which do not or only barely exist in other plant oils.

### 1.2.3 Carbohydrate

Peanut has complex carbohydrate components. From the aspect of structural composition, they are divided into monocarbohydrates, oligocarbohydrates, and polycarbohydrates, where the polycarbohydrates are further divided into homopolycarbohydrates and heteropolycarbohydrates based on the difference in composition. The homopolycarbohydrates are polycarbohydrates composed of the same glycosyls, including starch and cellulose; heteropolycarbohydrates are polycarbohydrates composed of two or more kinds of monocarbohydrate units. Peanut kernel contains 10–23% carbohydrate, where starch accounts for about 4%, and the remainder are free sugars, both soluble and insoluble. The soluble sugars, mainly comprising sucrose, fructose, and glucose, are mostly sucrose and also contain small amounts of oligocarbohydrates, such as stachyose, raffinose, and verbascose, etc. The insoluble free sugars mainly include galactose, xylose, arabinose, and glucosamine (Yao, 2005). The carbohydrates are mostly composed of sucrose, and the sweetness of uncooked peanuts mainly comes from sucrose. The special flavor (Newell et al., 1967) and nut fragrance (Mason et al., 1969) of baked peanuts mainly come from the Maillard reaction between the monocarbohydrates (glucose and fructose) and free amino acids during the baking process (Savage and Keenan, 1994). The studies on the

**TABLE 1.4** Fatty Acid Content (%) of Several Common Plant Oil

| Source | C14:0 | C16:0 | C17:0 | C18:0 | C20:0 | C22:0 | C24:0 | C16:1 | C18:1 | C20:1 | C22:1 | C18:2 | C18:3 |
|---|---|---|---|---|---|---|---|---|---|---|---|---|---|
| Almond oil | — | 2.65 | 0.02 | 0.18 | 0.02 | — | — | — | 86.79 | — | — | 9.82 | 0.03 |
| Tea oil | 1.79 | — | — | 3.55 | — | — | — | 11.99 | 74.41 | — | — | 8.26 | 0.97 |
| Cottonseed oil | 0.50 | 20.40 | — | 1.40 | — | — | — | 0.30 | 14.60 | — | — | 62.80 | — |
| Olive oil | 1.30 | 14.41 | — | 3.26 | 0.52 | — | — | — | 71.25 | — | — | 9.77 | 1.15 |
| Peanut oil | — | 10.90 | — | 2.70 | 1.10 | 1.70 | 0.60 | — | 46.50 | 0.70 | — | 35.40 | 0.10 |
| Soybean oil | — | 19.05 | — | 3.87 | — | — | — | — | 16.99 | — | — | 51.39 | 8.70 |
| Sesame oil | — | 8.77 | — | 5.30 | — | 0.10 | 0.31 | 0.15 | 39.70 | 0.19 | — | 44.69 | 0.96 |
| Sunflower seed oil | — | 4.44 | — | 3.82 | 0.28 | 1.06 | — | — | 38.11 | 0.16 | — | 51.82 | — |
| Corn oil | — | 11.60 | — | 1.30 | — | — | — | — | 30.60 | — | — | 55.80 | — |
| Wheat germ oil | — | 27.89 | — | 1.48 | 2.92 | — | — | — | 15.06 | — | — | 52.64 | — |
| Rice bran oil | 0.61 | 22.14 | 0.03 | — | — | 0.93 | — | — | 32.70 | 1.54 | — | 35.30 | — |
| Rapeseed oil | — | 4.50 | — | 1.40 | 0.50 | 0.30 | 0.20 | — | 32.00 | 4.40 | 24.80 | 23.60 | 8.10 |
| Basil oil | 0.21 | 3.94 | — | 1.62 | 0.31 | — | — | 0.61 | 12.14 | — | — | 15.32 | 62.91 |

Note: "—" indicates undetected.

relationship between the sweetness and taste of baked peanuts and the hardness and sugar content of peanut seeds indicate that higher seed hardness and sucrose content can ensure better taste and quality, and there is a closer relationship between the taste and quality of baked peanuts with sweetness than with hardness (Wan, 2007).

The chemical components of dietary fibers in peanuts mainly include cellulose, hemicellulose, pectin, and lignin, which mainly exist in the peanut shell, peanut kernel (cake), and peanut stem and leaf. The peanut shell has the most dietary fiber, accounting for 65–80% of the shell, the peanut cake has 4–6%, and the peanut stem and leaf contain about 21.8% of dietary fiber (Liao, 2004). Since the peanut has a complex composition of dietary fibers, a number of categorization methods have been developed. According to their different solubility in water, they are categorized into water-soluble dietary fiber and insoluble dietary fiber; according to the quality, they are categorized into common dietary fiber and high-quality dietary fiber; according to the different degrees of coliform bacteria fermentation, they are categorized into partially fermented and fully fermented dietary fiber. The dietary fibers in the peanut also have good water holding capacity, cation binding, exchanging and absorption effects, as well as other physiochemical properties. In addition, they also have physiological functions, such as regulating blood sugar, lowering cholesterol, lowering blood pressure, and preventing obesity, etc. The dietary fibers in the peanut can be added to dairy products, meat products, beverages, baked products, and other foods to improve both the sensory quality and nutritional functions of these foods, and they can also be made into snack foods or diet foods. With extensive sources and numerous physiological functions, the dietary fibers in the peanut are widely applied, having broad market prospects.

## 1.2.4 Vitamin

Peanuts are rich in vitamins, including niacin, vitamin E ($V_E$), vitamin B1, vitamin B2, vitamin B6, pantothenic acid, and folic acid. Among them, the content of niacin, $V_E$, and pantothenic acid is relatively high, accounting for above 85% of the total content of vitamins, followed by vitamin $B_1$, vitamin $B_2$, and vitamin $B_6$—the content of the three is 0.1–1 mg/100 g, accounting for about 10% of the total amount of vitamins. The peanut has a relatively low content of vitamin K, folic acid, and biotin—lower than 0.1 mg/100 g, accounting for less than 5% of the total content of vitamins.

The $V_E$ in the peanut mainly contains three kinds of isomers, namely, $\alpha$-$V_E$, $\gamma$-$V_E$, and $\delta$-$V_E$ 3, the content of which are respectively about 20 mg/100 g, 7 mg/100 g, and 0.5 mg/100 g. It is reported that the natural $V_E$ (especially $\alpha$-$V_E$) has effects in enhancing immunity, antiaging, reducing the incidence of cardiovascular diseases, and preventing cancer. $V_E$ has great influence on the oil stability, and Tong et al. (2009) adopted a fat oxidation stability tester to determine the differences in oxidation stability of small-packaged plant oil that does not contain antioxidants. The test suggests that the oxidation stability of corn oil is better than that of peanut oil because the deodorized corn oil contains $V_E$ of 80–120 mg/100 g, while the peanut oil has a significantly lower content of $V_E$, 26.8–51 mg/100 g, compared with the corn oil.

According to the domestic report, the total content of $V_E$ is different among peanuts of different varieties. On this basis, Zhang et al. (2012) collected three isomers of peanuts, namely, $\alpha$-$V_E$, $\gamma$-$V_E$, and $\delta$-$V_E$, of 45 varieties from five major peanut plantation areas (Shandong, Guangdong, Henan, Fujian, and Jiangsu) for determination and statistical analysis. The results reveal that among the 45 peanut varieties subject to analysis, the peanuts from Guangdong have the highest content of $\alpha$-$V_E$ but those from Fujian have the highest content of $\gamma$-$V_E$ and $\delta$-$V_E$. This study lays a theoretical foundation for selecting $V_E$-rich peanut varieties.

### 1.2.5    Other Nutrients

Besides containing rich protein, fat, and carbohydrates, the peanut also contains minerals, saponins, resveratrol, proanthocyanidins, flavonoids, and other bioactive components. The mineral content in peanut kernel is only 2–3%. However, from the nutritional perspective, the peanut is rich in zinc, potassium, phosphorus, and magnesium, but has a relatively lower content of calcium, iodine, and iron. $Zn^{2+}$, $Mg^{2+}$, and other metal ions are the components or activators of many metalloenzymes. The content of $Zn^{2+}$ of 100 g peanut oil has reached 8.48 mg, being 37 times that of salad oil, 16 times that of rapeseed oil, and 7 times that of soybean oil. The peanut root, followed by stem, red skin, leaves, and shell, contains large amounts of resveratrol, and the peanut kernel contains the smallest amount (Zhang et al., 2009). The luteolin is mainly distributed in the peanut shell, but the content varies within the range of 0.25–1.12% (Tang et al., 2005) depending on the origin, species, and degree of maturity. Peanut skin is rich in proanthocyanidins, accounting for 17% of the dry weight of peanut skin, and about 50% of proanthocyanidins are oligomers with high bioactivity (Karchesy and Hemingway, 1986).

### 1.2.6    Antinutritional Factors in the Peanut

The antinutritional factors in the peanut mainly include trypsin inhibitor, agglutinin, phytate, condensed tannin, and $\alpha$-amylase inhibitor (Ahmed et al., 1998; Jeanne et al., 2005).

Trypsin inhibitor is a substance that can bind with trypsin to inhibit the enzyme. The experiment conducted by Yang et al. (1998) reveals that the 2S protein component of the peanut has trypsin inhibitory activity, and the trypsin inhibitor was extracted from peanut kernel by means of purification, and various sulfhydryl reductants and proteases were adopted to passivate its activity under low temperature so as to improve the nutritional value of peanut protein products. The study by Tian et al. (2009) proves that the peanut contains at least three kinds of trypsin inhibitor components, the molecular mass is 30–70 kDa, and the isoelectric point is pH5.0–5.8. The content of trypsin inhibitor of the baked and shelled peanuts decreases from 16.31 TIU/mg to 12.6 TIU/100msg.

The agglutinin can mainly be found in grain legumes, peanuts, and peanut meal. Dev et al. (2006) state that peanut agglutinin is a kind of agglutinin with quaternary open structure, which is unique to bean-pods, and it is a kind of homotetramer protein, that is, a monomeric dominance group folded by three βs. The experiment carried out by Sun et al. (2011) revealed that the crude agglutinin extract of peanut kernels failed to agglutinate the ABO-type blood cells of human beings but could agglutinate the sialidase-processed ABO-type blood cells. The agglutination of red blood cells by agglutinin can be inhibited by four kinds of carbohydrates, namely, lactose, melibiose, raffinose, and D-galactose, and it is relatively stable within the pH range of 5.0–11.0. However, the activity disappears completely 10 minutes later under the constant temperature of 55°C. Among the six peanut varieties subject to testing, the agglutinin in the peanuts from Hebei Gaoyou was most active. At the early growth period of peanuts, the agglutinin exists in the cotyledons and roots.

Phytic acid (salt) is also named inositol 6-phosphate, and its phosphate radical can be chelated with several types of metal ions (such as $Zn^{2+}$, $Ca^{2+}$, $Cu^{2+}$, $Fe^{2+}$, $Mg^{2+}$, $Mn^{2+}$, $Mo^{2+}$, and $Co^{2+}$, etc.) to form corresponding insoluble compounds in the form of stable phytates, which are not easily absorbed by the intestine, thereby suppressing the intake of it by living bodies. The phytic acid content in the peanut is 1.05–1.76% (by dry weight) and the phytic acid content in the soybean is 1–2.22% (by dry weight). Compared with the unprocessed peanuts, the germinated, baked, and shelled peanuts have significantly lower content of peanut phytic acid, and the phytic acid content of germinated, baked, and shelled peanuts decreased from 14.66 to 8.45 mg/g.

Tannin, also known as tannic acid, is a water-soluble polyphenol, having a bitter taste. It is divided into polyflavonoid with antinutrition effect and hydrolyzable tannin with toxicity. The polyflavonoid is the condensation of flavonoids inside plants. It generally cannot be hydrolyzed but is water-soluble due to its strong polarities. The hydroxide radicals can react with trypsin and amylase or their substrates (protein or carbohydrate) to reduce the utilization rate of protein or carbohydrate. It can also bind with gastrointestinal mucosa protein to form insoluble complexes on the intestinal mucosal surface to damage the intestinal wall, interfere with the absorption of some minerals (such as iron ions), and influence the growth and development of animals. Tannin can combine with metal ions, such as calcium, iron, and zinc ions, to form deposits and can also form complexes with vitamin $B_{12}$ to reduce their utilization rate. The tannin content of the shelled and processed peanuts reduced from 0.09 to 0 mg/100 g.

α-Amylase inhibitor can mainly be found in wheat, kidney bean, taro, mango, unripe bananas, and other foods. If much of such foods is eaten uncooked or undercooked, the amylase inhibitor takes effect to make the starch contained in the food unable to be digested, absorbed, or utilized by the organic bodies, and most of the starch will be directly excreted. If this condition lasts for long, the nutrition absorption rate of the human body will decrease, and growth and development will be affected. The fully heated or processed beans or wheat foods can basically fully deactivate the digestive enzyme protein inhibitor. The content of α-amylase inhibitor of the germinated, baked, and shelled peanuts decreased from 62.5 to 18.0 AIU/g.

## 1.3    PROGRESS OF INTERNATIONAL STUDIES ON PEANUT PROCESSING

Peanut plantations can be found in 100 countries on five continents. In 2012, the world's peanut productivity reached 41,185,900 t, and is still increasing year by year. With the rapid development of the peanut processing industry, scientific research in the peanut processing field has been undertaken by people all over the world. Currently, different countries lay different emphases on the studies of peanut processing but most of them focus on the analysis of processing quality of raw peanuts, processing technique and flow, product flavor and nutrition, as well as the preparation and functional evaluation for active components. A lot of attention has been paid to the studies in the peanut processing field, especially the United States, China, India, and Argentina. The research team of the author keeps close track on the studies in relevant fields both at home and abroad. On this basis, the relevant research literatures are systematically arranged and compiled so as to give an in-depth analysis of the research news, focuses, and directions in the peanut processing field in the United States, India, Argentina, and China. The purpose is to keep up with the latest international developments, give full play to the features and advantages of the research team, and lead the sound development of the peanut processing industry and technology.

### 1.3.1    Research Progress in the United States

The United States' peanuts mainly comprise four types, namely, Virginia type, Valencia type, Runner type, and Spanish type, and the main production area is in the southeast United States: 47.6% in Georgia, 12.7% in Alabama, 12.4% in Florida, and 9.2% in Texas. In 2012, the peanut plantation area in the United States reached 1,638,000 ha and the yield reached 3,060,000 t.

The literature dated from 2000 to 2012 was retrieved from the US Dissertation Database and *Peanut Science*, etc. and categorization and compilation work was carried out according to the raw material quality analysis, product process optimization, product nutritional quality and flavor, product activity evaluation, and primary processing (drying, grading, and storage), etc. The results are as shown in Fig. 1.2. It can be seen that there has been increasingly more literature on the quality of raw peanuts in the United States since 2013, accounting for 42% of the retrieved literature, mainly including studies on

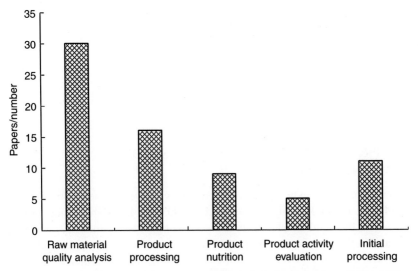

**FIGURE 1.2**   Classification of peanut study literatures in the United States from 2000 to 2012.

near-infrared rapid nondestructive testing techniques for the nutritional components in peanuts, such as fatty acid, vitamin E, and sterol, as well as technical studies on the processing features and quality evaluation for the special peanut varieties dedicated to the production of peanut butter and peanut candies. On one hand, the rapid retrieval of nutritional data of peanuts is required in the peanut trade in the US market, and therefore, it is essential to learn about the processing quality of different varieties or raw materials; on the other hand, these studies can provide data support for the peanut breeding experts so as to cultivate the peanut varieties with high nutrition and that are adaptable for different processing purposes. Therefore, the near-infrared rapid nondestructive testing technique for the nutritional and functional components in the peanuts, as well as the processing quality evaluation technology designed for the special peanut varieties used for producing peanut butter and peanut candies, have become the study focus over the past decade. The studies on peanut product processing technique optimization account for 22% of the retrieved literature, mainly involving the preparation of peanut protein and functional short peptides by means of protein meal enzymolysis, and the preparation techniques of resveratrol and polyphenolic active substances from peanut root, stem, and leaves, etc. The studies on product nutritional quality and flavor mainly include the improvement of product quality by adding functional factors to improve the nutrition and stability of peanut butter as well as the flavor and nutrition of snack foods, such as baked peanuts, etc. The studies on product activity evaluation mainly include the ACE inhibitory activity of protein hydrolytes and functional short peptides, antibiotic activity, and studies on the antioxidant activity of resveratrol and phenols from peanut root, stem, and leaves, etc. Studies on the primary processing of peanuts (drying and storage) mainly focused on the drying method and parameters of microwave processing, the low-oxygen and airtight storage method, parameters, warehouse structure, and so forth. However, there is relatively little such literature in recent years, mainly because the processing theory and practical techniques regarding peanut drying, grading and separating, and storage, which used to be the study focus in the United States in the 1990s, have already become very sophisticated nowadays.

## 1.3.2 Study Progress in India

The peanut plantation in India is widespread throughout the country, although it is mainly distributed in the north, southwest, middle, and southeast, the peninsula, and the south of India. The southwest of India mainly plants Virginia prostrate type peanuts, the northwest of India mainly plants upright type peanuts, and the middle and southwest of India plants a mixture of Virginia upright type and Spain upright type peanuts. In 2011, the peanut plantation area in India reached 4,190,000 ha and the productivity reached 6,933,000 t (according to the statistics of FAO).

On the basis of the retrieval of peanut-related literatures in India, classification and compilation have been carried out according to the study content, and refer to Fig. 1.3 for the results. It can be seen that from 2000 to 2012, the quantity of peanut study literature increased significantly, approximately doubling from 1990 to 2000. The peanut processing studies in India cover a wide range, including the quality of raw peanuts, peanut protein preparation process optimization, and structural analysis on peanut protein component and agglutinin, etc. With regard to the study on the quality of raw peanuts, the efforts are mainly directed to the analysis on the protein and oil content, fatty acid composition, and other basic components of large sample size peanut varieties so as to determine the quality evaluation technologies, methods and standards, and

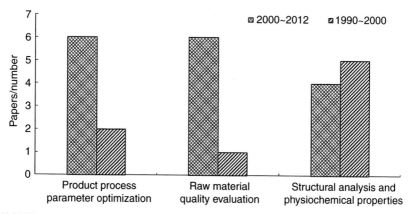

FIGURE 1.3    Analysis on peanut study literatures in India.

to select the varieties suitable for the production of peanut butter and candies. These studies have always been the focus of the field of peanut processing in India. With regard to the product process parameter optimization, the efforts are mainly directed to studying the preparation processes and techniques of agglutinin, phospholipase, peanut oil, and protein component, etc. With regard to the structural analysis and physiochemical properties, the efforts are mainly directed toward studying the structure and properties of peanut agglutinin and the components, structure, and functional evaluation of peanut protein.

### 1.3.3 Study Progress in Argentina

Cordoba Province is the main peanut production area in Argentina, accounting for about 95% of the peanut plantation area in the country. The main varieties include one to three Foluolanna type small and prostrate peanuts. In 2011, the peanut plantation area reached 230,900 ha and the yield reached 701,500 t (FAO statistics). A total of 18 studies on peanut processing in Argentina were retrieved, and it was found that the studies mainly focused on the basic component analysis of peanut varieties and analysis on the sensory and processing quality of raw and fried peanuts. From 1995 to 1999, the studies were mainly directed toward the basic components of peanut varieties in each area, including protein, fat, ash, sterol, and fatty acid, etc., accounting for 33% of the retrieved literature. From 2000 to 2012, the studies mainly focused on the comparison of the natures of the processed products made of different peanut varieties, the nutritional variation during the processing, and the influence of the posttreatment on the peanut products. The most representative is the influence on the product nutrition, flavor, and shelf life of the high fatty acid peanut variety and the common peanut variety, and this is because Argentina has always paid attention to peanut variety breeding, especially high fatty acid peanuts. These types of peanut products have high nutrition and long shelf life, and are the main type of peanuts for export from Argentina.

### 1.3.4 Study Progress in China

Peanuts are widely planted in China, and peanut plantations can be found all over China, except Tibet and Qinghai. In 2011, the peanut plantation area in China reached 4,673,400 ha and the yield reached 16,114,200 t, and the provinces with peanut yield exceeding 1 million tons include Henan (4,297,900 t), Shandong (3,385,800 t), Hebei (1,289,200 t), and Liaoning (1,265,400 t).

More than 100 papers were retrieved from the postgraduate and doctoral dissertations related to peanut processing. Statistical analysis was made for the retrieved literatures, as shown in Fig. 1.4. In recent decades, peanut processing studies mainly have focused on three aspects, namely, processing technique and flow, product flavor and nutrition, and activity evaluation, which account for 80% of the retrieved literature. Before 2007, the studies on peanut processing

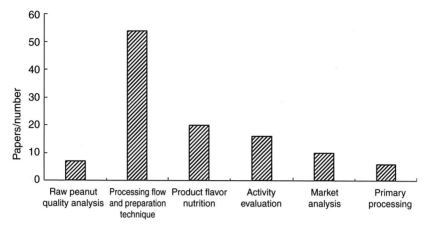

**FIGURE 1.4** Categorization and analysis of peanut study literatures from 2003 to 2012 in China (categorized based on study content).

mainly focused on the peanut oil and protein extraction techniques, and after 2007, especially since 2012, the volume of literature on peanut processing has increased significantly, accounting for 60% of the total number of retrieved literature, and most studies focus on the extraction and preparation of peanut protein, peptide, the functional components from by-products, and activity evaluation.

Categorical analysis was carried out according to the raw material of peanuts, peanut oil, peanut protein (peptide), snack food, and by-products (root, stem, and leaves), as shown in Fig. 1.5. The studies mainly focus on the utilization of peanut meal, protein and peptide, and by-products (root, stem, and

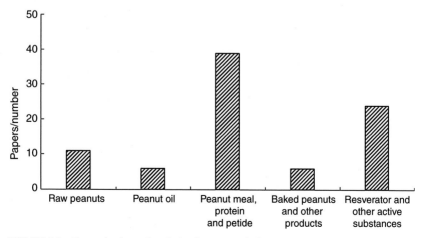

**FIGURE 1.5** Categorization and analysis of peanut study literatures from 2003 to 2012 in China (categorized based on objects of study).

leaves), which respectively account for 45% and 28% of the retrieved literature. With regard to the peanut meal, protein and peptide, the studies mainly focus on the preparation and modification of protein or peptide as well as functional activity evaluation; with regard to the comprehensive utilization of by-products, the studies mainly focus on the detection, extraction preparation, and activity evaluation of effective components (such as resveratrol, luteolin, flavone, and pigment, etc.).

Chapter 2

# Peanut Processing Quality Evaluation Technology

A. Shi, Q. Wang, H. Liu, L. Wang, J. Zhang, Y. Du, X. Chen
*Institute of Food Science and Technology, Chinese Academy of Agricultural Sciences, Beijing, China*

## Chapter Outline

*Peanuts: Processing Technology and Product Development.* http://dx.doi.org/10.1016/B978-0-12-809595-9.00002-8

## 2.1 PREAMBLE

As one of the main oil-bearing crops in the world, the peanut is planted in over 100 countries in the world. However, the sensory quality, nutritional quality, and processing characteristics are vastly different among the peanuts from different regions and of different varieties. For example, the peanuts in China are mainly used for oil production and export, and the oil content, fatty acid proportion, content of vitamin E, sterols, and other active components are the important quality indexes that determine the quality of peanut oil products. However, kernel shape, skin color, sugar content, and the oleic acid/linoleic acid (O/L) ratio are the main quality indexes that determine the quality of peanuts for export. For another example, the US peanuts are mainly divided into Runner, Valencia, Spanish, and Virginia types, and the different types of peanuts have different processing characteristics and qualities. Runner peanuts are greatly favored due to their unique pod size and high yield, and more than half of Runner peanuts are used for peanut butter production; Valencia peanuts are mainly used for baking, being sold in shells, and sauced-steaming; Spanish peanuts, due to their high oil content, have become the ideal varieties for oil production, and they are also used for producing desserts and peanut butter; Virginia peanuts have large kernels, and are mainly used for baking in shells or producing salted or flavored peanuts (Licher, 2005).

However, various peanuts are currently put into mixed use in China. There is not only a scarcity of special varieties dedicated to different processing purposes but also a shortage of studies on the processing characteristics of different varieties. There is also a lack of peanut processing quality evaluation technologies, methods, and standards, all of which greatly restrict the development of the peanut processing industry in China. Therefore, to analyze the quality and processing characteristics of raw peanuts and define the relationship between raw peanuts and product quality will be of great significance for scientifically and rationally utilizing the peanut resources and improving product quality.

## 2.2 RAW MATERIAL INSPECTION TECHNOLOGY FOR PEANUT PROCESSING

The processing characteristics of peanut varieties or raw material are closely related to the quality of peanut products, and the quality of varieties or raw materials determines the quality of processed products. Without proper processing of varieties or raw materials, it is hard to produce high quality peanut products. Therefore, the processing quality of peanut varieties has a direct influence on the product quality. It will effectively promote the sound development of the peanut processing industry in China to give a comprehensive analysis on the processing quality of peanut varieties or raw materials, study the processing quality inspection technology, method, and standard, and select the correct special varieties for processing. This section summarizes the inspection methods for the sensory quality, physiochemical nutritional quality, and processing quality of raw peanuts for processing to lay a foundation for the peanut processing quality evaluation technology.

### 2.2.1 Testing Method for Sensory Quality of Raw Peanuts

The sensory quality of peanuts mainly includes the quality attributes that have direct influence on the peanut value, such as unsound pod, unripe pod, crushed pod, worm-eaten pod, damaged pod, pod size, pod shape, pod waist, unsound kernel (worm-eaten kernel, spotted kernel, budding kernel, crushed kernel, unripe kernel, and damaged kernel), deteriorated kernel (moldy kernel, heat-damaged kernel, discolored kernel, oily kernel, spotted kernel, and other damaged kernels), abnormal seed rate, mass of 1000 grains, impurities, odor, whole or half of a peanut kernel, size of a peanut kernel, color, taste, and clamp-shaped kernel, and other appearance properties of peanut pods and kernels.

1. Unsound pod, unripe pod, crushed pod, worm-eaten pod, damaged pod, pod shape, pod waist, unsound kernel (worm-eaten kernel, spotted kernel, budding kernel, crushed kernel, unripe kernel, and damaged kernel): GB/T 5494—2008 Inspection of grain and oil—Determination of foreign matter and unsound kernels of grain and oilseeds.
2. Deteriorated kernel (moldy kernel, heat-damaged kernel, discolored kernel, oily kernel, spotted kernel, and other damaged kernels): SN/T 0798—1999 Inspection of import and export grain and oil—Inspection Terminologies.
3. Mass of 1000 grains: GB/T 5519—2008 Cereals and pulses—Determination of the mass of 1000 grains.
4. Impurity: GB/T 5494—2008 Inspection of grain and oil—Inspection of impurity and unsound kernel of grain and oilseeds.
5. Color: GB/T 5492—2008 Inspection of grain and oil—Methods for identification of color, odor and taste of grain and oilseeds.
6. Odor: GB/T 5492—2008 Inspection of grain and oil—Methods for identification of color, odor, and taste of grain and oilseeds.
7. Whole or half of a peanut kernel: GB/T 1532—2008 Peanut.
8. Abnormal variety rate: SN/T 0803.4—1999 Test method for the purity and intermixing degree of import and export of oilseeds.
9. Size of kernel: NY/T 1893—2010 Grade and specifications of peanuts for processing.

### 2.2.2 Method for Testing Physiochemical and Nutritional Quality of Raw Peanuts

Physiochemical and nutritional qualities refer to the varieties of various chemical components and nutrients essential to the human body, as well as the quantity and quality of various chemical components and nutrients, including crude protein, crude fat, crude fiber, moisture, ash, total carbohydrate, amino acids, fatty acids, vitamins, minerals, phytosterols, resveratrol, squalene, other nutritional components, and protease inhibitor, plant agglutinin, and other antinutritional factors.

1. Crude protein: GB 5009.5—2010 National food safety standard determination of food protein.
2. Crude fat: GB/T 5009.6—2003 Determination of food fat.

3. Crude fiber: GB/T 5515—2008/ISO 6865:2000 Grain and oil inspection Determination of crude fiber content in grain Medium filtering method.
4. Water: GB 5009.3—2010National food safety standard determination of water in food.
5. Ash: GB 5009.4—2012 National food safety standard Determination of ash in food.
6. Total carbohydrate
   a. Reducing sugar: GB/T 5009.7—2008 Determination of reducing sugar in food.
   b. Sucrose: GB / T 5009.8—2008 Determination of sucrose in food.
   c. Starch: GB/T 5009.9—2008 Determination of starch in food.
   d. Total carbohydrate: Phenol-sulfuric acid method.
7. Amino acid: GB/T 5009.124—2003 Determination of amino acid in food.
8. Fatty acid: GB/T 17376—2008 Animal and vegetable fat and oil—Preparation of Fatty acid methyl ester; GB/T 17377—2008 Animal and oil vegetable fat—Analysis by gas chromatography of methyl esters of fatty acid;
9. Vitamin E: NY/T 1598—2008 Determination of vitamin E component and content in edible vegetable oil—High Performance Liquid Chromatography.
10. Phytosterol: GB/T 25223—2010/ISO 12228:1999 Animal and vegetable fat and oil Determination of resveratrol in peanuts—High Performance Liquid Chromatography.
11. Resveratrol: GB/T 24903—2010 Grain and oil inspection Determination of resveratrol in peanuts—High Performance Liquid Chromatography.

## 2.2.3   Raw Peanuts Processing Quality Test Method

The processing quality (characteristics) mainly includes pure kernel yield, oil yield, protein extraction rate, O/L ratio, unsaturated fatty acid/saturated fatty acid value, arachin/conarachin value, and other quality attributes which are closely related to processing. The processing characteristics of peanuts bear close relationship to their main processed products.

1. Pure kernel yield: GB/T 5499—2008 Inspection of grain and oil—Test method of pure kernel yield of oilseeds in shells.
2. Oil yield: Extracted oil mass/Mass of oil in raw materials.
3. Protein extraction rate: Extracted protein mass/Mass of protein in raw materials.
4. O/L ratio: GB/T 17376—2008 Animal and vegetable fat and oil—Preparation of Fatty acid methyl ester; GB/T 17377—2008 Animal and oil vegetable fat—Analysis by gas chromatography of methyl esters of fatty acid.
5. Unsaturated fatty acid/saturated fatty acid value: GB/T 17376—2008 Animal and vegetable fat and oil—Preparation of Fatty acid methyl ester; GB/T 17377—2008 Animal and oil vegetable fat—Analysis by gas chromatography of methyl esters of fatty acid.
6. Protein subunit composition and relative content: SDS-PAGE analysis with optical density.

## 2.3 QUALITY EVALUATION TECHNOLOGY FOR PEANUTS SUITABLE FOR THE PRODUCTION OF GEL-TYPE PROTEIN

Gelation is one of the main functional properties of protein, and it is also a functional property with market development prospects in actual applications. The gelation of peanut protein is closely related to the composition, structure, and subunit content of protein, and these properties are determined by the peanut variety. The protein composition, structure, and the functional properties of protein vary in different peanut varieties. In recent years, the studies on the quality evaluation technologies, methods, and standards for peanuts suitable for the production of gel-type protein, as well as the composition, subunit content, and functional properties of protein of different peanut varieties have increasingly been emphasized. Wang (2012) and Wang et al. (2012) carried out studies on the quality of different raw peanuts and gel-type protein, and established the relational model for the quality of raw peanuts and gel-type protein in a bid to provide the basis for the screening of peanut varieties or raw material used for the production of gel-type peanut protein.

### 2.3.1 Quality Index Determination

Table 2.1 shows the determination of sensory quality, physiochemical, and nutritional qualities of peanuts, as well as the indexes of processing quality.

**TABLE 2.1 Indexes of Peanut Quality**

| Sensory Quality | Physiochemical and Nutritional Quality | Processing Quality |
|---|---|---|
| Pod shape, red skin, kernel shape, appearance, odor, 100-pod weight, 100-kernel weight | Moisture, crude fat, crude protein, total carbohydrate, ash, crude fiber, total amino acids, 17 kinds of amino acids (aspartic acid, threonine, serine, glutamic acid, proline, glycine, alanine, cysteine, valine, methionine, isoleucine, leucine, tyrosine, phenylalanine, lysine, histidine, and arginine), protein composition and subunit content (arachin, conarachin, conarachin I, conarachin II, 40.5, 37.5, 35.5, 23.5, 18, 17, and 15.5 kDa), 13 kinds of fatty acids (O/L ratio), vitamin E (total amount of $\alpha$-vitamin E, $\gamma$-vitamin E, $\delta$-vitamin E, vitamin E), sterols (campesterol, stigmasterol, $\beta$-sitosterol, and total sterols), squalene | Pure kernel yield, oil yield, protein extraction rate, O/L ratio, unsaturated fatty acid/saturated fatty acid value, arachin, and conarachin value |

**TABLE 2.2 Regression Significance Indexes of Peanut Quality and Gelation**

| No. | Index | P Value | No. | Index | P Value |
|---|---|---|---|---|---|
| 1 | Pod shape | 0.011 | 5 | Arginine | 0.005 |
| 2 | Crude protein | 0.018 | 6 | Conarachinl | 0.003 |
| 3 | Crude fiber | 0.010 | 7 | 23.5 kDa | 0.001 |
| 4 | Leucine | 0.009 | | | |

## 2.3.2 Prediction of Gelation

1. Screening of peanut quality evaluation indexes
   The significance analysis is carried out for the regression coefficients of individual indexes of peanut quality and protein gelation according to regression analysis, and seven indexes have significant correlation with gelation at the level of 0.01 (Table 2.2).
2. Peanut quality correlation analysis
   The seven quality indexes that have been selected are used for correlation analysis (Table 2.3), for the indexes that are significant above the level of 0.01, and for those whose correlation coefficient is larger than 0.600. One index is used to represent the other index, and therefore, the retained indexes include pod shape, crude protein, arginine, conarachin I, and 23.5 kDa protein subunit.
3. Modeling
   The dependent variables for establishing the regression equation should conform to normal distribution, but the gelation data does not fit into the normal distribution. Therefore, it is required to carry out Box-Cox conversion between each main component and gelation to make the variables conform to normal distribution before carrying out regression analysis. Throughout the regression analysis, all the regression coefficients of each main component and gelation are significant at the level of 0.5 (Table 2.4), and therefore, the relation between each main component and gelation is established, and the determination coefficient of the model is $R^2 = 0.768$. On this basis, the relation between each index and gelation is established; refer to Eq. (2.1) for the results.

$$\text{Gelation} = e^{\substack{(1.571 - 0.024\,74 \times \text{pod shape} + 0.007\,009 \times \text{crude protein} - 0.043\,51 \times \\ \text{arginine} - 0.005 \times \text{conarachin I} - 0.060\,57 \times 23.5\,\text{kDa})}} \tag{2.1}$$

The model is verified using the remaining 21 varieties to investigate the accuracy and promotional applicability of the model established. Through the outlier analysis, it is found that the gelation of "Huaguanwang" is an outlier, and it is then deleted. The five indexes (pod shape, crude protein, arginine, conarachin I, and 23.5 kDa) are substituted into Eq. (2.1) to calculate the gelation of 20 varieties. The model calculation results and gelation

**TABLE 2.3 Peanut Quality Correlation**

| | Pod shape | Crude Protein | Crude Fiber | Leucine | Arginine | Conarachin I | 23.5 kDa |
|---|---|---|---|---|---|---|---|
| Pod shape | 1.000 | | | | | | |
| Crude protein | 0.347* | 1.000 | | | | | |
| Crude fiber | 0.241 | 0.293 | 1.000 | | | | |
| Leucine | 0.029 | 0.699** | 0.252 | 1.000 | | | |
| Arginine | 0.236 | 0.732** | 0.409** | 0.649** | 1.000 | | |
| Conarachin I | 0.022 | −0.094 | −0.206 | −0.262 | −0.380* | 1.000 | |
| 23.5 kDa | 0.133 | 0.188 | 0.157 | 0.198 | 0.272 | −0.457** | 1.000 |

Note: "*" represents significance when $P$ is less than 0.05; "**" represents significance when $P$ is less than 0.01.

**TABLE 2.4** Significance of Regression Coefficients Between Gelation and Main Component (Actual Application)

| Variant | Coefficient | F Value | P Value |
|---|---|---|---|
| Gelation | −1.5192579 | 7.05 | 0.0100 |
| Main component 1 | −0.0571799 | 12.92 | 0.0010 |
| Main component 2 | 0.0116931 | 0.10 | 0.7501 |
| Main component 3 | −0.0812101 | 5.64 | 0.0233 |

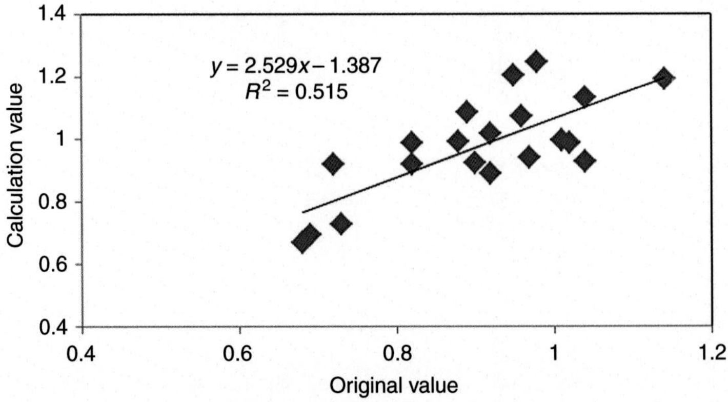

**FIGURE 2.1** Fitting figure for original value of gelation and practical application model.

comprehensive value are used for regression analysis, and the correlation coefficient of the two is 0.178 (Fig. 2.1).

4. Prediction of Gelation

The above model is adopted for the prediction of the protein gelation of 86 domestic peanut varieties, and Table 2.5 shows the protein gelation of 152 varieties (including prediction varieties and modeling varieties).

## 2.3.3 Determination of Peanut Varieties Suitable for the Production of Gel-Type Protein

1. Criteria of Evaluation Index

K-means cluster analysis is carried out for five qualities of each peanut variety, and each index is categorized into three classes, that is, Grade I (suitable), Grade II (basically suitable), and Grade III (not unsuitable), and the weighted value of each index is deemed as its highest score (Table 2.6), that is, Grade I, and the score is assigned to the index of each grade with the same method.

**TABLE 2.5 Gelation Ranking and Value of 152 Peanut Varieties**

| No. | Name of Variety | Gelation | No. | Name of Variety | Gelation | No. | Name of Variety | Gelation | No. | Name of Variety | Gelation |
|---|---|---|---|---|---|---|---|---|---|---|---|
| 1 | Quanhua464 | 1.27 | 15 | Pingdu Lijing | 1.16 | 29 | Yueyou No. 14 | 1.11 | 43 | Laiyang Dunhuasheng | 1.08 |
| 2 | Luhua11 | 1.27 | 16 | Yuhua15 | 1.16 | 30 | Fenghua4 | 1.10 | 44 | Jimo Paman | 1.08 |
| 3 | Shuangji No. 2 | 1.26 | 17 | Linqing Yiwohou | 1.16 | 31 | Yueyou40 | 1.10 | 45 | Qixia Bankangpi | 1.07 |
| 4 | Bianhua No. 3 | 1.24 | 18 | Weihua No. 6 | 1.16 | 32 | Huayu 19 | 1.10 | 46 | Tianfu No. 4 | 1.07 |
| 5 | Fenghua No. 1 | 1.23 | 19 | Haihua 1 | 1.15 | 33 | Luhua8 | 1.10 | 47 | Linhua No. 1 | 1.07 |
| 6 | Kainong 30 | 1.23 | 20 | Guihua166 | 1.15 | 34 | Qixia Laobaoji | 1.10 | 48 | Laiyang BanzhanmanLarge Peant D1-122 | 1.07 |
| 7 | Fenghua 3 | 1.22 | 21 | Honghua No. 1 | 1.14 | 35 | Juye Tuoyang | 1.10 | 49 | Hua 55 | 1.07 |
| 8 | Laiyang Banman | 1.22 | 22 | Laixi Liman | 1.13 | 36 | Hua 27 | 1.10 | 50 | Yueyou 52 | 1.06 |
| 9 | Zhongkaihua No. 4 | 1.21 | 23 | Huayue 22 | 1.13 | 37 | Hueyu No. 25 | 1.10 | 51 | Zhencheng Layang | 1.06 |
| 10 | Huaguawang | 1.20 | 24 | JI9814 | 1.12 | 38 | Luhua10 | 1.09 | 52 | 034-256-1 | 1.06 |
| 11 | Shanhua No. 7 | 1.18 | 25 | Laiyang Xuefang Banman | 1.12 | 39 | Hua 37 | 1.09 | 53 | Huayu No. 33 | 1.06 |
| 12 | Minhua No. 9 | 1.17 | 26 | Muping Yiwohou | 1.11 | 40 | Kunyu Dalidun | 1.08 | 54 | Qingdao Paman | 1.05 |

*(Continued)*

**TABLE 2.5 Gelation Ranking and Value of 152 Peanut Varieties (cont.)**

| No. | Name of Variety | Gelation | No. | Name of Variety | Gelation | No. | Name of Variety | Gelation | No. | Name of Variety | Gelation |
|---|---|---|---|---|---|---|---|---|---|---|---|
| 13 | Zhenhua82 | 1.17 | 27 | Muping Dunhuasheng | 1.11 | 41 | Luhua10 | 1.08 | 55 | Shanhua No. 9 | 1.05 |
| 14 | Penglai Zaohuasheng | 1.17 | 28 | Zhaoyuan Banman | 1.11 | 42 | Minhua No. 10 | 1.08 | 56 | Xixuan21 | 1.05 |
| 57 | Baisha101 | 1.05 | 81 | Huayu17 | 1.00 | 105 | Huayu8 | 0.95 | 129 | Zhonghua No. 15 | 0.89 |
| 58 | Cangshan Siguozi | 1.04 | 82 | Wendang Banman | 1.00 | 106 | Xiaopuyang | 0.95 | 130 | Yueyou86 | 0.88 |
| 59 | Laiyang Laobaoji | 1.04 | 83 | Qufu Doupeng | 1.00 | 107 | Ganhua No. 7 | 0.94 | 131 | Baihua | 0.88 |
| 60 | Huayu16 | 1.04 | 84 | Guihua771 | 1.00 | 108 | Yuhua11 | 0.93 | 132 | Yuhua9326 | 0.88 |
| 61 | Luhua No. 9 | 1.04 | 85 | Shanyou250 | 1.00 | 109 | Sishui Zhanyang | 0.93 | 133 | Quanhua551 | 0.88 |
| 62 | Ningyang Daliman | 1.04 | 86 | Qixia Buluoye Laobaoji | 1.00 | 110 | Heyou No. 4 | 0.93 | 134 | Yuhua No. 14 | 0.88 |
| 63 | Shuangji22 | 1.04 | 87 | Qingdao Mansheng dali | 0.99 | 111 | Yuanhua No. 8 | 0.93 | 135 | Puhua23 | 0.88 |
| 64 | Xuhua14 | 1.04 | 88 | Shuanghong No. 2 | 0.99 | 112 | Quanhua327 | 0.93 | 136 | Xuhua No. 5 | 0.86 |
| 65 | Huayu No. 32 | 1.04 | 89 | Qinglan No. 2 | 0.99 | 113 | Dongping Manhuasheng | 0.92 | 137 | Luhua12 | 0.85 |
| 66 | Hua 67 | 1.04 | 90 | Zhufeng No. 1 | 0.98 | 114 | Hongguan | 0.92 | 138 | Guihua35 | 0.84 |

| No. | Variety | Value | No. | Variety | Value | No. | Variety | Value | No. | Variety | Value |
|---|---|---|---|---|---|---|---|---|---|---|---|
| 67 | Pingdu Dalidun | 1.03 | 91 | Huayu No. 36 | 0.98 | 115 | Hua17 | 0.92 | 139 | Yueyou25 | 0.83 |
| 68 | Quanhua10 | 1.02 | 92 | Xuhua 13 | 0.98 | 116 | Zhenzhuhei | 0.92 | 140 | Xuhua15 | 0.82 |
| 69 | Laiwu Paman | 1.02 | 93 | Qinglai No. 8 | 0.98 | 117 | Qixia Daliman | 0.91 | 141 | Yuanza9307 | 0.82 |
| 70 | Kainong37 | 1.02 | 94 | Luhua No. 2 | 0.98 | 118 | Zhonghua No. 4 | 0.91 | 142 | Baisha1016 | 0.81 |
| 71 | Lufeng No. 2 | 1.02 | 95 | Guihua95 | 0.98 | 119 | Qingdao Liman | 0.91 | 143 | Yueyou551 | 0.81 |
| 72 | Yueyou92 | 1.02 | 96 | Silihong | 0.97 | 120 | Zhenzhuhong | 0.91 | 144 | Yuanza9102 | 0.80 |
| 73 | Pingdu Paman | 1.02 | 97 | Fenghua5 | 0.97 | 121 | Huayu20 | 0.91 | 145 | Yueyou45 | 0.79 |
| 74 | Xiangxiang | 1.02 | 98 | Huayu23 | 0.97 | 122 | Hua 93 | 0.90 | 146 | Wucai Huasheng | 0.78 |
| 75 | Yishui Dapaman | 1.01 | 99 | Zhengnong7 | 0.97 | 123 | Yuhua9327 | 0.90 | 147 | Huayu28 | 0.77 |
| 76 | Yexian Yiwohou | 1.01 | 100 | Jihua2 | 0.97 | 124 | Haiyu No. 6 | 0.90 | 148 | Huayu31 | 0.76 |
| 77 | Xuhua 3 | 1.01 | 101 | Wenshang Mansheng | 0.96 | 125 | Dingtao Banmanyang Huasheng | 0.90 | 149 | Fenghua No. 6 | 0.73 |
| 78 | Zhonghua No. 8 | 1.01 | 102 | Xintai Zhanyang | 0.96 | 126 | Zhenghua5 | 0.90 | 150 | Luhua15 | 0.72 |
| 79 | Luhua14 | 1.01 | 103 | Zhongnong108 | 0.96 | 127 | Heyou No. 11 | 0.89 | 151 | Zhonghua No. 4 | 0.69 |
| 80 | Yueyou39 | 1.01 | 104 | Lanhua No. 2 | 0.96 | 128 | Hei husheng | 0.89 | 152 | Xianghua509-77 | 0.68 |

**TABLE 2.6 Criteria of Each Index**

| Index | | Grade I | Grade II | Grade III |
|---|---|---|---|---|
| Pod shape– | Categorical value | ≤3.10 | 3.10–6.69 | ≥6.69 |
| | Score | 17 | 11 | 5 |
| Crude protein + | Categorical value | ≥27.42 | 24.27–27.42 | ≤24.27 |
| | Score | 21 | 15 | 9 |
| Arginine + | Categorical value | ≥3.70 | 3.00–3.70 | ≤3.00 |
| | Score | 16 | 11 | 6 |
| Conarachin I+ | Categorical value | ≥29.37 | 23.82–29.37 | ≤23.82 |
| | Score | 23 | 16 | 9 |
| 23.5 kDa– | Categorical value | ≤20.80 | 20.80–23.72 | ≥23.72 |
| | Score | 23 | 16 | 9 |

Note: " + " indicates that the higher the value, the greater the influence; "–" indicates that the lower the value, the greater the influence

2. The Final Scores and Categorization of Each Variety
   The sum of the scores of each character index is deemed as the final scores of each variety, and the final scores of each variety are categorized into three classes, that is, Grade I (suitable), Grade II (basically suitable), and Grade III (unsuitable) (Table 2.7).
3. Determination of Peanut Varieties Suitable for Gel-Type Protein Production
   The above evaluation standards are adopted to evaluate the gelation of 152 peanut varieties, and the peanut varieties suitable for gel-type protein production include 31 varieties, namely, "Quanhua 464," "Luhua 11," "Shuangji No. 2," "Bianhua No. 3," "Fenghua No. 1," "Kainong 30," "Fenghua 3," "Zhongkaihua No. 4," "Huaguanwang," "Shanhua 7," "Minhua No. 9," "Zhanhua 82," "Yuhua 15," "Haihua 1," and "Guihua 166."

## 2.4 QUALITY EVALUATION TECHNOLOGY FOR PEANUTS SUITABLE FOR HIGH-SOLUBLE PROTEIN PRODUCTION

The protein solubility refers to the soluble performance of a protein in water or salt solution, and it is closely related to the protein composition and structure. In the meantime, the solubility of protein has a direct influence on the quality of the processed products. Wang et al. (2012b) studied the relationship between the quality characteristics of peanut varieties and protein solubility, and adopted the method of supervised principal component regression analysis to establish the quality evaluation model for the high-solubility protein suitable for drinks production. Through this model, the high-soluble protein of the unknown peanut varieties can be predicted so as to find a special processing variety with good solubility, to improve product quality and increase enterprise efficiency.

**TABLE 2.7** Grading of Assignment Values of the Varieties

| Category | Criteria | Number of Samples | Name of Samples |
|---|---|---|---|
| Suitable | ≥74.5 | 13 | Luhua 11, Shuangji No. 2, Muhua No. 3, Bianhua No. 3, Fenghua No. 1, Honghua No. 1, Luhua No. 14, Kainong 30, Zhanhua 82, Minhua 82, Yueyou No. 14, Yuhua 15, Ji 9814, Shanhua No. 7. |
| Basically suitable | 64.5–74.5 | 32 | Huayu No. 6, Shanhua No. 9, Shanyou 250, Zhenzhuhong, Yuanhua No. 8, 034-256-1, Lufeng No. 2, Hongguan, Zhonghua No. 4, Fenghua No. 6, Haihua 1, Silihong, Quanhua 551, Huayu 16, Baihuasheng, Heihuasheng, Zhongnong 108, Qinglan No. 8, Yuhua 9326, Yuhua 9327, Guihua 771, Xuhua 13, Yueyou No. 11, Heyou No. 11, Kainong 37, Luhua No. 9, Hua 17, Huayu 20, Zhonghua No. 8, Huayu 8, Longhua 243 |
| Unsuitable | ≤64.5 | 16 | Yueyou 45, Fenghua 3, Huayu 23, Huayu 28, Baisha 1016, Xuhua No. 5, Xuhua15, Yuanze 9102, Zhonghua No. 15, Luhua 15, Zhengnong No. 7, Huayu 31, Xinaghua 509-77, Wucai Huasheng, Yuanza 9307, Fenghua 4 |

## 2.4.1　Determination of Quality Index

A total of 70 indexes are determined, including seven sensory quality indexes, 57 physicochemical and nutritional quality indexes, and six processing quality indexes of peanuts (Table 2.1).

## 2.4.2　Determination of Solubility

1. Selection of Peanut Quality Evaluation Index
   Significance of regression analysis is carried out for the protein solubility and peanut quality (Table 2.8), and the results reveal that 15 quality indexes have significant correlation with solubility at the level of 0.05.
2. Peanut Quality Correlation Analysis
   Correlation analysis is carried out for the peanut protein solubility and other indexes of peanut protein (Table 2.9). It can be seen that the protein solubility has an insignificant correlation with other protein components, such as ash, crude fat, crude fiber, and protein purity, and therefore, other protein

**TABLE 2.8 Significance of Regression Coefficient Between Solubility and Peanut Quality**

| No. | Index | Regression Coefficient | No. | Index | Regression Coefficient | No. | Index | Regression Coefficient |
|---|---|---|---|---|---|---|---|---|
| 1 | Crude fat | 0.001 | 6 | Arginine | 0.044 | 11 | 18 kDa | 0.002 |
| 2 | Crude protein | 0.007 | 7 | Conarachin I | 0.007 | 12 | 17 kDa | 0.023 |
| 3 | Total carbohydrate | 0.020 | 8 | 40.5 kDa | 0.023 | 13 | 15.5 kDa | 0.001 |
| 4 | Cystine | 0.002 | 9 | 37.5 kDa | 0.035 | 14 | Protein extraction rate | 0.010 |
| 5 | Methionine | 0.004 | 10 | 23.5 kDa | 0.025 | 15 | Pure kernel yield | 0.030 |

**TABLE 2.9** Correlation Between Solubility and Protein Quality

|  | Ash | Crude Fat | Crude Fiber | Protein Purity | Solubility |
|---|---|---|---|---|---|
| Ash | 1.000 | | | | |
| Crude fat | 0.043 | 1.000 | | | |
| Crude fiber | 0.024 | 0.016 | 1.000 | | |
| Protein purity | −0.307* | −0.309* | −0.006 | 1.000 | |
| Solubility | −0.196 | 0.111 | −0.126 | 0.218 | 1.000 |

Note: "*" indicates significance above the level of $P<0.05$.

components will not be analyzed. Therefore, we only take the protein solubility as the analysis index of soluble protein.

3. Establishment of Model

The dependent variable for establishing the regression equation should conform to normal distribution. Since the solubility does not conform to normal distribution, Box-Cox transformation is made between each principal component and solubility to enable the dependent variables to conform to normal distribution before carrying out regression analysis (Table 2.10).

According to the regression analysis, the regression coefficient between each index and solubility is significant at the level of 0.05. The regression equation for each index and solubility is established to obtain Eq. (2.2).

$$\text{Solubility} = (1.490\,17 \times \text{crude fat} - 3.377\,5 \times \text{Cystine} - 0.390\,96 \times \text{Conarachin I} + 6.016\,274 \times 15.5\ \text{kDa} + 266.736\,6)^{\frac{4}{5}} \quad (2.2)$$

To investigate the accuracy and promotional suitability of the model, 21 varieties are adopted for verification. The outlier analysis indicates that "Zhongnong 108" is the outlier of solubility, and therefore, the remaining 20

**TABLE 2.10** Significance of Regression Coefficient Between Solubility and Peanut Quality

| No. | Index | Regression | No. | Index | Regression |
|---|---|---|---|---|---|
| 1 | Crude fat | 0.001 | 5 | Conarachinl | 0.007 |
| 2 | Crude protein | 0.007 | 6 | 18 kDa | 0.002 |
| 3 | Cystine | 0.002 | 7 | 15.5 kDa | 0.001 |
| 4 | Methionine | 0.004 | 8 | Protein extraction rate | 0.010 |

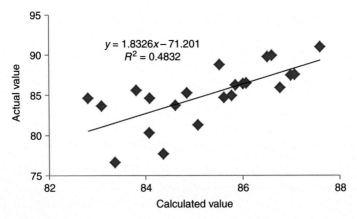

**FIGURE 2.2**   Relational diagram between original value of solubility and calculated value of actual application model.

varieties are used for analysis. Four indexes, namely, the crude protein, cysteine, conarachin I, and 15.5 kDa 4 are substituted into Eq. (2.2) to calculate the solubility of 20 varieties, and regression analysis is carried out between the calculation results of the model and the determined results, and the relational coefficient of the two is 0.699 (Fig. 2.2). Therefore, this method is adopted as the quality evaluation model for the peanuts suitable for soluble protein production.

4. Prediction of Solubility
   The above model is adopted to predict the protein solubility of 86 peanut varieties in China, and Table 2.11 shows the protein solubility of 152 varieties, including predicated varieties and modeling varieties.

### 2.4.3   Determination of Peanut Varieties Suitable for Soluble Protein Production

1. Evaluation Criteria
   a. Index Evaluation Criteria
      K-means cluster analysis is carried out respectively for four qualities of 54 peanut varieties, and each index is categorized into three classes, that is, Grade I (suitable), Grade II (basically suitable), and Grade III (unsuitable) (Table 2.12), and the weighted value of each index as determined in Section 2.3.2 is deemed as its highest score, that is, Class I, and the score is assigned to the index of each grade with the same method.
   b. Final Categorization of the Varieties
      The sum of the scores of each character index is deemed to be the final scores of each variety, and the final scores of each variety are categorized into three classes, that is, Grade I (suitable), Grade II (basically suitable), and Grade III (unsuitable) (Table 2.13). Through the comparative analysis

**TABLE 2.11 Solubility Ranking and Value of 152 Peanut Varieties**

| Ranking | Name of Variety | Solubility | Ranking | Name of Variety | Solubility | Ranking | Name of Variety | Solubility | Ranking | Name of Variety | Solubility |
|---|---|---|---|---|---|---|---|---|---|---|---|
| 1 | Shanhua No. 7 | 93.93 | 14 | Jimo Paman | 89.41 | 27 | Yueyou86 | 88.51 | 40 | Guihua166 | 87.37 |
| 2 | Yuhua9326 | 93.44 | 15 | Lanhua No. 2 | 89.18 | 28 | Heihuasheng | 88.45 | 41 | Qingdao Paman | 87.31 |
| 3 | Baisha1016 | 93.02 | 16 | Yuhua9327 | 89.03 | 29 | Dingdao Ban manyang Huasheng | 88.43 | 42 | Linhua No. 1 | 87.31 |
| 4 | Yhua15 | 91.72 | 17 | Laiwu Paman | 88.93 | 30 | Laiyang Banzhanman Dahuasheng D1-122 | 88.23 | 43 | Huayu No. 25 | 87.30 |
| 5 | Wucai Huasheng | 91.55 | 18 | Yuhua11 | 88.91 | 31 | Hua37 | 88.14 | 44 | Luhua12 | 87.26 |
| 6 | Ningyang Damanli | 91.20 | 19 | Qixia Buluoye Laobaoji | 88.84 | 32 | Yuanhua No. 8 | 87.97 | 45 | Weihua No. 6 | 87.23 |
| 7 | Hua67 | 91.19 | 20 | Wenshang Mansheng | 88.79 | 33 | Shuangji No. 2 | 87.72 | 46 | Jihua2 | 87.17 |
| 8 | Baihuasheng | 90.68 | 21 | Luhua No. 9 | 88.76 | 34 | Zhenzhuhei | 87.55 | 47 | Qingdao Mansheng Dali | 87.08 |
| 9 | Luhua15 | 89.97 | 22 | Laiyang Laobaoji | 88.71 | 35 | Huayu No. 33 | 87.51 | 48 | Minhua No. 10 | 87.05 |

(Continued)

**TABLE 2.11 Solubility Ranking and Value of 152 Peanut Varieties (cont.)**

| Ranking | Name of Variety | Solubility | Ranking | Name of Variety | Solubility | Ranking | Name of Variety | Solubility | Ranking | Name of Variety | Solubility |
|---|---|---|---|---|---|---|---|---|---|---|---|
| 10 | Luhua8 | 89.96 | 23 | Luhua11 | 88.65 | 36 | Xintai Zhanyang | 87.47 | 49 | Xiaopuyang | 87.03 |
| 11 | Guihua35 | 89.68 | 24 | Pingdu Pman | 88.62 | 37 | Zhufeng No. 1 | 87.44 | 50 | Luhua10 | 86.96 |
| 12 | Baisha101 | 89.53 | 25 | Zhencheng Layang | 88.57 | 38 | Kunyu Dalidun | 87.38 | 51 | Luhua10 | 86.94 |
| 13 | Yueyou25 | 89.41 | 26 | Zhaoyuan Banman | 88.51 | 39 | Laiyang Banman | 87.38 | 52 | Xuhua14 | 86.84 |
| **Ranking** | **Name of Variety** | **Solubility** | **Ranking** | **Name of Variety** | **Solubility** | **Ranking** | **Name of Variety** | **Solubility** | **Ranking** | **Name of Variety** | **Solubility** |
| 1 | Shanhua No. 7 | 93.93 | 14 | Jimo Paman | 89.41 | 27 | Yueyou86 | 88.51 | 40 | Guihua166 | 87.37 |
| 2 | Yuhua9326 | 93.44 | 15 | Lanhua No. 2 | 89.18 | 28 | Heihuasheng | 88.45 | 41 | Qingdao Paman | 87.31 |
| 3 | Baisha1016 | 93.02 | 16 | Yuhua9327 | 89.03 | 29 | Dingdao Ban manyang Huasheng | 88.43 | 42 | Linhua No. 1 | 87.31 |
| 4 | Yhua15 | 91.72 | 17 | Laiwu Paman | 88.93 | 30 | Laiyang Banzhanman Dahuasheng D1-122 | 88.23 | 43 | Huayu No. 25 | 87.30 |

| No. | Variety | Value | No. | Variety | Value | No. | Variety | Value | No. | Variety | Value |
|---|---|---|---|---|---|---|---|---|---|---|---|
| 5 | Wucai Huasheng | 91.55 | 18 | Yuhua11 | 88.91 | 31 | Hua37 | 88.14 | 44 | Luhua12 | 87.26 |
| 6 | Ningyang Damanli | 91.20 | 19 | Qixia Buluoye Laobaoji | 88.84 | 32 | Yuanhua No. 8 | 87.97 | 45 | Weihua No. 6 | 87.23 |
| 7 | Hua67 | 91.19 | 20 | Wenshang Mansheng | 88.79 | 33 | Shuangji No. 2 | 87.72 | 46 | Jihua2 | 87.17 |
| 8 | Baihuasheng | 90.68 | 21 | Luhua No. 9 | 88.76 | 34 | Zhenzhuhei | 87.55 | 47 | Qingdao Mansheng Dali | 87.08 |
| 9 | Luhua15 | 89.97 | 22 | Laiyang Laobaoji | 88.71 | 35 | Huayu No. 33 | 87.51 | 48 | Minhua No. 10 | 87.05 |
| 10 | Luhua8 | 89.96 | 23 | Luhua11 | 88.65 | 36 | Xintai Zhanyang | 87.47 | 49 | Xiaopuyang | 87.03 |
| 11 | Guihua35 | 89.68 | 24 | Pingdu Pman | 88.62 | 37 | Zhufeng No. 1 | 87.44 | 50 | Luhua10 | 86.96 |
| 12 | Baisha 101 | 89.53 | 25 | Zhencheng Layang | 88.57 | 38 | Kunyu Dalidun | 87.38 | 51 | Luhua 10 | 86.94 |
| 13 | Yueyou25 | 89.41 | 26 | Zhaoyuan Banman | 88.51 | 39 | Laiyang Banman | 87.38 | 52 | Xuhua14 | 86.84 |
| 53 | Huayu No. 36 | 86.81 | 78 | Qixia Bankangpi | 85.88 | 103 | Qixia Laobaoji | 84.82 | 128 | Quanhua10 | 81.92 |
| 54 | Haihua1 | 86.79 | 79 | Xiangxiang | 85.85 | 104 | Pingdu Dalidun | 84.75 | 129 | Yueyou92 | 81.66 |

(Continued)

**TABLE 2.11 Solubility Ranking and Value of 152 Peanut Varieties (cont.)**

| Ranking | Name of Variety | Solubility | Ranking | Name of Variety | Solubility | Ranking | Name of Variety | Solubility | Ranking | Name of Variety | Solubility |
|---|---|---|---|---|---|---|---|---|---|---|---|
| 55 | Shuangji22 | 86.70 | 80 | Laiyang Xuefang Banman | 85.84 | 105 | Huayu No. 32 | 84.64 | 130 | Huayu20 | 80.93 |
| 56 | Pingdu Lijing | 86.60 | 81 | Huayu8 | 85.80 | 106 | Quanhua327 | 84.63 | 131 | Fenghua No. 1 | 80.88 |
| 57 | Shanhua No. 9 | 86.56 | 82 | Xixuan 21 | 85.75 | 107 | Zhonghua No. 4 | 84.62 | 132 | Minhua No. 9 | 80.80 |
| 58 | Haiyu No. 6 | 86.53 | 83 | Zhongkaihua No. 4 | 85.71 | 108 | Yuanza9102 | 84.56 | 133 | Yueyou 52 | 80.66 |
| 59 | Qufu Doupeng | 86.49 | 84 | Xuhua 15 | 85.64 | 109 | Luhua No. 2 | 84.33 | 134 | Zheng-nong7 | 80.30 |
| 60 | 034-256-1 | 86.47 | 85 | Zhonghua8 | 85.63 | 110 | Hua 27 | 84.25 | 135 | Long-hua243 | 79.96 |
| 61 | Quanhua 551 | 86.45 | 86 | Qinglan No. 2 | 85.62 | 111 | Yueyou 14 | 84.25 | 136 | Shany-ou250 | 79.95 |
| 62 | Cangshan Siguozi | 86.42 | 87 | Heyou No. 4 | 85.60 | 112 | Fenghua No. 6 | 84.21 | 137 | Ji9814 | 79.60 |
| 63 | Qinglan No. 8 | 86.41 | 88 | Puhua23 | 85.55 | 113 | Yueyou 39 | 84.15 | 138 | Fenghua3 | 77.76 |
| 64 | Juye Tuoyang | 86.41 | 89 | Yexian Yiwohou | 85.53 | 114 | Lufeng No. 2 | 84.12 | 139 | Kainong30 | 77.71 |
| 65 | Zhenghua 5 | 86.40 | 90 | Linqing Yiwohou | 85.44 | 115 | Xuhua 13 | 83.95 | 140 | Huayu 28 | 77.70 |

| 66 | Wendang Banman | 86.35 | 91 | Penglai Zaohuasheng | 85.43 | 116 | Huayu16 | 83.75 | 141 | Yueyou40 | 77.65 |
|---|---|---|---|---|---|---|---|---|---|---|---|
| 67 | Laiyang Dunhuasheng | 86.33 | 92 | Sishui Zhanyang | 85.35 | 117 | Huaguanwang | 83.71 | 142 | Yueyou45 | 77.54 |
| 68 | Hua93 | 86.22 | 93 | Qingdao Liman | 85.34 | 118 | Laixi Liman | 83.69 | 143 | Heyou No. 11 | 77.35 |
| 69 | Bianhua3 | 86.20 | 94 | Hua17 | 85.28 | 119 | Hongguan | 83.60 | 144 | Huayu31 | 77.21 |
| 70 | Yishui Dapaman | 86.17 | 95 | Huayu17 | 85.24 | 120 | Hua55 | 83.60 | 145 | Huayu22 | 76.78 |
| 71 | Dongping Manhuasheng | 86.13 | 96 | Honghua No. 1 | 85.03 | 121 | Silihong | 83.48 | 146 | Yuan-za9307 | 76.66 |
| 72 | Xianghua 509-77 | 86.12 | 97 | Tianfu No. 4 | 84.95 | 122 | Yueyou551 | 83.39 | 147 | Xuhua No. 5 | 76.60 |
| 73 | Xuhua3 | 86.08 | 98 | Ganhua No. 7 | 84.94 | 123 | Quanhua464 | 82.82 | 148 | Huayu19 | 75.11 |
| 74 | Muping Yiwohou | 86.03 | 99 | Zhonghua No. 15 | 84.94 | 124 | Huayu23 | 82.65 | 149 | Zhen-zhuhong | 74.97 |
| 75 | Qixia Daliman | 86.01 | 100 | Yuhua No. 14 | 84.93 | 125 | Fenghua4 | 82.34 | 150 | Zhanhua82 | 73.61 |
| 76 | Zhong-nong108 | 85.99 | 101 | Muping Dunhuasheng | 84.85 | 126 | Guihua95 | 82.23 | 151 | Guihua 771 | 72.84 |
| 77 | Shuanghong No. 2 | 85.89 | 102 | Luhua14 | 84.84 | 127 | Kainong37 | 81.97 | 152 | Fenghua5 | 72.63 |

**TABLE 2.12 Criteria of Each Index**

| Index | | Grade I | Grade II | Grade III |
|---|---|---|---|---|
| Crude protein + | Categorical value | ≥27.58 | 24.49–27.58 | ≤24.49 |
| | Score | 18 | 12 | 6 |
| Cysteine − | Categorical value | ≤0.48 | 0.48–0.85 | ≥0.85 |
| | Score | 27 | 18 | 9 |
| Conarachin I− | Categorical value | ≤23.49 | 23.49–29.69 | ≥29.69 |
| | Score | 31 | 20 | 9 |
| 15.5 kDa− | Categorical value | ≤5.78 | 5.78–7.60 | ≥7.60 |
| | Score | 25 | 17 | 9 |

Note: " + " indicates that the higher the value, the greater the influence; "−" indicates that the lower the value, the greater the influence

**TABLE 2.13 Grading of Assigned Values of the Varieties**

| Category | Criteria of Categorization | Number of Samples | Name of Samples |
|---|---|---|---|
| Suitable | ≥74 | 18 | Yueyou86, Yuhua9326, Yuanhua No. 8, Wucai Huasheng, Yuhua 9327, Xuhua 14, Fenghua No. 3, Huayu 28, Ji 9814, Baihuasheng, Heihuasheng, Shuangji No. 2, Xuhua 15, Honghua No. 1, Huayu 16, Luhua 14, Qinghua No. 8, Huayu No. 8 |
| Basically suitable | 57–74 | 21 | Zhonghua No. 15, Zhengnong No. 7, Hua 17, Yuhua 15, Hongguan, Huayu 23, 034-256-1, Fenghua No. 1, Baisha 1016, Huaguanwang, Zhenzhuhong, Zhonghua No. 4, Luhua No. 9, Shanhua No. 9, Kainong 37, Xuhua 13, Huayu 30, Minhua No. 9, Lufeng No. 2, Xianghua 509-77, Fenghua 6 |
| Unsuitable | ≤57 | 15 | Haiyu No. 6, Luhua 15, Quanhua 551, Fenghua No. 5, Huayu 19, Guihua 771, Zhanhua 82, Heyou No. 11, Yuanze 9307, Xuhua No. 5, Huayu 22, Huayu 31, Kainong 30, Yueyou 40, Yueyou 45 |

of the results, it is found that the accuracy of suitability is 66.67%, the accuracy of basic suitability is 72%, and the accuracy of unsuitability is 80%.

2. Determination of Peanut Varieties Suitable for Soluble Protein Production
The above criteria are adopted to evaluate the solubility of 152 peanut varieties in China, and the peanut varieties suitable for soluble protein production include 50 peanut varieties, namely, "Shanhua No. 7," "Yuhua 9326," "Baisha 1016," "Yuhua 15," "Wucai Huasheng," "Hua 67," "Baihuasheng," "Luhua 15," "Luhua 8," "Guihua 35," "Baisha 101," "Yueyou 25," "Lanhua No. 2," "Yuhua 9327," and "Yuhua 11."

## 2.5 QUALITY EVALUATION TECHNOLOGY FOR PEANUTS SUITABLE FOR PEANUT OIL PRODUCTION

Characterized by high nutrition and a delightful odor, peanut oil is good for cooking and has become one of the major edible oils in China. With the improvement of people's lives, the quality of peanut oil has increasingly received a lot of attention, and the experts in the industry have become more and more interested in studies on the relationship between the quality of raw peanuts and peanut oil.

Worthington et al. (1972) determined the fatty acid composition and oil stability for 82 peanut varieties of different genotypes from the United States, and established the relationship between the stability of peanut oil and fatty acid composition. Yan (1998) studied the quantitative relationship between the oxidative stability index (OSI) and fatty acid composition (FAC): OSI(h) = 7.5123 + %C16:0 × (0.2733) + %C18:0 × (0.0797) + %C18:1 × (0.0159) + %C18:2 × (0.1141) + %C18:3 × (0.3962), where %C16:0 represents the C16:0 percentage in the fat. The quantitative relationship was adopted to predict the OSI value in soybean oil, corn oil, sunflower seed oil, rapeseed oil, and olive oil, and the results revealed that the error between the predictive value and the experimental value is less than 10%. The diversification of peanut varieties results in the great difference in peanut quality as well as the peanut oil produced from the peanuts. Currently, there are relatively few studies on the influence of the quality of different peanut varieties on the peanut oil produced from them. We can only find that Zhang et al. (2012) carried out studies on the relationship between the quality of 45 peanut varieties and the peanut oil produced from them, and established the relational model between the quality of raw peanuts and the peanut oil. The model will be helpful in predicting the quality difference of peanut oils produced with different peanut varieties, facilitating the production of excellent peanut oil, and laying down a basis for the selection of peanut varieties suitable for oil production.

### 2.5.1 Determination of Quality Index

A total of 70 indexes are determined, including seven sensory quality indexes, 57 physicochemical and nutritional quality indexes, and six processing quality indexes of peanuts (Table 2.1).

**TABLE 2.14** Relation Between Peanut Quality and Peanut Oil Quality

| No. | Index | Significance |
| --- | --- | --- |
| 1 | Crude fat | 0.004 |
| 2 | O/L ratio | 0.034 |
| 3 | Unsaturated fatty acids | 0.043 |

## 2.5.2 Establishment of Peanut Oil Quality Evaluation Model

1. The Peanut Quality Evaluation Selection
   The regression analysis method is adopted to carry out regression coefficient significance analysis for the individual indexes of peanut quality as well as the peanut oil quality, three indexes have significant correlation with gelation at the level of 0.01 (Table 2.14).
2. Transformation Peanut Oil Quality Indexes
   For some index values, the higher the value, the better the oil quality; however, for other index values, the lower the value, the better the oil quality. Therefore, for the convenience of the follow-up calculation, the evaluation index values of the peanut oil of the 34 varieties are all transformed so that the higher that index value, the better the oil quality, and the results are shown in Table 2.15.
   The quality of the peanut oil of 34 peanut varieties is transformed so that the higher the value, the better the quality; then a standard treatment is carried out so as to reduce the influence of unit differences on the data. The coefficient before each index in the quality evaluation model for peanuts suitable for oil production is used as the weighted value (Table 2.16).
3. Modeling
   The dependent variables for establishing the regression equation should conform to normal distribution, but the integrated value of peanut oil does not fit into normal distribution. Therefore, it is required to carry out Box-Cox conversion between each principal component and the integrated value of peanut oil to make the variables conform to normal distribution before carrying out regression analysis (Table 2.17).

$$Y_1 = 5.999\,998\,2 - 1.394\,673\,8 \times \text{standardized data of fatty acid content}$$
$$+ 0.647\,305\,4 \times \text{standardized data of O/L ratio} \tag{2.3}$$
$$- 1.930\,320\,4 \times \text{standardized data of unsaturated fatty acid content}$$

Change $Y_1$ in the equation (2.3) into $Y$, and convert the standardized data into original data to obtain $Y = -0.412\,547 \times$ crude fat $+ \quad$ (2.4)
$40.560138 \times$ oleic/linoleic $- 0.618\,986 \times$ Unsaturated fatty acids

**TABLE 2.15 Evaluation Index of Oil Quality of Different Peanut Varieties**

| No. | Name | Sensory Quality | | Physicochemical and Nutritional Quality | | Processing Quality | | | | |
| --- | --- | --- | --- | --- | --- | --- | --- | --- | --- | --- |
| | | Color | | Water and Volatile | Unsaponifiables | Induction Time | Peroxide Value | Acid Value | Iodine Value | Saponification Value |
| | | Red | Yellow | | | | | | | |
| 1 | Zhonghua No. 8 | −0.65 | −11.00 | −0.125 | 4.24 | 4.19 | 3.2310 | 0.411 | 88.13 | 177.32 |
| 2 | Shanhua No. 7 | −0.40 | −3.50 | −0.045 | 19.76 | 5.76 | −2.367 | −0.358 | −96.87 | −176.17 |
| 3 | Silihong | −0.60 | −2.80 | −0.080 | 12.10 | 5.19 | −2.972 | −0.280 | −97.62 | 170.64 |
| 4 | Luhua11 | −0.70 | −5.50 | −0.105 | 9.74 | 4.67 | −3.081 | −0.273 | −93.15 | 175.63 |
| 5 | Bianhua No. 3 | −0.90 | −9.25 | −0.085 | 13.23 | 4.61 | −2.201 | −0.643 | −93.21 | 148.47 |
| 6 | Haihua1 | −0.65 | −4.40 | −0.090 | 10.24 | 5.23 | −4.040 | −0.389 | −90.17 | 177.33 |
| 7 | Shuangji No. 2 | −0.40 | −5.35 | −0.050 | 9.67 | 4.45 | 2.242 | −0.271 | −92.62 | −170.97 |
| 8 | Shanhua No. 9 | −0.85 | −3.35 | −0.145 | 11.70 | 4.39 | −2.699 | −0.364 | −98.06 | −174.84 |
| 9 | Fenghua 5 | −0.45 | −4.30 | −0.200 | 17.91 | 3.74 | −6.131 | −2.574 | −93.14 | 176.72 |
| 10 | Baihuasheng | −0.60 | −3.95 | −0.080 | 10.95 | 4.62 | −3.281 | −0.328 | −97.24 | 175.48 |
| 11 | Fenghua No. 1 | −0.60 | −4.40 | −0.040 | 8.95 | 3.91 | −4.573 | −0.307 | −96.24 | −175.40 |
| 12 | Fenghua3 | −0.70 | −6.50 | −0.080 | 6.08 | 3.91 | −4.953 | −0.235 | −91.78 | −177.17 |
| 13 | Fenghua4 | −1.70 | −13.00 | −0.080 | 9.14 | 4.37 | −3.203 | −0.563 | −88.68 | 177.17 |
| 14 | Xuhua No. 5 | −0.70 | −5.40 | −0.070 | 7.70 | 3.81 | −3.428 | −0.429 | −93.03 | −166.78 |
| 15 | Yuanhua No. 8 | −0.80 | −6.50 | −0.090 | 9.00 | 3.73 | −3.717 | −0.361 | −100.91 | −186.26 |
| 16 | Xuhua13 | −0.65 | −7.35 | −0.050 | 11.28 | 3.82 | −4.269 | −0.362 | −101.69 | −188.27 |
| 17 | Xuhua 14 | −0.70 | −6.40 | −0.055 | 11.57 | 4.27 | −2.326 | −0.230 | −101.84 | −186.26 |

*(Continued)*

**TABLE 2.15 Evaluation Index of Oil Quality of Different Peanut Varieties (cont.)**

| No. | Name | Sensory Quality | | Physicochemical and Nutritional Quality | | Processing Quality | | | | |
|---|---|---|---|---|---|---|---|---|---|---|
| | | Color | | | | | | | | |
| | | Red | Yellow | Water and Volatile | Unsaponifiables | Induction Time | Peroxide Value | Acid Value | Iodine Value | Saponification Value |
| 18 | Huayu 19 | -0.55 | -3.35 | -0.070 | 11.13 | 4.71 | -3.069 | -0.381 | -98.63 | -183.92 |
| 19 | Huayu20 | -0.45 | -3.25 | -0.050 | 10.66 | 4.63 | -3.232 | -0.406 | -98.26 | -181.98 |
| 20 | Huayu22 | -0.60 | -5.70 | -0.040 | 8.01 | 4.79 | -2.192 | -0.668 | -94.15 | -186.53 |
| 21 | Huayu23 | -0.60 | -6.20 | -0.075 | 6.34 | 5.63 | -3.774 | -1.048 | -99.14 | -183.26 |
| 22 | Huayu28 | -0.60 | -13.00 | -0.065 | 9.70 | 3.94 | -2.892 | -0.526 | -92.50 | -181.51 |
| 23 | Huayu31 | -0.60 | -5.35 | -0.100 | 13.61 | 4.44 | -2.724 | -0.622 | -100.53 | -185.63 |
| 24 | Baisha1016 | -0.55 | -5.30 | -0.055 | 18.83 | 4.36 | -4.076 | -0.540 | -108.98 | -187.57 |
| 25 | Wucai Huasheng | -0.80 | -5.20 | -0.080 | 10.05 | 4.97 | -2.276 | -0.356 | -103.60 | -181.13 |
| 26 | Heihuasheng | -0.55 | -5.10 | -0.075 | 17.21 | 4.81 | -1.686 | -0.477 | -103.95 | -185.91 |
| 27 | 034-256-1 | -0.70 | -6.45 | -0.060 | 9.47 | 5.23 | -1.613 | -3.020 | -96.15 | -203.69 |
| 28 | Ji 9814 | -0.80 | -10.45 | -0.105 | 11.76 | 4.63 | -2.727 | -0.388 | -108.61 | -190.03 |
| 29 | Yuhua15 | -0.70 | -5.50 | -0.085 | 8.22 | 3.59 | -0.806 | -1.201 | -107.34 | -217.13 |
| 30 | Yuhua9326 | -0.55 | -4.35 | -0.080 | 10.66 | 4.04 | -3.411 | -0.375 | -116.14 | -179.48 |
| 31 | Yuhua9327 | -0.90 | -6.05 | -0.070 | 12.23 | 4.31 | -1.413 | -0.372 | -98.60 | -203.19 |
| 32 | Kainong30 | -0.50 | -9.90 | -0.045 | 10.02 | 4.05 | -3.128 | -0.377 | -107.07 | -188.67 |
| 33 | Kainong37 | -0.80 | -4.00 | -0.105 | 17.32 | 4.49 | -2.454 | 0.422 | -113.39 | -193.59 |
| 34 | Yuanza9102 | -0.50 | -5.40 | -0.045 | 18.65 | 3.70 | -1.037 | -0.316 | -103.25 | -186.92 |

**TABLE 2.16** Weighted Value of Each Index

| No. | Index | Coefficient | Weighted Value |
|---|---|---|---|
| 1 | Crude fat | 1.394 673 8 | 35 |
| 2 | O/L ratio | 0.647 305 4 | 16 |
| 3 | Unsaturated fatty acids | 1.930 320 4 | 49 |

**TABLE 2.17** Table of Regression Coefficient

| Variable | Coefficient | F Value | P Value |
|---|---|---|---|
| Coefficient | 5.999 9982 | 173.27 | 0.000 10 |
| Crude fat $X_1$ | 1.3946738 | 8.35 | 0.007 20 |
| O/L ratio $X_2$ | 0.6473054 | 1.87 | 0.018 22 |
| Unsaturated fatty acids $X_3$ | 1.9303204 | 16.45 | 0.000 30 |

The crude fat, O/L ratio, and unsaturated fatty acid content, as well as the peanut oil quality of the other 11 peanut varieties were determined. The peanut oil qualities of the 11 varieties are transformed so that the higher the value, the better the quality. These data are standardized, equally weighted, and added, and the added result is $Y_3$. Then, three indexes, namely, crude fat, O/L ratio, and unsaturated fatty acid content, are substituted into Eq. (2.4) to obtain the calculated value $Y_2$, and the suitability of the equation is determined by analyzing the correlation coefficient of $Y_3$ and $Y_2$. The two values are then fitted, and the fitting results are as shown in Fig. 2.3, and the correlation coefficient $R = 0.70$ of the two can be used to predict the integrated value of any kind of peanut.

## 2.5.3 Determination of Peanut Varieties Suitable for Peanut Oil Production

1. Evaluation Index Standard

   K-means cluster analysis is carried out respectively for three indexes of 45 peanut varieties for model establishment and verification, and each index is categorized into three classes according to Eqs (2.3) and (2.4), that is, Grade I (suitable), Grade II (basically suitable), and Grade III (unsuitable), and the weighted value of each index is deemed as its highest score (Table 2.18), that is, Class I, and the score is assigned to the index of each grade with the same method.

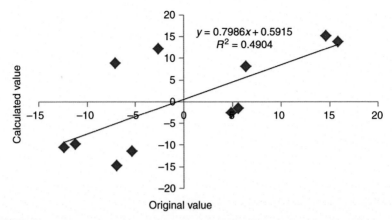

**FIGURE 2.3**    Relational diagram between actual integrated value of peanut oil and theoretical value.

**TABLE 2.18 Evaluation Index Standard**

| Index | | Grade I | Grade II | Grade III |
|---|---|---|---|---|
| Crude fat+ | Categorical value | ≥53.68 | 47.70–53.68 | ≤47.70 |
| | Score | 35 | 25 | 15 |
| O/L ratio+ | Categorical value | ≥1.48 | 1.00–1.48 | ≤1.00 |
| | Score | 16 | 11 | 6 |
| Unsaturated fatty acids− | Categorical value | ≤37.63 | 37.63–44.66 | ≥44.66 |
| | Score | 49 | 34 | 19 |

Note: " + " indicates that the higher the value, the greater the influence; "−" indicates that the lower the value, the greater the influence

2. Categorization of Final Score of Each Variety

K-means cluster analysis (Table 2.19) is carried out for the sum of the assigned value of each variety, and the comparison of the results indicates that the accuracy of suitability is 90%, the accuracy of basic suitability is 63.6%, and the accuracy of unsuitability is 78.6%. These data can be used as the quality evaluation criteria for peanuts used for oil production.

3. Determination of Peanut Varieties Suitable for Oil Production

The above standard is adopted to evaluate the oil production suitability of 45 peanut varieties and the peanut varieties suitable for oil production include 20 varieties, namely, "Yuanza 9102," "Luhua No. 9," "Xuhua 14," "Haihua 1," "Lufeng No. 2," "Yuhua 9326," "Baisha 1016," "Yuhua 15," "Shuangji No. 2," "Shanhua No. 7," "Zhonghua No. 15," "Luhua 11," "Bianhua No. 3," "Huayu 31," "Zhongnong 108," "Baihuasheng," "Xuhua 15," "Huayu 19," and "Zhonghua No. 8."

**TABLE 2.19** Suitability of Peanuts Used for Oil Production

| Category | Criteria | Number of Samples | Name of Samples |
|---|---|---|---|
| Suitable | ≥80 | 19 | Yuanza 9102, Luhua No. 9, Xuhua 14, Haihua 1, Lufeng No. 2, Yuhua 9326, Baisha 1016, Yuhua 15, Shuangji No. 2, Shanhua No. 7, Zhonghua No. 15, Luhua 11, Bianhua No. 3, Huayu 31, Zhongnong 108, Baihuasheng, Xuhua 15, Huayu 19, Zhonghua No. 8 |
| Basically suitable | 70–80 | 11 | Huayu 28, Huayu 23, Shanhua No. 9, Luhua 14, Kainong 37, Luhua 15, Huayu 20, Yuhua 9327, Zhonghua No. 4, Kainong 30, Fenghua No. 1 |
| Unsuitable | ≤70 | 13 | Silihong, 034-256-1, Huaguanwang, Xuhua No. 5, Huayu 22, Ji 9814, Fenghua No. 6, Yuanhua No. 8, Fenghua No. 4, Huayu 16, Fenghua No. 3, Fenghua No. 5, Xuhua 13 |

## 2.6   QUALITY EVALUATION TECHNOLOGY FOR PEANUTS FOR EXPORTS

Peanuts are an important oil-bearing crop in China and an important agricultural export. Over many years, the peanut and peanut products in China have played an important part in the worlds' peanut exports (Yang, 2009). According to the FAO statistics, from 2000 to 2005, the exports of shelled peanuts, peanuts in shells, and processed peanuts reached 29.8%, 33.9%, and 44.6%, respectively, of the total peanuts and peanut products in the world (Qu et al., 2008).

The peanut variety is closely related to the peanut quality. Since the peanut-importing countries and regions mainly use the peanuts for food processing instead of oil production, they often favor peanuts of particular varieties (Holley & Hammons, 1968; Worthington et al., 1972). In addition, with the world's economic development, the requirements of the peanut-importing countries regarding the hygiene, physiochemical, and nutritional indexes are gradually improving, and their attention on such indexes as size, color, and water content in the past is switching to such indexes as aflatoxin content, sugar content, and O/L ratio, etc., and they are placing higher and higher requirements on these indexes. This situation restricts the peanut export industry of China to a certain degree (Chen & Duan, 1994; Jiang et al., 1998).

Therefore, it is necessary to establish a quality evaluation model for exported peanut quality for the peanut kernels and peeled kernels so as to improve the quality control level of the raw peanut export enterprises and to promote the sound development of the peanut export industry (Chen et al., 2007a; Yin et al., 2011; Li et al., 2011).

## 2.6.1   Determination of Quality Indexes

There are a total of 70 indexes for peanut quality determination, including seven sensory quality indexes, 57 physicochemical and nutritional quality indexes, and six processing quality indexes (Table 2.1).

## 2.6.2   Predication of Suitability

1. Prediction of Peanuts Suitable for Export
   a. Selection of Peanut Quality Evaluation Index
      The results of regression analysis for the indexes and export peanut quality (Table 2.20) indicate that the three indexes, namely, 100-kernel weight, O/L ratio, and aflatoxin content, have significant relationships with the quality of peanuts for export at the level of 0.01, and it is suggested that these are important indexes influencing the quality of peanuts for export.
   b. Inter–Index Relationship
      Analysis of the correlation among the three indexes shows that the correlation coefficient of all these indexes is 0.2 or below (Table 2.21), indicating that the three indexes are mutually independent and can individually reflect their respective relationship with the quality of peanuts for export.
   c. Modeling
      A multiple linear regression equation is established by means of the stepwise multiple regression method, where the independent variables include 100-kernel weight $X_1$, kernel yield $X_2$, fat content $X_3$, protein content $X_4$, content of polysaccharides $X_5$, polysaccharide content $X_6$, O/L ratio $X_7$, water content $X_8$, unsound kernel $X_9$, and aflatoxin content $X_{10}$, and the dependent variable ($Y$) is the total score of the quality of the peanuts for export.

**TABLE 2.20 Correlation Analysis for 100-Kernel Weight, O/L Ratio, and Aflatoxin Content**

| Index | 100-Kernel Weight | O/L Ratio | Aflatoxin Content |
|---|---|---|---|
| 100-kernel weight | 1.000 | | |
| O/L ratio | 0.037 | 1.000 | |
| Aflatoxin content | 0.120 | 0.090 | 1.000 |

**TABLE 2.21** Correlation Analysis for 100-Kernel Weight, O/L Ratio, and Aflatoxin Content

| Index | 100-Kernel Weight | O/L Ratio | Aflatoxin Content |
|---|---|---|---|
| 100-kernel weight | 1.000 | | |
| O/L ratio | −0.037 | 1.000 | |
| Aflatoxin content | 0.120 | −0.090 | 1.000 |

The specific steps of stepwise multiple regression include the following: first, respectively calculate the contribution of each independent variable $X$ on the dependent variable $Y$, select the independent variable with the greatest contribution according to the descending order and substitute it into the equation, then recalculate the contribution of each independent variable $X$ to the dependent variable $Y$, and resubstitute it into the equation. In the meantime, investigate whether the variables in the equation lose statistical significance due to the introduction of new variables. If they do, delete the first dependent variable. Repeat the steps until there is no detectable variable in the equation and no variable that can be substituted into the equation. Then three indexes are retained, including 100-kernel weight, O/L ratio, and aflatoxin content, to obtain the regression equation.

$$Y = 26.291 + 0.234 \times X_1 + 14.609 \times X_7\, 2.964 \times X_{10} \qquad (2.5)$$

The effectiveness of the regression model for the peanuts for export is determined by the determination coefficient ($R^2$) and $F$ test. Therefore, the model summary (Table 2.22) and variance analysis (Table 2.23) are carried out. The square root ($R^2$) of the determination coefficient of the custom regression model above is 0.573, indicating a good fit between the model and the data. The statistics of the regression equation is $F = 22.366$, and the significance of the significance test results is 0.000, which is lower than 0.05, indicating a significant linear relationship of the multiple regression equation. It is obvious that it is effective to use the multiple linear regression model to predict the composite score of the quality of peanuts for exports.

**TABLE 2.22** Summary for Quality Model of Peanuts for Export

| Model | R | $R^2$ | Adjustment $R^2$ | Estimated Standard Error (%) |
|---|---|---|---|---|
| 1 | 0.757 | 0.573 | 0.547 | 5.599 71 |

**TABLE 2.23** Variance Analysis Table

| Model | Sum of Squares | DOF | Average Variance | F | Significance |
|---|---|---|---|---|---|
| Regression | 2103.959 | 3 | 701.320 | 22.366 | 0.000 |
| Residual | 1567.836 | 50 | 31.357 | | |
| Total | 3671.795 | 53 | | | |

Sixty-five varieties are adopted for modeling to obtain the model of three indexes (Eq. (2.6)), 10 varieties are adopted to verify the model, and it is shown that the correlation coefficient between the predicted value and the actual value is 0.87. It is indicated that this equation provides better prediction results, and this model can be used to determine the quality evaluation indexes for peanuts for export, to predict the peanut quality of unknown varieties, and to provide means for the processing enterprises to select peanut varieties suitable for export processing.

$$\text{Quality of peanuts for export} = 26.291 + 0.234 \times 100 - \text{kernel} \\ \text{weight} + 14.609 \times O/L \text{ ratio} - 2.964 \times \text{aflatoxin content} \tag{2.6}$$

d. Prediction of Export Suitability

The difference between the composite score for the peanuts for export and the predicted value of the model is only about 10%, the correlation coefficient is 0.839, and the significance is 0.002, indicating a highly significant correlation at the level of 0.01 (Table 2.24).

There is a degree of error between the predicted values of the 10 varieties and the actual composite score. The error for all the varieties is lower than 3%, except for "Yuanza 9307" with a relative error reaching 3.92%. This indicates that the regression prediction model for the quality of peanuts for export can effectively predict the quality of the known peanut varieties for export.

2. Prediction of Export Suitability of Peeled Kernels

a. Refer to 2.5.1 for peanut quality evaluation index selection and peanut quality evaluation index selection.

b. Correlation Analysis of Peanut Qualities

As shown in Table 2.25, it is known that there is a highly significant correlation between the 100-kernel weight and aflatoxin content and the composite score, while there is a slightly lower correlation between the O/L ratio and the composite score. However, the significance of the latter is also below 0.1. It is indicated that the three indexes are important factors that influence the quality of peeled peanut kernels. The significance of the correlation coefficient among 100-kernel yield, O/L ratio,

**TABLE 2.24** Verification and Analysis of Quality Predication Model for Peanuts for Export

| Variety | Predicated Value | Compos-ite Score | Relative Error (%) | Correlation Coefficient | Significance |
|---|---|---|---|---|---|
| Longhua243 | 60.29 | 58.94 | 0.95 | 0.839 | 0.002 |
| Yueyou45 | 51.96 | 49.41 | 1.81 | | |
| Yueyou52 | 59.24 | 60.88 | 1.16 | | |
| Yueyou86 | 50.44 | 48.79 | 1.17 | | |
| Zhanhua82 (oil) | 51.60 | 50.13 | 1.04 | | |
| Heyou11 | 52.13 | 48.06 | 2.88 | | |
| Yuhua15 | 61.19 | 58.12 | 2.17 | | |
| Yuhua9326 | 58.61 | 57.85 | 0.54 | | |
| Guihua771 | 55.11 | 53.02 | 1.48 | | |
| Yuanza9307 | 53.70 | 59.24 | 3.92 | | |

**TABLE 2.25** Correlation Analysis for 100-Kernel Weight, O/L Ratio, and Aflatoxin Content

| | 100-Kernel Weight | O/L Ratio | Aflatoxin Content | Total Scores |
|---|---|---|---|---|
| 100-kernel weight | 1.000 | 0.037 | 0.120 | 0.396** |
| Significance of correlation coefficient | | 0.781 | 0.354 | 0.001 |
| O/L ratio | 0.037 | 1.000 | 0.090 | 0.229 |
| Significance of correlation coefficient | 0.781 | | 0.501 | 0.084 |
| Aflatoxin content | 0.120 | 0.090 | 1.000 | 0.276* |
| Significance of correlation coefficient | 0.354 | 0.501 | | 0.030 |
| Scores | 0.396** | 0.229 | 0.276* | 1.000 |
| Significance of correlation coefficient | 0.001 | 0.084 | 0.030 | |

Note: "*" indicates significance at the level of $P<0.05$. "**" indicates significance at the level of $P<0.01$.

**TABLE 2.26** Table of Regression Model Coefficient for the Quality of Peeled Peanut Kernels

| Model | Nonstandard Coefficients | | Standard Coefficient | | |
|---|---|---|---|---|---|
| | Regression Coefficient | Standard Error (%) | Trial | $t$ | Significance |
| Constant | 30.839 | 4.110 | | 7.503 | 0.000 |
| 100-kernel yield | 0.091 | 0.032 | 0.341 | 2.796 | 0.007 |
| O/L ratio | 5.648 | 2.584 | 0.266 | 2.186 | 0.034 |
| Aflatoxin content | 1.909 | 0.921 | 0.253 | 2.073 | 0.043 |

and aflatoxin content are all larger than 0.3, indicating that these indexes are mutually independent and can individually reflect their respective relationship with the quality of peeled peanut kernels.

c. Modeling

Multiple linear regression equation is established by means of the step-wise multiple regression method, where the independent variables include 100-kernel weight ($X_1$), kernel yield ($X_2$), fat content ($X_3$), protein content ($X_4$), content of polysaccharides ($X_5$), polysaccharide content ($X_6$), O/L ratio($X_7$), water content ($X_8$), unsound kernel ($X_9$), and aflatoxin content ($X_{10}$), and the dependent variable ($Y$) is the total quality score of the peeled peanut kernels. All the three indexes, 100-kernel weight ($X_1$), O/L ratio ($X_7$), and aflatoxin content ($X_{10}$), are included (Table 2.26) to finally form the regression equation, as follows:

$$Y = 30.839 + 0.091X_1 + 5.648X_7 - 1.909X_{10} \qquad (2.7)$$

As shown in the regression coefficient table (Table 2.26), the level of significance of the 100-kernel weight, O/L ratio, and aflatoxin content are all lower than 0.05, indicating the significant influence of the independent and dependent variables and the existence of significant linear regression relationships. The effectiveness of the regression model for the quality of the peeled peanuts is determined by F test. Therefore, the model is subjected to variance analysis. According to the variance analysis (Table 2.27), the statistics of the regression equation is $F = 5.967$, and the significance of the significance test results is 0.001, which is lower than 0.05, indicating a significant linear relationship of the multiple regression equation. It is clear that it is effective to use the multiple linear regression model to predict the composite score of the quality of peeled peanut kernels.

**TABLE 2.27** Variance Analysis Table

| Model | Sum of Squares | DOF | Average Variance | F | Significance |
|---|---|---|---|---|---|
| Regression | 401.317 | 3 | 133.772 | 5.967 | 0.001 |
| Residual | 1120.925 | 50 | 22.419 | | |
| Total | 1522.242 | 53 | | | |

Ten varieties are adopted to verify the Eq. (2.8), and it is shown that the correlation coefficient between the predicted value and the actual value is 0.741. This indicates that this equation provides good prediction results, and this model can be used to determine the quality evaluation indexes for peanuts for export, to predict the peanut quality of unknown varieties, and to provide means for the processing enterprises to select peanut varieties suitable for peeled peanut kernels for export.

$$\text{Quality of peeled peanut kernels for export} = 30.839 + 0.091 \times 100 - \text{kernel weight} \times \text{O/L ratio} - 1.909 \times \text{aflatoxin content} \quad (2.8)$$

**d.** Prediction of Suitability of Peeled Peanut Kernels for Export

The difference between the composite score (Table 2.28) for the peeled peanut kernels for export and the predicted value of the model (Table 2.29) is only about 10%, the correlation coefficient is 0.741, and the significance is 0.014, indicating a significant correlation at the level of 0.05 (Table 2.30).

**TABLE 2.28** Predicted Scores of 10 Peanut Varieties

| Peanut Variety | 100-Kernel Weight (g) | O/L Ratio | Aflatoxin Content (µg/kg) | Predicated Score |
|---|---|---|---|---|
| Longhua243 | 68.10 | 1.29 | 0.25 | 43.83 |
| Yueyou45 | 63.70 | 0.84 | 0.49 | 40.42 |
| Yueyou52 | 76.50 | 1.04 | 0.05 | 43.58 |
| Youyou86 | 63.55 | 0.86 | 1.09 | 39.38 |
| Zhanhua82 (Oil) | 50.55 | 1.03 | 0.54 | 40.24 |
| Heyou11 | 65.85 | 1.02 | 1.51 | 39.71 |
| Yuhua15 | 99.30 | 1.19 | 1.95 | 42.90 |
| Yuhua9326 | 88.00 | 1.10 | 1.47 | 42.26 |
| Guihua771 | 65.20 | 0.96 | 0.16 | 41.89 |
| Yuanza9307 | 74.90 | 0.99 | 1.55 | 40.29 |

**TABLE 2.29** Actual Scores of 10 Peanut Varieties

| Peanut Variety | Water Content (%) | Moisture Score | Proportion of Unsound Kernel (%) | Score of Unsound Kernel | Proportion of Half Kernel (%) | Score of Half Kernel | Aflatoxin Content (µg/kg) | Score | Grade | O/L Value | Score of O/L Value | Red Skin Residual Rate % | Score | Score of Red Skin Residual | Total Score |
|---|---|---|---|---|---|---|---|---|---|---|---|---|---|---|---|
| Longhua 243 | 4.28 | 2.16 | 0.88 | 10.60 | 13.35 | 3.33 | 0.25 | 10.68 | 41.85 | 9 | 1.29 | 9.71 | 1.59 | 3.075 | 56.87 |
| Yueyou45 | 4.23 | 2.31 | 0.97 | 10.15 | 13.55 | 3.23 | 0.49 | 7.68 | 44.74 | 9 | 0.84 | 10.16 | 1.51 | 3.675 | 56.56 |
| Yueyou52 | 4.29 | 2.13 | 0.73 | 11.35 | 13.65 | 3.18 | 0.05 | 8.08 | 37.25 | 13 | 1.04 | 9.96 | 1.62 | 2.850 | 62.27 |
| Yueyou86 | 4.55 | 1.35 | 1.03 | 9.85 | 12.95 | 3.53 | 1.09 | 7.44 | 44.85 | 9 | 0.86 | 10.14 | 1.74 | 1.950 | 51.46 |
| Zhanhua82 (Oil) | 4.48 | 1.56 | 1.12 | 9.40 | 12.78 | 3.61 | 0.54 | 9.36 | 56.38 | 5 | 1.03 | 9.97 | 1.69 | 2.325 | 49.70 |
| Heyou 11 | 4.19 | 2.43 | 0.99 | 10.05 | 13.98 | 3.01 | 1.51 | 10.20 | 43.28 | 9 | 1.02 | 9.98 | 1.42 | 4.350 | 52.78 |
| Yuhua15 | 4.15 | 2.55 | 1.07 | 9.65 | 14.23 | 2.89 | 1.95 | 6.44 | 28.70 | 15 | 1.19 | 9.81 | 1.43 | 4.275 | 56.37 |
| Yuhua 9326 | 4.13 | 2.61 | 0.95 | 10.25 | 13.87 | 3.07 | 1.47 | 8.36 | 32.39 | 11 | 1.10 | 9.90 | 1.16 | 6.300 | 57.24 |
| Guihua 771 | 4.10 | 2.70 | 0.88 | 10.60 | 13.69 | 3.16 | 0.16 | 8.68 | 43.71 | 9 | 0.96 | 10.04 | 1.61 | 2.925 | 57.78 |
| Yuanza 9307 | 4.14 | 2.58 | 1.05 | 9.75 | 13.89 | 3.06 | 1.55 | 8.20 | 38.05 | 13 | 0.99 | 10.01 | 1.43 | 4.275 | 56.47 |

**TABLE 2.30** Verification of Predication Model of Peeled Peanut Kernel Quality

| Variety | Predicted Value | Composite Score | Relative Error% | Correlation Coefficient | Significance |
|---|---|---|---|---|---|
| Longhua 243 | 43.83 | 56.87 | 9.22 | 0.741 | 0.014 |
| Yueyou45 | 40.42 | 56.56 | 11.41 | | |
| Yueyou52 | 43.58 | 62.27 | 13.21 | | |
| Yueyou86 | 39.38 | 51.46 | 8.54 | | |
| Zhanhua82 (Oil) | 40.24 | 49.70 | 6.69 | | |
| Heyou11 | 39.71 | 52.78 | 9.24 | | |
| Yuhua15 | 42.90 | 56.37 | 9.52 | | |
| Yuhua9326 | 42.26 | 57.24 | 10.59 | | |
| Guihua771 | 41.89 | 57.78 | 11.23 | | |
| Yuanza9307 | 40.29 | 56.47 | 11.44 | | |

## 2.6.3    Determination of Peanuts Suitable for Export

Seven percent of peanuts in our country are for export, which ranks first in world trade volume. Although the peanut quality indexes are different among different countries or regions, the requirements for the quality of the peanut kernels by the main export destinations, such as the EU and Japan, mainly include six indexes, namely, skin color, water content, aflatoxin content, O/L ratio, unsound kernel content, and grade. The quality of peeled peanut kernels for export mainly involves water content, aflatoxin content, O/L ratio, unsound kernel content, red skin residual, half kernel content, and grade. This section will place an emphasis on the evaluation standard for peanuts suitable for export processing and the evaluation standard for peeled peanut kernels suitable for export processing in an effort to lay down a basis for the selection of peanut varieties or raw peanuts for export.

1. The method of K-means clustering analysis is adopted to determine the peanut varieties suitable for export, and 65 peanut varieties are categorized into three grades (Table 2.31).
2. The method of K-means clustering analysis is adopted to determine the peeled peanut kernel varieties suitable for export, and 65 peanut varieties are categorized into three grades (Table 2.32).
3. Determination of Peanut Varieties Suitable for Export

   The above evaluation criteria are adopted to evaluate the export suitability of 65 peanut varieties, and there are 15 peanut varieties suitable for export,

**TABLE 2.31 Criteria of Suitability of Peanuts for Export**

| Category | Number of Varieties | Name of Varieties |
|---|---|---|
| Suitable | 15(>66) | Huayu28, Luhua14, Lufeng2, 9616, Baisha1016, Zhonghong 108, Bianhua No. 3, Shuangji No. 2, Yuanhua No. 8, Xuhua 13, 034-256-1, 600-6, 365-1, L03-329-336, L03-601-604 |
| Basically suitable | 25(60-66) | Huayu19, Huayu 22, Luhua9, Fenghua3, Fenghua No. 5, Shanhua No. 7, Xuhua No. 14, 818, 780-15, Huayu No. 8, Huayu16, Huayu20, Huayu23, Huayu25, Huayu31, Luhua11, 25(60–66) Luhua15, Fenghua No. 1, Fenghua No. 4, Fenghua No. 6, Variety060, Baisha101, Zhonghua No. 4, Zhonghua No. 8, Zhonghua No. 15 |
| Unsuitable | 25(<60) | Shanhua No. 9,Haihua 1, Haiyu No. 6, Qinglan No. 8, Honghua No. 1, Xuhua No. 5, Xuhua 15, Hongguan, Wucai Huasheng, Kainongbai No. 2, Baihuasheng, Zhongyu No. 1, Heihuasheng, Huaguanwang, Ji 9814, 060-18-w, Fuhua No. 1, Silihong, 8130, Hua 17, Xianghua509-77, Zhengong No. 7, Yuanza 9102, 606, L03-257-294 |

**TABLE 2.32 Criteria of Suitability of Peeled Peanut Kernel Varieties for Export**

| Category | Number of Varieties | Name of Varieties |
|---|---|---|
| Suitable | 13 (>49) | Huayu No. 8, Huayu16, Huayu31, Baisha101, Shanhua No. 7, Shanhua No. 9, Haihua1, Honghua No. 1, Xuhua14, 034-256-1, Ji9814, Zhengnong No. 7, Yuanza9102 |
| Basically suitable | 30 (45–49) | Huayu19, Huayu22, Huayu23, Huayu25, Huayu28, Luhua9, Luhua14, Fenghua No. 1, Fenghua No. 3, Fenghua No. 5, Lufeng No. 2, Pinzhong060, 9616, Baisha1016, Zhongnong108, Hiayu No. 6, Qinglan No. 8, Bianhua No. 3, Shuangji No. 2, Yuanhua No. 8, Xuhua 13, Zhongyu No. 1,Heihuasheng, Huaguanwang, 818, 600-6, 780-15, 365-1, 03-329-336, L03-601-604 |
| Unsuitable | 20 (<45) | Huayu 20, Luhua11, Luhua15, Fenghua No. 4, Fenghua No. 6, Zhonghua No. 4, Zhonghua No. 8, Zhonghua No. 15, Xuhua No. 5, Xuhua15, Kainongbai No. 2, Baihuasheng, 060-18-w, Fuhua No. 1, Silihong, 8130, Hua17, Xianghua509-77, 606, L03-257-294 |

including "Huayu 28," "Luhua 14," "Lufeng 2," "9616," "Baisha 1016," "Zhongnong 108," "Baihua No. 3," "Shuangji No. 2," "Yuanhua No. 8," and "Xuhua 13," etc.; and there are 15 peeled peanut kernel varieties suitable for export, including "Huayu No. 8," "Huayu 16," "Huayu 31," "Baisha 101," "Shahua No. 7," "Shanhua No. 9," and "Haihua 1," etc.

Chapter 3

# Peanut Oil Processing Technology

Q. Wang, H. Liu, H. Hu, R. Mzimbiri, Y. Yang, Y. Chen

*Institute of Food Science and Technology, Chinese Academy of Agricultural Sciences, Beijing, China*

## Chapter Outline

## 3.1 PRETREATMENT TECHNOLOGY

Peanut pretreatment refers to a series of treatment processes before oil production, such as cleaning, shelling, grading and selection, drying, cooling, thermal conditioning, pressing, and cooking. Through these treatments, the impurities can be removed, so that the materials have a certain structural performance to meet the requirements of the oil production processes.

## 3.1.1 Cleaning

Impurities might be blended in during the harvest, transport, and storage processes of the peanuts. Therefore, after the peanuts are brought into the production workshops, further cleaning is required to meet the requirements of production processes and ensure product quality.

The impurities in the raw peanuts can be categorized into three types:

1. Inorganic impurities: dust, sand, metal, and so on
2. Organic impurities: stems and leaves, hemp, shell, and so on.
3. Oil-bearing impurities: Moldy kernel, worm-eaten kernel, and unsound kernel, etc.

These impurities will reduce the oil yield and the quality of fat and residual cake, shorten the service life of equipment, and may even cause production accidents, etc.

The methods of screening, winnowing, magnetic and gravity separation are usually adopted to remove the impurities from peanuts in combination with their corresponding equipment. After cleaning, the peanuts are not allowed to contain large impurities, such as stone, iron contamination, and hemp rope, etc. The limit for the impurity content in the cleaned raw peanuts is 0.1%.

Cleaning efficiency $\eta$ can be adopted to evaluate the impurity cleaning effect of various kinds of cleaning equipment. In actual production, Eq. (3.1) is usually adopted to calculate the impurity removal efficiency (Liu, 2009; Hu, 2011).

$$\eta = \frac{a \times b}{a} \times 100\% \qquad (3.1)$$

where $\eta$ is the impurity removal efficiency (%), $a$ is the content of the designated impurity in the feeding machine (%), and $b$ is the content of the designated impurity in the discharging machine (%).

### 3.1.2   Shelling

The peanuts are shelled to improve the oil yield from the peanuts and reduce the fat loss. Manual shelling and mechanical shelling can be adopted. Manual shelling is mainly adopted for peanuts for export and seeds, which have relatively high requirements for peanuts, but this method is not only labor-consuming but also inefficient. Mechanical shelling is both time-saving and labor-saving, but tends to cause mechanical damage to the peanuts. For the peanuts used for oil production, there is not a high requirement for undamaged peanut kernels, and therefore, mechanical equipment can be used for shelling.

### 3.1.3   Grading and Selection

During the processing, methods such as magnetic separation, gravity separation, winnowing, and color sorting are adopted in combination with manual selection to remove impurities and unsound kernels from peanuts and ensure that all the peanut kernels are full, uniform, and free from shriveled, moldy, rotten, or worm-eaten kernels, or other incomplete or unsound kernels. Especially, the introduction of a color sorting process can effectively remove the aflatoxin-contaminated peanuts (the efficiency can reach above 99.0%).

### 3.1.4   Drying

The cleaned peanut kernels are first dried and then subjected to air cooling to reduce the water content of the peanut kernels and enable the red skin of the

**TABLE 3.1** Water Content and Peeling Rate of Peanut Kernels at Different Drying Temperature and Time

| Temperature (°C) | 40 | 50 | 60 | 70 | 80 |
|---|---|---|---|---|---|
| Time (h) | 10.0 | 9.0 | 7.0 | 4.0 | 3.0 |
| Water content (%) | 5.21 ± 1.05 | 5.05 ± 1.62 | 4.58 ± 0.98 | 4.61 ± 1.32 | 4.85 ± 0.85 |
| Peeling rate (%) | 90.2 ± 1.0 | 93.5 ± 0.7 | 96.0 ± 1.2 | 98.9 ± 1.0 | 98.2 ± 0.8 |

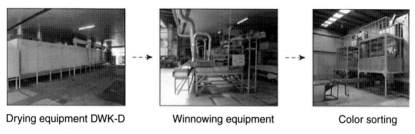

Drying equipment DWK-D          Winnowing equipment          Color sorting

**FIGURE 3.1**    Figure of drying-peeling and color sorting production equipment.

kernels to be crisp and easy to peel. The drying temperature is controlled between 40 and 80°C, and the duration is controlled according to the water content of the peanuts. Generally speaking, the peeling efficiency is relatively high when the water content of the peanut kernels is reduced below 5%. As shown in Table 3.1, when the drying temperature exceeds 60°C, the water content of the peanut kernels is reduced significantly, the peeling rate can reach above 95%, and the peeling effect is good at 70°C. Refer to Fig. 3.1 for the drying-peeling and color sorting production equipment.

## 3.2   PEANUT PRESSING TECHNOLOGY

Among the total peanuts for consumption in China, more than 50% of peanuts are used for oil production, which has become the main pattern of peanut consumption in China. Currently, there are two pressing methods: high-temperature pressing and cold pressing. More than 90% of oil production in China adopts the traditional technique of high-temperature pressing, and the peanut oil produced with this method has a strong fragrant flavor and is therefore greatly favored by consumers. However, the peanut oil produced with the high-temperature pressing method has a poor sensory quality and a heavy loss of vitamin E, sterol, wheat germ phenol, phospholipid, and other nutritional factors, and what is more, the oil has poor stability. Only less than 10% of peanut oil is produced with cold pressing technique, but the peanut oil produced with this technique maintains the original nutritional quality of peanuts, and peanut protein powder with low variability can be produced during the pressing process so that both the peanut utilization rate and the economic benefits can be improved. Therefore, the peanut oil produced with cold pressing method has more promising market prospects.

## 3.2.1 High-Temperature Pressing

Before pressing, it is necessary to dry the peanuts repeatedly with a hot air drier to control the water content within the range of 5–6%, and then bring down the temperature of the oil quickly with cool air to a temperature below 40°C. After crushing, the small-channel kernels will be heated to 180–200°C in a stir-fry furnace for hot-air frying so as to improve the oil yield and enhance of the fragrance of the peanuts. The large-channel kernels will be pressed into flakes through the flaking machine to destroy the cellular tissue of the peanuts. The flakes produced by the pressing are called uncooked flakes, which after being steamed and roasted, are called cooked flakes. The steaming and roasting can further destroy the cells, cause coagulation and denaturation of protein, as well as the segregation and combination of phospholipids and gossypol, so as to enable the oil to be extracted from the flakes and improve the quality of the crude oil (Fig. 3.2).

## 3.2.2 Cold Pressing

Refer to Fig. 3.3 for the production flow of the cold pressing process of peanut oil.

1. Determination of Temperature and Time for Thermal conditioning
   As shown in Fig. 3.4, during the process the temperature rises from 60 to 90°C, the oil/residual oil ratio of the system decreases gradually, reaching

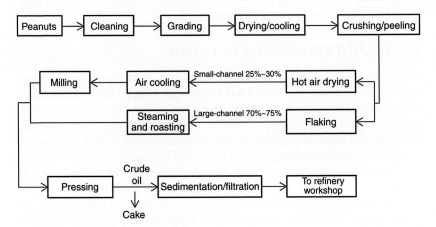

**FIGURE 3.2** High-temperature peanut oil production flow chart.

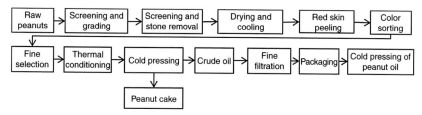

**FIGURE 3.3** Flow chart of cold pressing of peanut oil.

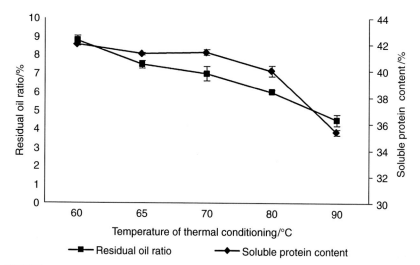

**FIGURE 3.4** Influence of the temperature of thermal conditioning on oil residual oil ratio and soluble protein content of residual cake.

4.5% at 90°C. However, during the process of the temperature rise, the content of the soluble protein of the residual cake decreases slightly at 60–70°C, and when the temperature exceeds 70°C, the content of the soluble protein of the residual cake decreases rapidly and reaches merely 35–38% at 90°C. Therefore, the best temperature for thermal conditioning is about 70°C.

As shown in Fig. 3.5, within 30–110 min of thermal conditioning, the residual oil ratio plummets, that is, with the prolonging of the time of thermal conditioning, the residual oil ratio decreases constantly. Within the period of 50–70 min of thermal conditioning, the changes in the soluble protein content tend to level off. With further time, the soluble protein content begins to decrease again. It is clear that with the prolonging of the time of thermal conditioning, the soluble protein of the peanut protein powder decreases rapidly, indicating a serious denaturation of peanut protein powder. To ensure that the peanut protein powder is not denatured and the residual oil ratio of the peanut protein is not higher than 6.5%, the period for thermal conditioning is determined to be 70 min.

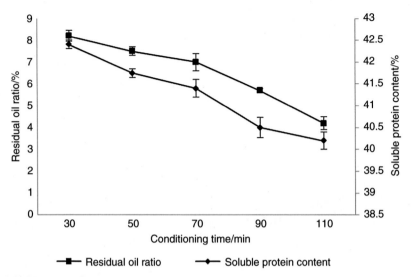

**FIGURE 3.5**    Influence of conditioning time on residual oil and soluble protein content of residual cake.

### 2. Determination of Pressing Moisture

During the steaming and roasting process of conventional production, the protein is greatly denatured. In the meantime, premature oil exudation easily occurs during the steaming and roasting process, which is very detrimental to the reduction of the residual oil ratio. Through repeated trials, it has been found that, during the actual production process, when the peanut kernels have the proper pressing moisture (6–8%) and temperature (60–90°C), the residual oil ratio can be reduced while ensuring low protein denaturation. Through thermal conditioning, the materials can be softened and become plastic, and the cell structures can be easily damaged, squeezed, and oil extracted. Water can be added for conditioning in the case of a heavy loss of water.

As shown in Fig. 3.6, when the pressing moisture of peanut kernels is 2–10%, both the residual oil ratio and soluble protein content of the residual cake are on rise. The soluble protein content rises quickly when the pressing moisture is maintained between 2 and 6%, and levels off when the pressing moisture is maintained between 6 and 8%, and then rises slightly. During the actual production, an overly low pressing moisture of the peanut kernels can increase the energy consumption as well as the denaturation degree of the protein, and therefore, it is meaningless to keep the pressing moisture at an overly low level. Taking into consideration the residual oil rate and soluble protein content, the pressing moisture of 5% has been finally selected as the optimal moisture content of materials. In practical production, the pressing moisture can be controlled at 4–8%.

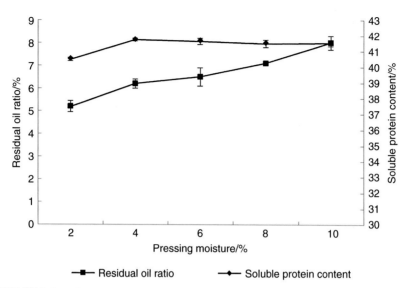

**FIGURE 3.6** Influence of pressing moisture on residual oil ratio and soluble protein content of residual cake.

3. Pressing Condition

   Two abnormalities easily occur during the cold pressing process: first, premature oil exudation, making the materials unable to be fed into the press; and second, the cake does not take shape, or although the cake takes shape, very little oil is extracted. This is greatly related to the pretreatment. Through adjusting the pressing temperature and adjusting the spindle speed and oil passage slot of the press during the actual production, the residual oil ratio in the residual cake can be effectively reduced. By analyzing the results of pressing experiments, it can be observed that, simply from the perspective of oil production, the residual oil ratio can be kept at a relatively lower level (about 6.5%) under the conditions of pressing temperature of 70°C, pressing moisture of 5%, and rotating speed of press at 10r/min when ZX10 type press is adopted for continuous pressing. From the perspective of ensuring the functional properties of proteins, only the factor of temperature needs to be taken into consideration, and it is better to keep the pressing temperature at 70°C (Tables 3.2–3.4).

4. Physical Fine Filtration

   A frame filter is adopted to press and filter three times under 30°C to obtain the product oil. Food-grade filter cloth is adopted (interception size is 200 mesh) to filter the crude oil at least twice, and the filter temperature is controlled between 10 and 50°C (30°C is preferred); the pressured required during filtration is 0.1–0.4 MPa (0.3 MPa is preferred). The obtained product is cold extracted peanut oil (Fig. 3.7).

**TABLE 3.2** Changes of Residual Oil Ratio of Single Pressing at 60–90°C

| No. | Control Temperature (°C) | Pressing Moisture (%) | Residual Oil Ratio of Peanut Meal (%) | No. | Control Temperature (°C) | Pressing Moisture (%) | Residual Oil Ratio of Peanut Meal (%) |
|---|---|---|---|---|---|---|---|
| 1 | 60 | 6.9 | 7.4 | 6 | 70 | 6.8 | 6.5 |
| 2 | 60 | 7.2 | 7.1 | 7 | 80 | 6.3 | 6.6 |
| 3 | 65 | 7.8 | 8.2 | 8 | 80 | 6.1 | 6.5 |
| 4 | 65 | 8.0 | 7.9 | 9 | 90 | 5.9 | 5.7 |
| 5 | 70 | 6.5 | 6.4 | 10 | 90 | 5.4 | 5.5 |

**TABLE 3.3** Influence of Rotational Speed of Press on Residual Oil Ratio of Peanut Meal

| No. | Main Spindle Speed (r/min) | Residual Oil Ratio of Peanut Meal (Dry Basis) (%) |
|---|---|---|
| 1 | 10 | 6.4 |
| 2 | 20 | 7.0 |
| 3 | 30 | 7.8 |
| 4 | 50 | 7.9 |

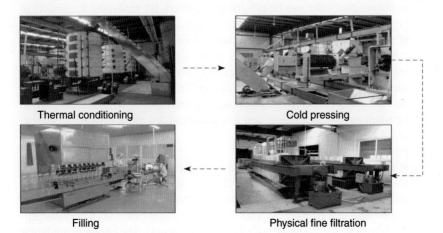

**FIGURE 3.7** Production process of cold pressing of peanut oil.

**TABLE 3.4** Changes of Soluble Protein After Single Pressing at 60–90°C

| No. | Pressing Temperature (°C) | Meal Protein (Dry Basis of Meal) (%) | Soluble Protein Content (%) | Soluble Protein (Meal or Finished Product) (%) | Remarks |
|---|---|---|---|---|---|
| 1 | 60 | 54.49 | 43.59 | 41.64 | Pilot test |
| 2 | | 54.83 | 43.45 | 41.56 | Pilot test |
| 3 | 65 | 54.63 | 43.70 | 41.36 | Pilot test |
| 4 | | 55.12 | 43.68 | 41.23 | Pilot test |
| 5 | 70 | 56.54 | 43.86 | 41.43 | Normal production sample of three shifts |
| 6 | | 57.14 | 43.84 | 41.25 | Production sample of three shifts |
| 7 | 80 | 54.78 | 45.30 | 41.12 | Pilot test |
| 8 | | 54.45 | 45.12 | 41.09 | Pilot test |
| 9 | 90 | 55.74 | 44.59 | 35.34 | Pilot test |
| 10 | | 55.38 | 44.52 | 35.26 | Pilot test |

## 3.3    PEANUT OIL EXTRACTION TECHNOLOGY

The leaching method, also named the extraction method, is a method that uses certain organic solvents that can dissolve fat to spray and immerse the oil-bearing materials so as to eventually separate the fat from the materials. Compared with the pressing method, the leaching method has the advantages of high oil yield, low residual oil ratio of the peanut meal, large production scale, and low production cost, etc. The disadvantages include that high capital investment is required, and the leaching agents are mostly flammable and explosive toxic substances, threatening production safety.

### 3.3.1    Separation

The leaching technology can be categorized in several ways. The leaching method of oil production can be categorized into intermittent and continuous leaching by the mode of operation. The leaching technology mainly includes processes such as feeding the raw materials into the leacher, discharging the meal from the leacher, injecting fresh agent, and extraction of thick blending oil, etc. Intermittent leaching is carried out in batches, briefly and periodically, while continuous leaching is carried out continuously. The leaching method of oil production can be categorized into immersion type leaching, spraying type leaching, and mixed type leaching. For immersion type leaching, the raw materials are directly put into the leacher and immersed into the extraction solvent to complete the leaching process. For the spraying type leaching, the solvent is sprayed on the raw materials to achieve sufficient contact and complete the leaching process. The mixed type leaching is a combination of spraying type and immersion type leaching methods, and both the flat-turn leacher and the rotary-turn leacher are representative of the equipment for mixed type leaching.

The leaching method of oil production can be categorized into direct leaching and prepressing leaching by the production method. For direct leaching, the oil-bearing materials are directly used for leaching after pretreatment, and this process is applied to the oil-bearing materials with low oil content. For prepressing leaching, the raw materials are first pressed to exude part of the oil, and then the cakes with relatively high oil content are subject to leaching. This prepressing leaching process is applied to the oil-bearing materials with a relatively high oil content, and can increase the equipment productivity and lower the production cost.

### 3.3.2 Leaching Process

After pretreatment of the peanuts, the leaching process mainly includes the following four steps: leaching, wet meal desolventizing, mixed oil evaporation and stripping, and solvent recovery. The peanut oil produced with the leaching method is generally needed to be brought to the refinery workshop for refining treatment before being canned and packaged.

1. Leaching Procedure

   The leaching procedure mainly makes use of the solid-liquid extraction principle to select certain fat-dissolving organic solvents to extract the peanut oil. Generally speaking, the leaching solvent should meet the following conditions: good fat solubility, stable physiochemical properties, easy separation from the fat and meal, and be safe, cost-effective, and easy to get. In fat production both in China and abroad, the aliphatic hydrocarbons, hexane mixture in particular, have been widely used. Currently, the most widely used leaching solvent in China is "No. 6 Solvent Oil," that is, light gasoline, which has a wide distillation range (60–90°C). In actual production, it is necessary to increase the temperature for wet meal desolventizing, evaporation, and stripping (usually 110°C), which can result in the decline of the fat and meal quality. Some other countries adopt commercial hexane, which has fewer components and a smaller distillation range, making it easy for meal-fat separation and reducing the protein denaturation.

The main leaching equipment includes the flat-turn type leacher and rotary-turn leacher used for continuous production.

2. Wet Meal Desolventizing Procedure

The meal from the leacher, generally containing about 25% solvent, is called wet meal, and the process of removing the solvent from the wet meal is called desolventizing. The purpose of desolventizing is to maximally remove the solvent so as to reduce the solvent consumption and ensure the safety of the meal during the production process, and passivate and destroy harmful toxins and antinutrients in the meal, and to improve the meal quality.

The method of thermal desorption is usually adopted for desolventizing, and the separation of solvent and meal is realized by increasing the temperature. The desolventizing process comprises two steps: first, the solvent is removed from the surface of the meal granule by means of surface vaporization; second, the solvent is enabled to have mass transfer from the interior to the surface, and is then removed through interior diffusion. To improve the desolventizing efficiency, measures, such as direct steaming, vacuuming, and mixing, etc., are usually adopted to assist the removal. To ensure that the solvent can be maximally removed from the meal, it is usually necessary to improve the desolventizing temperature, which will result in the protein denaturation in the meal. During the desolventizing process, the antinutrients in the meal can be passivated through proper heat-moisture treatment to improve the meal quality.

Liu et al. (2008) dried the graded peanut kernels at low temperature ($\leq 60°C$), the peeled kernels were directly fed into the low-temperature screw type press for prepressing, and the low-temperature prepressed cake was properly crushed to 10–12 mm and fed into the leacher after moisture conditioning. No. 6 solvent oil was used to extract the fat, and wet meal was fed into the hot mixing type low-temperature desolventizing device for low-temperature desolventizing at a temperature lower than 80°C. Next, ultrafine grinding was carried out to obtain defatted peanut protein powder so as to guarantee effective protein utilization during the peanut oil extraction.

3. Mixed Oil Evaporation and Stripping Procedure

The mixed oil produced from leaching comprises volatile solvent, nonvolatile oil dissolved therein, and 0.4–1.0% solid meal powder. The purpose of the procedures is to separate the solvent from the mixed oil and remove the solid meal powder to obtain relatively pure crude oil. The process of separating and removing the solid meal powder is called the purification of mixed oil, and usually, the methods of filtration, gravity sedimentation, and hydrocyclone separation are adopted for this purpose. The solvent separation is usually realized through evaporation and stripping procedures. During the evaporation process, most solvent is removed by means of boiling below a certain temperature. However, the high-density mixed oil has a very high boiling point, and it is impossible to remove all the solvent in the oil in the actual production process. Therefore, steam stripping is required to remove the residual solvent.

4. Solvent Recovery Procedure

The leaching solvent used in the leaching procedure is recycled for use, and most solvents will be returned to the production line for reuse except for a small amount of components with low boiling points that are discharged into the air along with the exhaust and a small amount of components with high boiling points that remain in the meal and crude oil. The solvent recovery includes condensation and cooling of solvent steam, separation of solvent and water, recovery of solvent from wastewater, as well as recovery of solvent from free gas, etc. The solvent vapor from the evaporator is cooled by the condenser and returns directly to the production line. Generally, the solvent content of the wastewater discharged from the water segregator should not exceed 0.01%, and the wastewater becomes milky if the wastewater contains a large amount of meal powder; in this case, a wastewater boiling tank will be used to recover the solvent. The exhaust comprises air, solvent, and water vapor and freezing, solid adsorption, and liquid absorption methods, etc., are usually adopted to recover the solvent from the exhaust (Liu, 2009; Hu, 2011).

5. Refining Procedure

The leached crude peanut oil will be brought to the refinery workshop for degumming, deacidification, bleaching, and deodorization, etc. The degumming procedure is mainly to remove the colloidal impurities from the oil, and the common degumming process includes hydration degumming and acid degumming. The deacidification procedure is mainly to remove the free fatty acids from the oil, and the main deacidification methods include distillation and alkali refining methods. The bleaching procedure is mainly to remove the pigments from the oil, so as to ensure the normal appearance and stability of the oil, and the common bleaching method adopted in industry is the absorption bleaching method. The deodorization is mainly to remove the unpleasant odor from the oil, improve the smoke point, increase the oil stability, and improve the color and quality of the oil, and the most widely applied deodorization method in the world is the vacuum steam deodorization method, which has good effects.

## 3.4 PEANUT OIL PRODUCTION LINE AND RELEVANT EQUIPMENT

### 3.4.1 Production Line Process

1. Cold-Pressed Peanut Oil

First, the sheller is used to shell the peanuts, and then the peanut kernels are transported to be dried in the low-temperature drying oven after being subjected to precleaning, cleaning by the gravity/magnetic separation destoner,

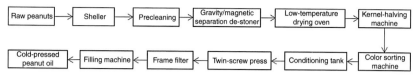

**FIGURE 3.8**    Flow chart of equipment for cold pressing of peanut oil.

and grading. Then, the dried peanut kernels are rapidly fed into the kernel-halving machine to peel the red skin, and then transported to the color sorting machine by a conveyor so as to effectively remove the yellow-spotted, moldy, worm-eaten, or incompletely peeled peanuts. After color sorting, the peanuts are put into the conditioning tank, where both the pressing moisture and temperature are conditioned to improve the oil yield. The conditioned peanut kernels are transported by a conveyor to the twin-screw press for cold pressing; the cold-pressed crude oil and cold-pressed peanut meal with low denaturation will be obtained. After the cold-pressed crude oil is filtered with frame filter, product oil is obtained, which will be packaged by a filling machine to form cold-pressed peanut oil products (Fig. 3.8).

2.  **Superfine Fragrant Peanut Oil**
    First, a sheller is used to shell the peanuts, and then the peanut kernels are transported to the low-temperature drying oven to be dried after being subjected to precleaning, gravity/magnetic separation de-stoner cleaning, and grading. Then, the dried peanut kernels are rapidly fed into the kernel-halving machine to peel the red skin, and then transported to the color sorting machine by a conveyor so as to effectively remove the yellow-spotted, moldy, worm-eaten, or incompletely peeled peanuts. 25–30% of peanuts are transported to the hot air roasting converter for roasting and then rapidly cooled by means of air suction. After that, the peeled kernels are crushed with a roller-type crusher. 70–75% of peanuts are transported to the twin-roller flaking mill for flaking, and then the flakes are steamed and roasted with a steaming and roasting cauldron. The large-channel and small-channel kernels are respectively transported to the screw press for oil pressing. After being mixed at a certain ratio, the crude oil is put into the sedimentation tank for sedimentation. After sedimentation, the crude oil is pumped into the frame filter for filtration, and the filtered product oil will be filled and packaged by a filling machine to form superfine fragrant peanut oil products (Fig. 3.9).

3.  **Peanut Oil (Leaching Method)**
    The prepressed peanut cake is fed into the scraper after its feeding volume is conditioned at the buffer bin, and then the peanut cake is elevated to be fed into the enclosed auger. After that, the materials are fed into the leacher for leaching. The materials are circulated in the leacher from the feeding port to the discharging port, sprayed by the mixed oil with a concentration gradient, and then drained. After that, fresh solvent is used for spray-leaching. The materials

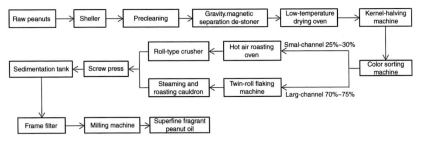

**FIGURE 3.9**    Flow chart of equipment for superfine fragrant peanut oil production.

are redrained and continuously transported to the wet meal scraper. After large impurities of the thick mixed oil are filtered out through the first tent filter in the leacher, the materials enter the second mixed oil filter, where the fiber impurities in the mixed oil are removed. The impurities from the filtration are pumped to the residue pump by the mixed oil circulation pump and then returned to the leacher, and the impurities are filtered through the bed of materials. The filtered mixed oil is then pumped into the third-layer hydrocyclone separator through the mixed oil pump, and after the starch impurities are removed, the oil will enter the mixed oil tank for sedimentation and evaporation.

After the sedimentation of the mixed oil in the mixed oil tank, the feeding pump of the first long tube evaporator sends the oil to the first long tube evaporator for evaporation. After climbing film evaporation, the solvent is separated from the mixed oil in the separator under the negative pressure of 0.5–0.6 MPa. The mixed oil condensed by the first long tube evaporator comes to the second long tube evaporator for evaporation through the heat exchanger, and the separated mixed oil is transported to the lower part of the separator for liquid seal. The feeding pump of the stripping tower, through the mixed oil heater, adjusts the oil to the prestripping temperature, and then the oil enters the steam stripping section. After that, the mixed oil enters the oil separation tank in the tower from the top of the steam stripping tower, and the evenly distributed oil enters the disc layer inside the tower for diffusion, where the residual solvent is removed through secondary steam stripping. The crude oil is pumped out from the steam stripping pump, and the crude oil, which reaches the output temperature through heat exchange, is transported to the crude oil tank. After being measured, the crude oil is then transported to the refining section by the crude oil pump. The crude oil is pumped to the acid-refining tank for degumming, and the degummed oil, obtained through sedimentation and separation, is then pumped to the neutralizing pot for alkali refining and deacidifying. Afterward, the disc-type centrifuge is used for oil-soap separation, and the alkali-refining oil obtained from the separation is heated by the heater and then blended with the meter-output clay in the vacuum discoloration tower for discoloration. The mixture of discolored oil and clay is exported to the vertical blade filter for filtration and then to the bag-type filter for fine filtration to remove the residual clay. Finally, the discolored

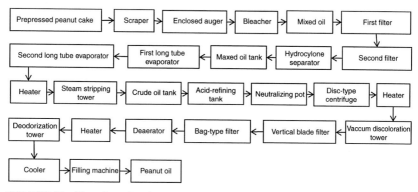

**FIGURE 3.10** Flow chart of equipment for peanut oil production with leaching method.

oil, free from pigment, residual soap, and metal oxides, is obtained. After the discolored oil is heated by the heater, it enters the deaerator for air removal. After being heated by the conductive oil in the heater to reach the deodorization temperature, the oil enters the deodorization tower for deodorization, and the deodorized oil is then cooled by the cooler and filled by the filling machine to form peanut oil products (Fig. 3.10).

## 3.4.2 Introduction of Key Equipment

1. Peanut Sheller

   The peanut shelling equipment is mainly categorized into two kinds: open drum (steel beating rod or beating plate)—fixed recessed grid that mainly features beating and rubbing; and enclosed rubber drum—rubber floating concave grid that mainly features squeezing and rubbing. The former type of peanut sheller is more widely used in production in China due to its higher efficiency and more reasonable structure.

   The open drum-recessed plate type sheller mainly comprises a feeding hopper, shelling drum (including beating rod and rotary bracket, etc.), recessed grid, tilt shaker, fan, and chassis (Fig. 3.11). During operation, the raw peanuts enter the shelling chamber through the feeding hopper, where the peanuts are subject to repeated beating and compacting by the high-speed rotary beating rod as well as the friction and rubbing jointly generated by the beating rod and recessed auger. Under these effects, the shells are crushed while the peanut kernels and crushed shells go through the recessed grid under the rotating air pressure and beating of the beating rod, and are then blown out of the machine by the fan. At the same time, the peanut kernels and the heavy impurities fall onto the shaker for preliminary cleaning, and are then discharged from the peanut outlet.

2. Peanut Color Sorter

   The peanut color sorter is a rapid food sorting equipment that integrates light, machinery, and electricity as a whole. It is mainly used for separating

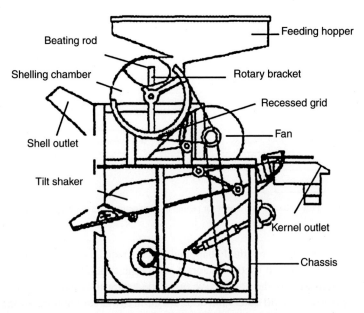

**FIGURE 3.11** Structure chart of open drum-recessed plant type sheller.

out from the milky kernels the kernels with different colors, including impurities and incompletely peeled, moldy, and yellow-spotted peanuts. The equipment makes use of the peanut surface color characteristics and the photoelectric detection principle to detect the outer color information of the objects subject to detection to effectively remove not only the impurities (such as stone, glass, and metal, etc.) from the peanut products but also the incompletely peeled, yellow-spotted, moldy, and worm-eaten peanut kernels to improve the food quality. The peanut color sorter, in place of manual sorting, can eliminate the interference of artificial factors, not only improving productivity but also reducing production cost (Zhao, 2010).

The structural components of peanut color sorter are as follows:

**a.** Feeding system

This system is for adjusting materials and maintaining relative stability of the individual posture and flow rate of materials so as to ensure the quantitative, equal, and continuous outflow process of materials. It is also used to ensure that materials are observed and detected one by one while passing through the viewing zone of optical system. The common feeding mechanism of a peanut color sorter includes a chute, vibrating feeder, roller, and vibrating screen. The feeding system is mainly composed of an adjustable door, hopper, and one/two/three items of feeding mechanism.

**b.** Optical system

The optical system is the key component of a peanut color sorter. It is for observing and detecting the materials so as to obtain accurate sorting information. The observation method, receiver category, and optical path structure will directly influence the performance feature, cost, and lifetime of the whole machine. In the research field of peanut color sorter, the optical system mainly includes an optical source, background plate, monochromator, and receiver. The receiver is the key device of the optical system. The commonly used receiver is a high-speed linear CCD/CMOS image sensor. It can make a rapid detection of the peanut moving at high speed and collect partial surface information which will then be rapidly processed and transferred. To reach such an effect of rapid detection, process, and transmission, the current receiver will generally, by using hardware circuits, assemble an image processing and transmission module in the receiver unit to generate functions including smoothing and noise reduction of signal, regeneration and amplification, digital image processing, etc. Such receivers are expensive.

**c.** Sorting system

The sorting system is used for sorting and rejecting the defective peanuts detected by the optical system. Under the control of a microprocessor, the run duration of peanuts from detecting point to the sorting point should match with the delay time for the optical system to detect the peanuts and transfer the signal to the sorting system which will conduct sorting afterward. The commonly used sorting system in the current market adopts a compressed air ejection method. When the peanuts move to the sorting area, qualified peanuts will move along the normal orbit and drop into the receiving mouth for quality goods while defective peanuts will be blown by the pulsed compressed air shot from nozzle and deviate from the normal orbit and thus drop into the receiving mouth for defective goods. The sorting is then completed.

**d.** Electronic control system

The function of the electronic control system is not only to control the operation of each component but also, most importantly, to receive detection signals of optical system and amplify the photoelectric pulse signal to generate pulse power, to initiate a pneumatic electromagnetic valve (or other sorting mechanisms), and to complete the sorting operation. Fig. 3.12 is the structure drawing of a color sorter SortexZ series. The design of the hopper ensures that materials are evenly and continuously transferred to the vibrating plate. The blowing direction of the ejector is the opposite of the moving direction of peanuts. This is because when the peanuts leave the chute, the ones that come into contact with the edge of chute will remain relatively stable, while the ones in the air will be relatively unstable. To be as close to the materials as possible, the nozzle holder is installed in the back of the materials. In such a case, the materials will be naturally separated when the nozzle blows. The closer the nozzle holder approaches the materials, the more accurate the ejection is and the lesser quality goods will be the rejected goods. This equipment is equipped with a channel chute with 64 channels. Each channel corresponds to an ejector and each ejector corresponds to a peanut. In such a case, the carryover rate will be extremely low and it is especially suitable

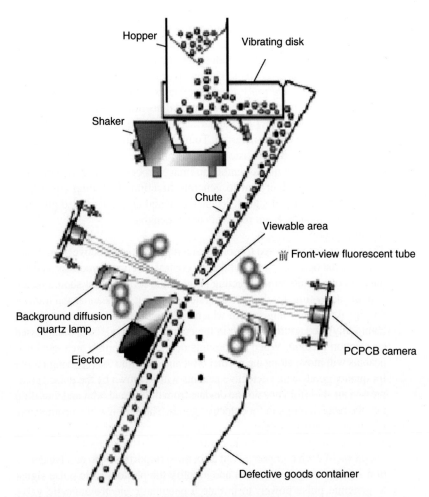

**FIGURE 3.12**    Structure drawing of SortexZ series color sorter.

for repeated screening. Compressed air is ejected to the materials and defective goods are rejected. The output will be enormous. Between the nozzle valve and the nozzle, there is a solid air pipe network which can accurately send a huge amount of air from valve to nozzle. The caliber of the nozzle is extremely small, merely 2–3 mm. In horizontal ejection, the gas diffusion will be well controlled. Vertical ejection is controlled by the nozzle valve whose specially made switching frequency will reach hundreds of nanoseconds (Tracey and Zang, 2003)

3. Twin-screw press
   The power driven squeezing screw shaft continuously rotates in the press cage and squeezes materials to generate oil. It is suitable for cold pressing. This equipment uses a twin screw and thus immensely increases the

Feeding port

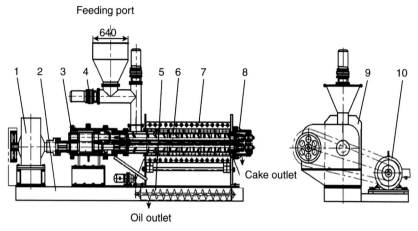

Oil outlet

**FIGURE 3.13** SLZ-30 twin-screw press for high oil containing seeds drawing.

propulsion capability and fundamentally prevents the materials from slipping or fixing on the rotation axle while pressing. Cold pressing results in high oil yield and good quality oil. In addition, cold pressing is conducted in indoor temperature which avoids damaging effective amino acids of protein, which may occur at high temperature, and thus high quality meal will be produced (Lin et al., 2006)

The SLZ-30 twin-screw press for high oil-containing seeds is mainly composed of a base, feeding device, transmission device, torque divider, squeezing screw shaft, press cage, and cake outlet device. The squeezing screw shaft, which adopts a twin screw structure and six levels of squeezing, is the key component of the machine. The twin screws are composed of squeezing axles of nine sections and six bevel gears. They are assembled in the "8" shaped chamber of the press cage with different bore diameters in the longitudinal direction. They are parallel-mounted with one rotating to the left with the other rotating to the right. In the feeding section, the top and bottom spirals adopt partial mesh form with openness in the vertical direction and closure in the horizontal direction. In the squeezing section, a nonintermeshing form is adopted with openness in both the vertical and horizontal directions. The vertical part of the press cage can expand and close around the hinges in both sides. On one hand, it is easy to assemble and maintain, and on the other hand, it will prevent the meal crumbs from gathering in the top of the press cage and thus increase the oil draining area. A dismountable grate type filter unit is assembled at the bottom of the feeding mouth which is beneficial to the timely discharge of air and hydrops in the feeding section and can prevent the slipping and detonation of materials in the chamber (Fig 3.13).

# Chapter 4

# Peanut Protein Processing Technology

H. Liu, A. Shi, L. Liu, H. Wu, T. Ma, X. He, W. Lin, X. Feng, Yuanyuan Liu
*Institute of Food Science and Technology, Chinese Academy of Agricultural Sciences, Beijing, China*

## Chapter Outline

Peanuts: Processing Technology and Product Development. http://dx.doi.org/10.1016/B978-0-12-809595-9.00004-1

## PREAMBLE

Peanut protein refers to the powdered peanut products, whose protein content is $\geq 50\%$, produced from low-denatured and defatted peanut meal. By adopting different production and processing methods, different types of peanut protein products can be obtained, namely, peanut protein powder, peanut tissue protein, peanut protein concentrate, peanut protein isolate, peanut globulin/conarachin component, and peanut protein film, etc.

## 4.1 PEANUT PROTEIN POWDER PROCESSING TECHNOLOGY

Traditional cold pressing technology solves the problem of serious denaturation of the protein that occurs during the high-temperature pressing process of the peanut oil. With this technology, the peanut oil yield is generally 41–43%, the peanut protein powder yield is generally 38–42%, and the residual oil ratio is generally 8–12%. It is obvious that the cold pressing technology adopted for the oil and protein extraction results not only in low oil yield but also a large oil content of the peanut protein powder, and this kind of protein powder not only has poor storage stability but also limits its application as a base stock or additive in food. Although low-denatured peanut protein powder can be obtained during oil production with this method, the oil and the protein can still not be completely separated efficiently. Currently, the industrialized technology that combines cold peanut oil pressing and functional peanut protein extraction as a whole can effectively solve the above problem. Through optimizing the parameters of drying, thermal conditions, and pressing and other processes and adopting efficient extraction technique with short-chain alkanes, the oil yield greatly increased and the peanut protein is less (or not) denatured.

### 4.1.1 Principle

Cold pressing and defatting technology has been adopted to produce peanut oil and low-denatured peanut protein powder. The production process uses kernel-peeling and cold pressing technology to produce the peanut oil product that reaches the edible vegetable oil that meets the national Grade I standard. Since the process is fully carried out at low temperature, various nutrients can be maximally retained, and the produced peanut oil is light-colored, and rich in unsaturated fatty acid while low in phospholipid, free from cholesterol, afla-toxin, or any toxic chemicals and residue, and generates less cooking fumes. In the meantime, the peanut protein powder obtained with the efficient cold extraction technique with alkanes can basically retain the protein nature in the peanut protein powder, where the residual oil ratio is 1.5% and the protein con-tent is 55.8%. This technique realizes the continuous and scaled production of the whole production process. A 50 t/d cold peanut oil pressing production line, built on the basis of this technological process, has been successfully put into operation.

## 4.1.2 Production Process

1. Efficient Extraction and Preparation of Peanut Meal With Low Oil Content and Denaturation

   The production process of peanut meal with low oil content and denatur-ation is as shown in Fig. 4.1. C3 or C4 short-chain alkanes are used for efficient extraction using the cold-pressed peanut cakes produced from the cold oil pressing. While keeping the oil extraction rate and protein nature stable, consideration should also be given to the actual production cost. In practice, the energy-saving device independently developed by our research team is designed for solvent recovery. The energy-saving device is designed to be a cylindrical empty tank container to overcome the system resistance. The collected condensate liquid is ejected from the top to make contact heat exchange with the solvent vapor that enters from below so that most of the solvent vapor is condensed, and the discharged liquid can be directly put into use after water separation at a temperature of 55°C without the need for heating.

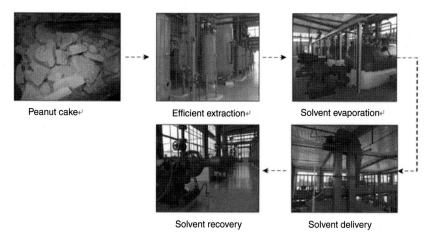

Peanut cake          Efficient extraction          Solvent evaporation

Solvent recovery          Solvent delivery

**FIGURE 4.1**   Production flow chart of efficient extraction.

To obtain odorless and pure peanut oil, it is also required to desolventize the mixed oil after the extraction, that is, the principle of heat recycling is utilized during the solvent recovery process. When the solvent evaporated from the mixed oil becomes high-temperature and high-pressure solvent gas after being compressed by the compressor, the heat generated during this process can replenish the heat required by solvent evaporation, and thus energy investment is reduced. In the meantime, the peanut meal is desolventized by means of depressurization and evaporation, and due to the decrease of pressure, part of the solvent evaporates from the meal. During the solvent evaporation, heat needs to be absorbed, and therefore a small amount of heat replenishmnent is required during this process. The desolventizing temperature is generally kept at about 45°C, which is insufficient to cause protein denaturation in the meal, and the solvent is recycled through recovery.

The production practice proves that the residual oil ratio in the peanut protein powder can be reduced below 1.5% and the protein content in the meal can reach 55–57% by stirring and extracting the pressed cake in the extraction tank five times using the extraction agent—short-chain alkanes at a temperature of 10–80°C (40°C optimally).

2. Ultrafine Grinding

To obtain quality peanut protein powder and meet different technology requirements, it is necessary to grind the wholly defatted cakes after the extraction of the pressed cake as described above. The powder-making involves repeated milling and screening until the fineness reaches the required indexes. The research uses a nonmesh-type airflow vortex mill to grind the wholly defatted meal to a fineness of above 200 meshes, and the air speed can be regulated according to the product requirements during this process. The finer the peanut protein powder, the greater the solubility and the greater the application value of the food. The defatted peanut powder obtained by milling is light-colored and similar to flour, and can therefore be added into flour or used in food, such as bread, dried noodles, pastries, and sausage, etc. In addition, the low-temperature defatted peanut meal can also be used for the preparation of protein isolates and protein concentrates.

3. Microwave Sterilization

   a. Single-Factor Experiment of Microwave Sterilization

      - Determination of Microwave Power

        The experiment further increases the microwave power to 300 W to study the influence of microwave power on the sterilization rate and the nitrogen solubility index of peanut protein powder. The results indicate (Figs. 4.2 and 4.3) that within the microwave power of 300–500 W, the sterilization rate rises significantly, and the nitrogen solubility index of peanut protein powder also rises with the rise of the microwave power. When the microwave power reaches 500 W, the sterilization rate and the nitrogen solubility reach 99.93% and

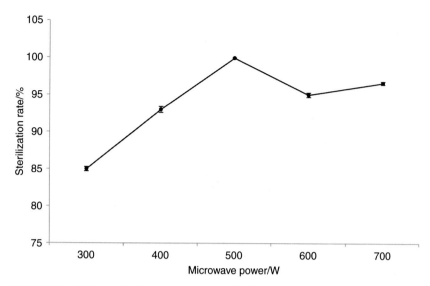

**FIGURE 4.2**    Influence of microwave power on sterilization rate.

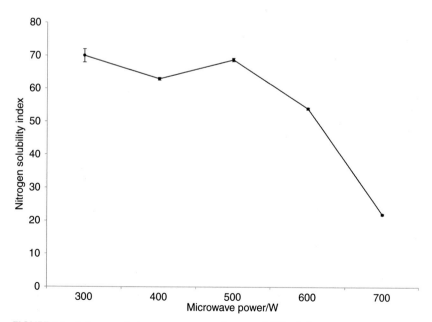

**FIGURE 4.3**    Influence of microwave power on nitrogen solubility index.

68.79%, respectively. The sterilization rate reaches its highest value and almost all the microorganisms are killed. When the microwave power exceeds 500 W, the changes of the sterilization rate level off but the nitrogen solubility index decreases dramatically. When the microwave power exceeds 700 W, the sterilization rate reaches 96.58% but the nitrogen solubility index is only 21.96%, reaching its lowest level. At the initial stage of microwave treatment, the microorganisms at the surface of the peanut protein are first killed, and due to the low power, the nature of the peanut protein powder basically remains unchanged, and almost all the micromechanisms are killed when the power rises to 500 W. At the further increase of the power, the nitrogen solubility index of the peanut protein powder decreases dramatically, which is a clear indication of protein denaturation. By comprehensively evaluating the two indexes, namely, the sterilization rate and the nitrogen solubility index, it is determined that the microwave value of 500 W is the optimal value.

- Determination of Sterilization Time

As shown in Figs. 4.4 and 4.5, within a sterilization period of 5–20 s, the sterilization rate shoots up, that is, with the prolonging of sterilization time, the sterilization rate rises constantly. When the sterilization time reaches 20 s, the sterilization rate and nitrogen solubility index are respectively 99.95% and 50.13%, and when the sterilization time increases from 20 to 25 s, the sterilization rate, which is 99.97%, basically remains unchanged, but the nitrogen index decreases greatly to 20.45%. It is obvious that with the prolonging of the sterilization time, the nitrogen solubility index of the peanut protein powder decreases rapidly, indicating a serious denaturation of

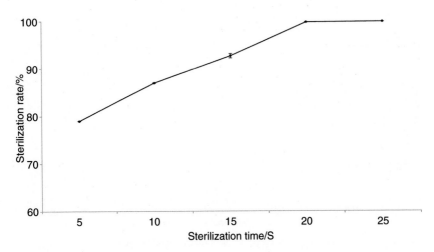

**FIGURE 4.4**   Influence of sterilization time on sterilization rate.

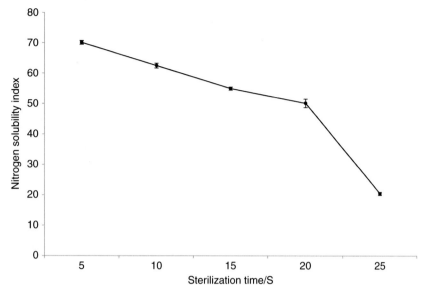

**FIGURE 4.5**   Influence of sterilization time of nitrogen solubility index.

the peanut protein powder. To ensure that the microorganism index of the peanut protein powder fully meets the requirements and the peanut protein powder is not denatured, the optimum sterilization time is determined to be 20 s.

- Determination of Material Concentration

Theoretically speaking, an over-large substrate concentration of a system will cause the concentration of the effective water in the system to be over-low, consequently reducing the material diffusion and movement and suppressing the microwave sterilization. The over-low material concentration will increase the collision probability of the protein molecules and microwave radiation rays to reduce the nitrogen solubility index. The experimental results (Figs. 4.6 and 4.7) show that 80% material concentration is insufficient to bring down the collision probability of the protein molecules and microwave radiation rays, that is, within the range of 80–90% material concentration, microwave sterilization is suppressed due to high concentration but the microwave radiation rays are not suppressed due to the insufficient concentration. In actual production, the over-low protein concentration will increase energy consumption and reduce equipment utilization rate, and therefore it is useless to keep the material concentration at an over-low level. As a result, the material concentration is not further brought down in this experiment. Finally, the experiment selects material concentration of 85% as an optimal material concentration, which is more practical in production

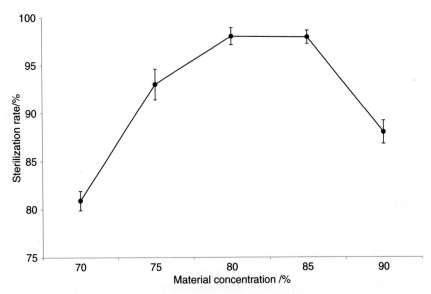

**FIGURE 4.6**   Influence of material concentration on sterilization rate.

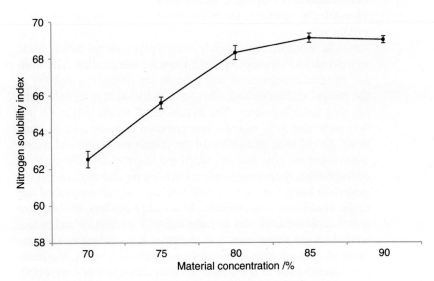

**FIGURE 4.7**   Influence of material concentration on nitrogen solubility index.

**TABLE 4.1** Components of Peanut Protein Powder (%)

| Protein (Dry Basis) | Water Content | Fat (Dry Basis) | Ash | Total Carbohydrate | Other |
|---|---|---|---|---|---|
| 56.1 | 9.2 | 1.2 | 4.2 | 26.7 | 2.6 |

compared with the material concentration of 70–90% as provided by Liang (1999) when determining the suitable material concentration for microwave sterilization of pepper powder.

**b.** Orthogonal Combination Experiment of Microwave Sterilization

On the basis of single-factor experiments, the relationship between the sterilization rate and nitrogen solubility index and the various factors are obtained through nine combined treatment experiments with four-factor and three-level orthogonal combinations. The results are analyzed and treated with DPS work platform, and the optimal process conditions are thus obtained through analysis. The analysis results show that, for the sterilization range, the optimal process combination for high sterilization rate is: microwave power 500 W, sterilization time 20 s, and material concentration 85%. Seen from the analysis results for the nitrogen solubility index range, the optimal process combinations for best nitrogen solubility index are A1B2C2, that is, microwave power 450 W, sterilization time 20 s, and material concentration 85%. Giving a comprehensive consideration to the sterilization rate and index nitrogen solubility index, it is obtained that the best process conditions for microwave sterilization are: microwave power 450 W, sterilization time 20 s, and material concentration 85%.

### 4.1.3   Product Quality Index

Quality indexes of peanut protein powder: color, milky; odor, normal fragrance of peanuts, not abnormal odor; solvent residue, not detected; fineness (100 mesh) (%), 99.76; lead (mg/kg), 0.05; arsenic (mg/kg), 0.05.

Microorganism index: total bacterial count (cfu/g), $\leq 2.2 \times 10^3$; coliform group (MPN/100 g), not detected, <30; pathogens, not detected.

Refer to Table 4.1 for the components determined from the peanut protein powder that is prepared.

## 4.2   PEANUT TISSUE PROTEIN PROCESSING TECHNOLOGY

### 4.2.1   Production Principle

The technology, equipment, and product quality indexes related to peanut tissue protein are all the same as those of soybean tissue protein, and the only difference lies in the composition and nature of the raw materials. The raw materials for peanut tissue protein production include low-denatured prepressed leached meal, mixed defatted peanut powder (mixed with peanut protein concentrate), and cold-pressed peanut meal (powder), etc. The main raw material for the production of peanut tissue protein extrusion is defatted peanut powder. The typical components of defatted peanut powder include: protein (N × 6.25) 56%, carbohydrates 32.0%, fat 6.0%, and water and ash 5.0%. Compared with defatted soybean powder, the defatted peanut powder has several advantages, such as white color, inexpensiveness, having no bean smell and low content of sulfur amino acids, and easier extrusion. Peanut tissue protein refers to the protein that is denatured after the peanuts are processed, where the protein molecules are reoriented to form a new organizational structure. As an important by-product of oil production, the defatted peanut powder has a protein content as high as 56%, and the amino acids contained in the protein are short of lysine and sulfur amino acids. The peanut tissue protein products, mainly produced from defatted peanut powder, will form a certain tissue structure or fiber structure similar to that of meat after being processed. The peanut tissue protein products, whose structure is characterized by water and fat absorption, can be added to other raw materials as food additives.

The deep-processing methods of defatted peanut powder mainly include the production of peanut tissue protein, peanut protein concentrate, and peanut protein isolate. Since complicated processes are involved during the production of peanut protein concentrate and peanut protein isolate, and many processes, such as acid and alkaline hydrolysis, washing separation and drying, etc., are required, there will be a heavy loss of peanut components and great energy consumption. However, the production process of peanut tissue protein is relatively simple, and all the components of the defatted peanut powder can be used. The quality tissue protein has a fiber structure similar to that of meat, and it can be added to meat products as a substitute for meat to improve the protein content and reduce production cost.

A single-screw extruder can be used to produce peanut tissue protein, and the high-pressure and high-shearing function of the single-screw extruder can be fully utilized to extrude the peanut tissue protein into flake-shaped products under extremely high shearing forces, and this kind of flake-shaped peanut tissue protein is fibrous.

## 4.2.2 Production Process

During production, the low-temperature defatted peanut meal and the additives are added into the mixer through the bunker after being measured. The materials are fed into the mixer for an even mixture, and then transported into the extruder for extrusion. The extruder is composed of a main shaft, enclosure, orifice plate, and cutter. The enclosure can be heated with steam pumped inside. There are toothed grooves of different shapes inside the enclosure, the helical-toothed main shaft can push the materials forward helically for extrusion. When the temperature of the materials in the enclosure rises to 160°C, the pressure reaches 1.96–3.92 MPa (20–40 kgf/cm²). Under high temperature and high pressure, the protein molecules are denatured. In the meantime, many granular proteins form objects with neat and uniform molecular arrangements. Because of the cavities formed at the outlet due to the evaporation caused by depressurization, the granular protein forms fibrous and meat-like food. Although the processing temperature is high, the time involved is extremely short, that is, 2–3 min. The extruded granular particles will go through wind transportation, degassing, and drying. After being dehydrated and cooled by a drying cooler, the food will be finely ground, graded, metered, and packaged.

When the double-screw extruder is used for producing peanut tissue protein, the supply amount of raw materials, rotating speed of screw, quantity of kneading disks, and the amount of water added will exert great influence on the product performance, and the supply amount of raw materials and the amount of water added exert the greatest influence on the products. The amount of water added has significant influence on residence time, and the more water is added the longer the residence time. The peanut protein in the cylinder is molten and then stripped after being cooled via the depressurization device. The peanut tissue protein has a special peanut flavor, and is light-colored and delicate. The main performance parameters of the peanut tissue protein produced from common leached meal are: color, brown; taste, special peanut flavor; absorbency, water absorption per 100 g dry products is 134–170 g, and after water absorption, the protein will be sponge-like and elastic; capacity, 133–169 g/L; water, about 9%; fat, 2.67% (dry basis); total protein, 56.89%; ash, 5.83%; residual leaching agent, <50 mg/kg; aflatoxin, not detected.

Taste and flavor are important parameters of peanut tissue protein. For the protein tissue of defatted peanuts, one important index is to evaluate whether there is fibrous structure inside and how does the extruded products taste. Therefore, at the time of extrusion, it is essential to remove the peculiar smell of the products, eliminate the antinutritional factors in the peanuts, and improve the taste of the peanut tissue protein. In addition, food base, vegetable oil, lecithin, salt, spices, and pigment, etc., will also be added as required.

## 4.3  PROCESSING TECHNOLOGY OF PEANUT PROTEIN CONCENTRATE

Currently, most of the peanuts in China are used for edible oil production, and peanut protein has not yet been put into wide use. In the domestic market, the peanut protein products mostly exist in the form of peanut protein flour and are used as the basic raw material in food processing. Since the protein content in the peanut protein flour is relatively lower than that in protein concentrates, it has many disadvantages, such as poor solubility, emulsifying activity, foaming property, and gelation, etc. Therefore, on the basis of peanut protein powder, it has become a top priority to further develop peanut protein concentrates of different functional properties and uses. Currently, the domestic studies on the preparation, solubility, and gelation of protein products mostly focus on soybean and rapeseed protein. There have been no reports of studies on the preparation, solubility, and gelation of peanut protein concentrates in China. The author's research team has adopted an alcohol precipitation process to prepare peanut protein concentrate and has employed a biological enzyme method to modify the protein. The preparation of peanut protein concentrate products with high solubility and gelation has promising market prospects.

### 4.3.1   Technological Process

The preparation and modification of peanut protein concentrates are as shown in Fig. 4.8.

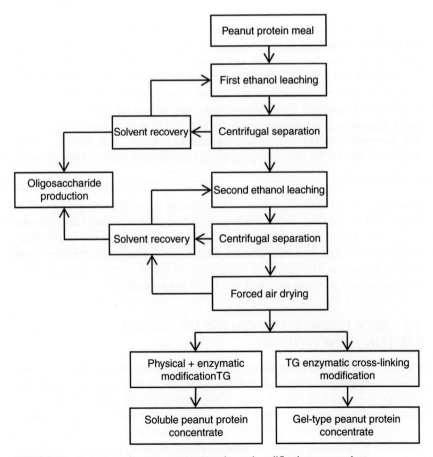

**FIGURE 4.8**   Peanut protein concentrate preparation and modification process chart.

## 4.3.2  Single-Factor Experiment of First Leaching

**1.** Determination of Solid–Liquid Ratio

As shown in Fig. 4.9, within the solid–liquid ratio of (1:6) to (1:12), the content of peanut protein concentrate (PPC) increases gradually, from 64.46% to 72.11%, increasing by 18.14% relative to peanut protein powder (protein content of dry base is 61.04%). Within (1:12) to (1:14), the PPC content changes insignificantly. Through variance analysis, the PPC content remains basically unchanged at a solid–liquid ratio of 1:12 and 1:14. However, there is a significant difference compared with the protein content at other levels. At the solid–liquid ratio of 1:12, the protein content reaches its maximal level of 72.11%. Theoretically speaking, the gradual increase of solid–liquid ratio will boost the protein leaching, remove the substances such as soluble sugar more effectively, and increase the protein content in the final products. However, when the solid–liquid ratio exceeds a certain value, an increase of the solid–liquid ratio will not cause significant change to the protein content. In the actual production, the over-high solid–liquid ratio will increase the cost of solvent used, prolong the leaching cycle, and reduce the working efficiency. Therefore, it is of little significance to have an over-high solid–liquid ratio. In this light, this experiment has not further raised the solid–liquid ratio and the final experiment has selected the solid–liquid ratio of 1:12 as the primary solid–liquid ratio.

**2.** Determination of Ethanol Concentration

The ethanol concentration has a significant influence on the protein content of PPC products (Fig. 4.10). The optimal ethanol concentration was about 70%, with which the obtained PPC products have the highest protein content, reaching 70.49%. When the ethanol concentration reaches 80%, the protein content is on the decrease, but with the continuous increase of ethanol concentration, the protein content does not change significantly. The underlying

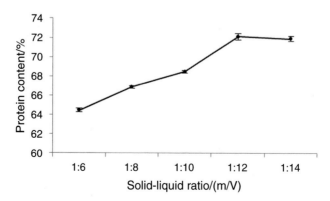

**FIGURE 4.9**  Influence of different solid-liquid ratio on PPC content.

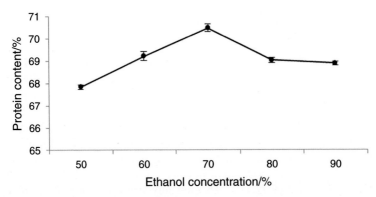

**FIGURE 4.10**    Influence of different ethanol concentration on PPC content.

mechanism might be that when the ethanol–water solution is about 70%, the protein structure of the peanut protein changes under the effect of alcohol denaturation, changing from the previous compact spherical shape into a loose structure. Consequently, the contact between ethanol–water solution and the nonprotein substances such as saccharide and pigments combined with protein. When parts of the soluble substances were extracted by the ethanol–water solution, protein content in PPC products was improved. What is more, with the further increased of ethanol concentration, some nonprotein-soluble substances (such as small molecular sugar) were precipitated in the ethanol–water solution, causing the relative decrease of protein content in the PPC products.

3. Determination of Leaching Time, Leaching Temperature, and Stirring Speed

The variance analysis indicates that the protein content of PPC products does not change significantly at different levels of leaching time, leaching temperature, and stirring speed. As shown in Fig. 4.11, with the prolonging of leaching time, the protein content of the PPC products first increases and then decreases slowly, increasing from 68.8% at 20 min of leaching to the maximal content of 68.9% at 40 min of leaching. After the leaching time is prolonged, the protein content decreases and the protein content is 68.28% at 100 min of leaching. As seen from the numerical value, the leaching time does not exert significant influence on the protein of PPC products, and therefore it is determined that the maximal leaching time is 40 min in the process parameter optimization. As seen from Fig. 4.12, the protein content of PPC products first increases and then decreases with the increase of leaching time, and when the temperature increases from 20 to 40°C, the protein content increases from 68.25% to 69.66%, and the protein content of PPC products reaches maximal value at 40°C; when the temperature increases from 40 to 60°C, the protein content decreases from 69.66% to 68.76%, and

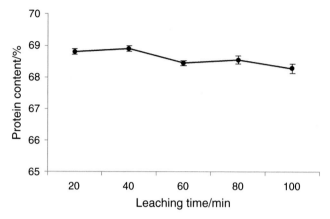

**FIGURE 4.11**  Influence of leaching time on PPC content.

therefore it is determined that the leaching temperature is 40°C in the process parameter optimization. The reason might be that, with the rise of temperature, the dissolution rate of the soluble substances of the peanut protein in the ethanol–water solution increases, which in turn boosts the leaching effect, speeds up the leaching rate of the soluble substances, and increases the protein content in the products. However, over-high temperature might cause the protein to denature to a certain degree and reduce its solubility, which will in turn retard the leaching of soluble substances and reduce the protein content in the products. In Fig. 4.13, the stirring speed does not exert significant influence on the protein content, which increases from 66.10% to 66.10% at the stirring speed of 70–130 r/min, and the protein content reaches the highest value at 70–130 r/min. In the following period, with the

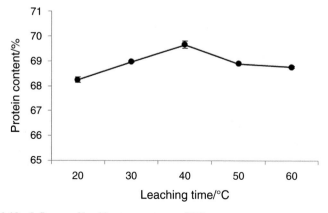

**FIGURE 4.12**  Influence of leaching temperature on PPC content.

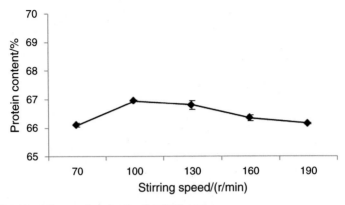

**FIGURE 4.13**    Influence of stirring speed on PPC content.

increase of the stirring speed, the protein content of the products does not change significantly. To reduce energy consumption, it is determined that the optimal stirring speed is 130 r/min in production.

## 4.3.3    Single-Factor Experiment of Secondary Leaching

As shown from the experimental results of the first single-factor experiment, stirring speed does not have a great influence on the protein content of products, and therefore the stirring speed of the secondary leaching is also determined to be 130 r/min.

The variance analysis shows that the solid–liquid ratio and ethanol concentration have significant influence on the protein content of products, but the leaching time and temperature do not have significant influence on the protein content of products. Fig. 4.14 shows that the protein content of products gradually increases to the maximal value and then decreases with the increase of the solid–liquid ratio and then the changing trend levels off. When the solid–liquid ratio changes from 1:5 to 1:15, the protein content of the products increases from 63.88% to 72.18%; when the solid–liquid ratio is 1:15, the protein content reaches the maximal value, with 18.25% more protein content than that of the original peanut protein powder (protein content of the dry basis is 61.04%). When the concentration of ethanol–water solution increases (Fig. 4.15), from 55% to 75%, the protein content of the products increases from 66.62% to 71.72%; however, the protein content does not change significantly with the continuous increase of ethanol concentration. This is because high-concentration ethanol will cause changes to the protein structure and increase the degree of protein denaturation, which not only prevents the combination between the nonprotein-soluble substances that combines with the protein and the ethanol but also decreases the functional properties of the products. From the perspective of production, the increase of ethanol concentration directly leads to

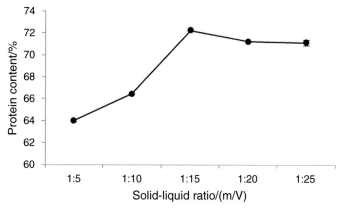

**FIGURE 4.14**  Influence of solid-liquid ration on PPC content.

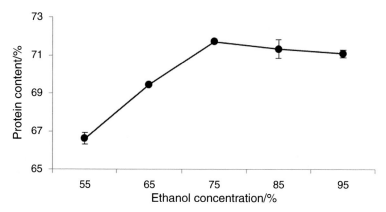

**FIGURE 4.15**  Influence of ethanol concentration on PPC content.

the increase of production cost, and therefore it is advised to adopt an ethanol concentration of 75%. When the leaching time is extended from 10 to 50 min (Fig. 4.16), the protein content does not change significantly, and therefore the optimal leaching time is determined to be 30 min. As shown in Fig. 4.17, when the leaching temperature is extended from 40 to 60°C, the protein content does not change significantly, with the average value being 69.51%, and the protein content reaches the highest value of 69.78% at 50°C. Therefore, the leaching temperature is determined to be 50°C.

## 4.3.4  Orthogonal Rotational Combination Experiment to Prepare Peanut Protein Concentrate Through Secondary Leaching

From the single-factor experiment results, it can be seen that the difference mainly lies on the solid–liquid ratio and ethanol concentration between the first

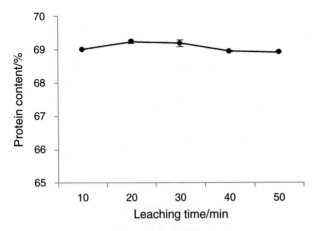

**FIGURE 4.16**    Influence of leaching time on PPC content.

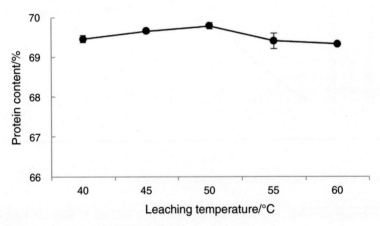

**FIGURE 4.17**    Influence of leaching temperature on PPC content.

leaching and secondary leaching, and other factors basically remain the same, such as leaching temperature, leaching time, and stirring speed. Therefore, according to the single-factor experimental results, the point where the highest protein content is obtained is selected to determine the parameters of the three factors as follows: for the first leaching, the leaching temperature is 40°C, the leaching time is 40 min, and the stirring speed is 130 r/min; for the secondary leaching, the leaching temperature is 50°C, the leaching time is 30 min, and the stirring speed is 130 r/min. Quaternary quadratic orthogonal rotational experiment is carried out with the solid–liquid ratio ($X_1$) and ethanol concentration

($X_2$) of the first leaching and the solid–liquid ratio ($X_3$) and ethanol concentration ($X_4$) of the secondary leaching. DPS working platform is used to process and analyze the results, establishing the mathematical model for the relationship between protein content of the products and various factors so as to obtain the optimal process conditions through model analysis.

Quadratic polynomial stepwise regression analysis is carried out for the experimental results. With the correlation coefficient $R$ of the regression model as an indicator, the insignificant phases will be deleted semiautomatically to finally obtain the regression model, as follows:

$$Y = 74.989\,392\,2 - 2.111\,483\,244\,7X_1 - 1.729\,715\,4426X_2 + 2.208\,726\,3124X_3$$
$$+0.774\,527\,440\,1X_4 - 2.276\,854\,3575X_1 \times X_1 - 0.926\,280\,4452X_2 \times X_2$$
$$- 1.933\,907\,5787X_3 \times X_3 - 2.420\,043\,4765X_4 \times X_4 - 0.831\,250\,000\,0$$

$$(4.1)$$

The correlation coefficient of the regression model is $R = 0.9510$ and the Durbin–Watson statistic is $d = 2.2759$, indicating that the residual errors are mutually independent. Therefore, the regression model established by the experiment is warranted.

The optimal combination to ensure the highest protein content according to the regression model is $X_1 = -0.6679$, $X_2 = -0.6340$, $X_3 = 0.8627$, and $X_4 = 0.1600$, that is, for the first leaching, the solid–liquid ratio is 1:10, the ethanol concentration is 63.7%; for the secondary leaching, the solid–liquid ratio is 1:19 and the ethanol concentration is 76.6%. The protein content (dry basis) of products determined under these experimental conditions is 77.26%. The optimal combination is basically consistent with the results of the single-factor experiments.

As seen from the model analysis results, within the range selected by the experiment, with the increase of the solid–liquid ratio and ethanol concentration, the protein content of the products decreases gradually. Therefore, by selecting the correct solid–liquid ratio and ethanol concentration during production, the energy consumption can be reduced and the equipment utilization can be improved. The single-factor experimental results indicate that when the solid–liquid ratio changes from 1:10 to 1:14 in the first leaching, the protein content of the products does not change significantly, and therefore the ratio of 1:12 is selected as the optimal solid–liquid ratio; when the ethanol concentration is 70%, the protein content changes significantly, but the protein content does not change significantly with the continuous increase or decrease of ethanol concentration. For the secondary leaching, when the solid–liquid ratio changes from 1:15 to 1:20, the protein content does not increase significantly, and therefore the ratio of 1:15 is selected as the optimal solid–liquid ratio. Within the range of 70% to 80% of ethanol concentration, the protein content changes significantly, and the protein content of the products reaches its highest level at about 75% but the protein content does not change significantly with the continuous increase or decrease of ethanol concentration. Studies show that the increase of ethanol concentration can help to remove the lipids, flavor precursors, and pigments that combine with the protein in the peanut protein powder, making their content significantly reduced in the alcoholic-leached peanut protein concentrate (for these substances can be dissolved by ethanol), and therefore the alcoholic-leached protein powder can be used to remove the peculiar smell and lighten the color.

Other studies reveal that the mechanism of ethanol-induced protein denaturation is different from thermal denaturation. In thermal denaturation, the protein becomes loose and disorderly, but the ethanol-induced denaturation results in the restructuring of the protein molecule to form a structure more orderly than that of the natural peanut protein, and driven by entropy changes, the protein aggregation particles are formed with the self-aggregation cycle, and the protein aggregation particles have relatively larger rigidity as well as greater and tighter conformational force, and the force that maintains such compact conformation is the secondary bonds that have relatively low bond energy.

By summarizing the practical and theoretical analyses, the author's research group proposes that the processes of the preparation of peanut protein concentrate through secondary leaching are as follows: for the first leaching, the solid–liquid ratio is 1:12, the ethanol concentration is 70%, the leaching time is 40 min, the leaching temperature is 40°C, and the stirring speed is 130 r/min; for the secondary leaching, the solid–liquid ratio is 1:15, the ethanol concentration is 75%, the leaching time is 30 min, the leaching temperature is 50°C, and the stirring speed is 130 r/min, and the content of the obtained peanut protein concentrate reaches 72.11%.

## 4.4 PROCESSING TECHNOLOGY OF PEANUT PROTEIN ISOLATE

Removing nonprotein component from low-temperature peanut meal, we obtain peanut protein isolate, which is a powdery high protein peanut product. In comparison with peanut protein concentrate, the soluble saccharides and insoluble glycan are removed from the peanut protein isolate, which also has a higher protein content (more than 90%, on dry basis). The preparation technology of peanut protein isolate includes alkali-solution and acid-isolation, separation membrane, etc. Nowadays, the production of peanut protein isolate both in China and abroad mainly relies on alkali-solution and acid-isolation, which, according to the degree of automation, can be divided into three types, that is, continuous, semicontinuous, and intermittent. The intermittent type, with advantages such as good maneuverability and lower construction cost, is commonly used in many small and medium-sized enterprises in China. The upfront investment for continuous and semicontinuous automation is higher, thus they are normally used in large-scale production lines. This section will describe the specific technology of alkali-solution and acid-isolation.

## 4.4.1 Process Principle

The main principle of alkali-solution and acid-isolation is that the protein in defatted peanut flour adequately dissolves in strong acid and neutral alkali and thus separates from other insoluble components. When the pH is about 4.5 (the isoelectric point of peanut protein), the solubility of protein is low and the protein will smoothly and completely precipitate out of the filtrate. In this case, the extraction and separation of peanut protein is completed.

After leaching in dilute lye, filtration or centrifugal separation, the insoluble substances, mainly polysaccharides or protein residuals, are removed from low-temperature defatted peanut flour. After adding acid in the supernatant and adjusting the pH to the isoelectric point, the protein will agglutinate and precipitate. After separation and drying of the protein precipitate, we get peanut protein isolate. Strong acid and strong base may cause excessive denaturation. Therefore, the ideal pH of the extracting solution is normally 7–9. Nowadays, when precipitating and concentrating protein in the filtrate, a NaOH solution of pH 7–8 is employed to extract peanut protein and hydrochloric acid is employed to adjust pH to about 4.5.

## 4.4.2 Requirements of Raw Materials and Auxiliary Materials

The quality and yield of peanut protein isolate depend on the quality of defatted peanut meal. The defatted meal, which is used to produce peanut protein isolate, should have no mildew, low levels of peanut skin, low impurity, high protein content, and high NSI. The specific requirements are as follows.

The peanut meal shall be milk white, powdery, normal smell, no mildew, or impurity; the protein content (dry base) should be ≥50%, fat content ≤1%, moisture content ≤8%, fiber content ≤4%, NSI >70%, and hygienic index should be met.

## 4.4.3 Production Process

The production process of peanut protein isolate can refer to the production process of soybean protein isolate. However, the specific technological parameter in each link is different. See Fig. 4.18 for the specific technological process.

1. Grinding and Leaching

   A dry grinding method is commonly employed in the production of peanut protein isolate. The peanut meal should be pulverized to 40–60 mesh in advance and then water is added for leaching. Secondary wet grinding, which has a significant effect on the dissolution rate and leaching efficiency, could also be employed. In secondary wet grinding, the defatted peanut meal is extracted in water for a period of time and then a colloid mill is used for secondary grinding and leaching.

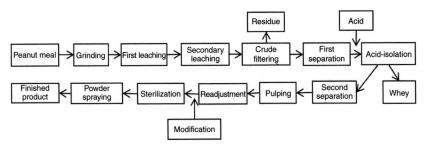

**FIGURE 4.18**  Technological process of peanut protein isolate production.

Li et al. (2004) defined the optimum condition for extraction of alkali liquor and compared the extraction ratio between the first leaching and second extraction. They found that the protein yield in the second extraction was far lower than that in the first leaching. In this case, we find that most of the protein has been extracted after two sets of leaching.

The factors which influence the dissolution rate and leaching efficiency include particle size of raw material, grinding method, pH, water addition, leaching temperature, and time.

The more extracting solution is used, the higher the dissolution rate and leaching efficiency will be. However, extra extracting solution will cause acid precipitation difficulty and higher cost. In addition, there will be greater difficulty in disposing and discharging of industrial waste water. Generally, the water added in the production of protein isolate is 8–20 times as much as the raw material.

Leaching temperature has a relatively high influence on the extraction ratio of peanut protein isolate. The extraction ratio will increase as the temperature increases. However, if the temperature exceeds 60°C, the extraction ratio will decrease. And if the temperature exceeds 65–70°C, the denaturation of peanut protein will begin. Therefore, too high a temperature will decrease the dissolution rate and solubility as well as other functions. In addition, the energy consumption of high temperature extraction is relatively high. Generally, the extraction temperature is 30–60°C.

Leaching time will influence the dissolution rate of protein. Under certain conditions, the dissolution rate will increase as the time increases. However, after a certain time, the dissolution rate will reach a stable value. Considering all kinds of indexes, the leaching time is usually less than 120 min.

We have discussed the influence of pH on the solubility of protein. When the pH is greater than 7, the dissolution rate will increase as the pH increases. However, strong alkaline conditions will cause "cystine lysine response" and amino acids will change into toxic compounds. In addition, the flavor and edible value of the product will be influenced. Therefore, the pH shall not be too high. Generally, it is 7. 0–8.5.

The first leaching and secondary leaching of peanut protein isolate refer to the extraction of soybean protein isolate and generally occur as follows: the defatted peanut flour is placed into the leaching tank; 8–16 times as much water as the mass of defatted flour is added; the pH of the leaching solution is adjusted to about 8.0 by NaOH solution; the solution is stirred at the speed of 30–35 r/min; the extraction liquor is poured out through the filter cartridge for secondary leaching. For the second leaching 4–10 times as much water as the mass of the residual of the first leaching is added; the pH is adjusted to about 8.0 by NaOH solution; the leaching is continued for 60–90 min under the same conditions as the first leaching.

2. Prefiltration and Primary Separation

   The aim of prefiltration and primary separation is to remove insoluble residuals. When the secondary leaching solution is poured through the filter cartridge then the prefiltration is completed. The extraction solutions of the first and second extraction are put together and then the thin residual is removed through centrifugal separation. The size of the filter cartridge used in the prefiltration is normally 60–80 mesh and the size of the filter cartridge used in centrifugal machine is generally 100–140 mesh.

3. Acid-Isolation

   The extraction solution is pumped into an acid-isolation tank, which is stirred continuously and 10–35% of the edible hydrochloric acid solution is slowly added, adjusting the pH to 4.4–4.6, and the peanut protein will precipitate under these conditions. When adding acid, the pH should be monitored. If the pH reaches isoelectric point, stirring is stopped for 20–30 min and the protein particle will precipitate. The pH has influence on the yield and purity of peanut protein isolate. The speed of protein particle formation and precipitate should be as fast as possible. The speed of acid addition and the stirring speed are the key influencing factors. Inappropriate handling will cause the following consequences: although the pH reaches isoelectric point, the agglutination and sedimentation speed are slow, the moisture content of the particle is high, and the supernatant is muddy. In this case, the yield will be low and it will have negative effects on the later processes. The optimum stirring speed should be 30–40 r/min. If the above-mentioned phenomenon happens because of inappropriate handling, you can reduce the pH at the isoelectric point and keep stirring. After the peanut protein is completely dissolved, add strong alkali liquor again and adjust the pH to the isoelectric point and the protein will rapidly agglutinate and precipitate.

4. Secondary Separation and Washing

   Secondary separation refers to the procedure of centrifugal dewatering of precipitate and removing of supernatant. After acid-isolation and centrifugation, a large amount of hydrogen ions will remain in the protein curd. Wash twice with 50–60°C warm water and the hydrogen will be removed. The pH of protein solution, after washing, will increase to about 6. In addition, hydrous ethanol can be used for washing. The advantage of this is avoiding the loss of protein and the disadvantage is a certain denaturation of protein and an influence on the product quality.

5. Pulping, Readjustment, and Modification

The protein precipitates generated from secondary separation and centrifugal machine are mostly lumpy and curdy. Before spray drying, we add moderate water and then whip or grind to make an even serosity. To improve the dispersibility of curd protein and the practicability of the end product, a 5% NaOH solution is to be employed to adjust the pH to 6.5–7.0, the stirring speed should be 85 r/min. To improve the functional characteristics of peanut protein and to meet the additional requirements of various food systems, modification, including physical, chemical, and emzymatic, is employed during pulping and/or after readjustment. In this case, one or several functional characteristics of peanut protein can be adjusted or reinforced purposefully.

6. Sterilization and Drying

The peanut protein isolate after pulping, readjustment, and modification is dried through sterilization handling at 140°C for 15 s. Spray drying process is normally used in the production of peanut protein isolate powder. The stock consistency is a key influence factor to the drying effect during spray drying. If the density and viscosity of stock entering the spray drying tower is high, the spray nozzle may be blocked and the atomization effect may be influenced. This may cause the instability of the spray tower. On the contrary, if the density of the stock is low and the particles are small while the mass volume is big, it is not suitable for application and transportation. A reasonable density should be 12–20% and we also need to consider the viscosity. Pressure spraying is normally used in producing peanut protein isolate. In this case, stock density and product quality can be improved. The temperature should be 150–170°C during powder-spraying, the tower body temperature should be 95–100°C, and the moisture removal temperature should be 85–90°C.

Alkali-solution and acid-isolation are normally used in the industrial production of peanut protein isolate. It is a simple and practicable method; however, there are several defects and deficiency, including halfway removing of soluble component, high consumption of acid–base of water, low yield of protein, high production cost, and waste water treatment pressure, etc.

## 4.5 PEANUT PROTEIN COMPONENT PREPARATION TECHNOLOGY

The peanut protein is divided into two categories: 90% are salt-soluble protein and 10% are water-soluble protein (Lusas, 1979). The salt-soluble protein mainly includes arachin (14S), conarachin I (2S), and conarachin II (7.8S), and their content is respectively 73% arachin, 6% conarachin I, and 21% conarachin

II (Huang and Fu, 1992). The arachin and conarachin are abundant and important components in peanuts, and their nature can influence various functional properties of peanut protein (Li, 2008). Currently, the arachin and conarachin are mainly extracted by the ammonium sulfate precipitation method (Tombs, 1965). The extraction of the peanut protein component with the ammonium sulfate precipitation method is based on the salting-out feature of the protein, through which the three components of arachin, conarachin I, and conarachin II can be divided. Due to the conditional restriction, the ammonium sulfate precipitation method only applies to the laboratory study of small-scale extraction instead of industrial production. The cold precipitation method refers to the method to separate the peanut protein component taking advantage of the characteristic that arachin can precipitate at low temperature. The cold-precipitated protein is thermodynamically reversible, that is, the protein can aggregate and precipitate at low temperature; however, when the temperature rises, the protein that has previously precipitated due to low temperature will be redissolved. This cold precipitation phenomenon can also be found in soybean protein, and many researchers have searched for the process for the separation and extraction of glycinin (11S) and conglycinin (7S). However, for the cold-precipitated protein of peanuts, the limited research mainly date to the 1970s and 1980s, during which, Basha and Pancholy (1982) separated and cold-precipitated the protein with Sephacryl S-300 column chromatography and discussed the influence of cold precipitation time, ionic strength, solution pH value, and other factors on the protein, and they also separated and prepared cold-precipitated protein I and cold-precipitated protein II.

The research team of the author has used ammonium sulfate precipitation method and cold precipitation method to separate and extract arachin and conarachin, respectively, and has determined the process conditions for separating and preparing peanut protein component with the ammonium sulfate precipitation method and the cold precipitation method so as to provide theoretical and methodological guidance for the industrialized preparation of the two components.

### 4.5.1 Fractional Separation With Ammonium Sulfate Precipitation Method

1. Comparison Among Different Peanut Total Protein Leaching Methods
   The peanut protein is mainly composed of peanut storage protein (Fig. 4.19), including arachin (40.5, 37.5, 35.5, 23.5 kDa), conarachin II (61.0 kDa), and conarachin I (15.5, 17.0, and 18.0 kDa).
   As shown in Fig. 4.20, the protein supernatant extracted with phosphate buffered saline (PBS) is precipitated with the ammonium sulfate of 40% saturation to obtain relatively pure arachin. Through optical density analysis, the component purity can reach 85.53 ± 0.86%. The supernatant collected with such precipitation is continuously precipitated with the ammonium sulfate of 65% saturation. As shown in Fig. 4.20, the main components are

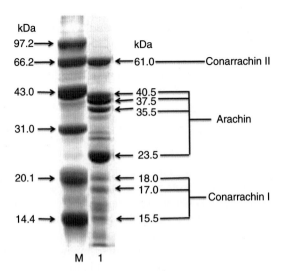

**FIGURE 4.19** Peanut protein component SDS-PAGE Atlas M. Standard protein; 1. Peanut protein isolate.

the mixture of arachin and conarachin. Through optical density analysis, the component purity is $73.91 \pm 0.58\%$. After the mixture obtained with ammonium sulfate of 40–65% saturation is removed, the ammonium sulfate of 85% saturation is used to precipitate the supernatant to separate and extract relatively pure conarachin, whose purity can reach $75.81 \pm 1.02\%$.

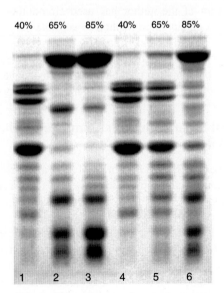

**FIGURE 4.20** SDS-PAGE atlas of peanut protein component prepared with two different extraction methods.

In Fig. 4.20, 1, 2, and 3 are ammonium sulfate precipitation components of different saturations extracted with phosphate buffered saline; 4, 5, and 6 are ammonium sulfate precipitation components of different saturations extracted with Tris-HCI method.

Tris-HCI method is used to extract protein supernatant, and the ammonium sulfate of different saturations is adopted for precipitation to obtain the protein component of corresponding purities, all of which are lower than those obtained with PBS extraction method through optical density analysis. Therefore, the experiment ultimately determines that the PBS extraction method should be used for the total protein leaching during the fractional separation of peanut protein components. The electrophoretic band obtained with this method basically has no hybrid band pollution, indicating a high purity of the arachin and conarachin, which can be used for the follow-up experiment.

2. The Influence of Temperature on Fractional Preparation of Peanut Protein Component With Ammonium Sulfate Precipitation Method

   As shown in Fig. 4.21, the temperature does not have significant influence on the fractional separation of peanut protein component. At a temperature of 25 and 4°C, the main component obtained with precipitation using the ammonium sulfate of 40% saturation is arachin, and through optical density analysis, the purity is respectively 83.97 ± 0.62% and 85.24 ± 0.99%. Accordingly, at a temperature of 25 and 4°C, the main component obtained with precipitation using the ammonium sulfate of 85% saturation is conarachin, and through optical density analysis, the purity is respectively 76.25 ± 0.83% and 75.58 ± 0.57%. Since the difference of the purity of the components obtained under the two conditions is minimal, we select fractional preparation of the peanut protein component at room temperature (25°C) in the follow-up experiment.

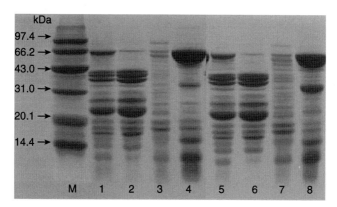

**FIGURE 4.21**   The influence on the ammonium sulfate fractionation effect for peanut protein component. 1,5, 0 saturated ammonium sulfate; 2,6, 40% saturated ammonium sulfate; 3,7, 65% ammonium sulfate; 4,8, 85% ammonium sulfate, where, the strip 1, 2, 3, 4 are under the temperature of 25°C; while the strip 5,6,7,8 are under the temprature of 4°C.

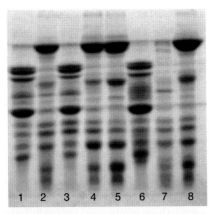

**FIGURE 4.22** The influence of NaCl salt ionic concentration on ammonium sulfate fractionation method. 1. 32% (the extracting solution is added with 0.5 mol/L of NaCI); 2. 80% component (the extracting solution is added with 0.5% NaCI); 3. 40% component (the extracting solution is added with 10% NaCI); 4. 65% component (the extracting solution is added with 10% NaCI); 5. 85% component (the extracting solution is added with 10% NaCI); 6. 40% component (the extracting solution is added with NaCI of 0.5 mol/L); 7. 70% component (the extracting solution is added with NaCI of 0.5 mol/L); 8. 85% component (the extracting solution is added with NaCI of 0.5 mol/L).

3. The Influence of NaCI Ionic Concentration on Peanut Protein Components Fractionally Prepared With Ammonium Sulfate Precipitation Method

   Fig. 4.22 shows that, by using 0.01 mol/L PBS (pH7.9) added with 0.5 mol/L NaCl for fractional separation, we can obtain the arachin component, as represented by No. 6 strip, reaching 85.14± 0.81% through optical density analysis. The conarachin, as represented by No. 8 strip, can reach a purity of 76.47 ± 0.67%. However, among the components obtained by using 0.01 mol/L PBS (pH7.9) added with 10% NaCl, the No. 3 strip represents the arachin component, whose purity is 76.59 ± 0.73%, lower than the purity of No. 6 strip; No. 5 strip represents the conarachin component, whose purity is 65.59 ± 0.88%, which is lower than that represented by No. 8. Therefore, the 0.01 mol/L PBS (pH 7.9) added with 0.5 mol/L NaCl is adopted for fractional separation of the arachin and conarachin components.

4. Peanut Protein Component Fractionation Process

   Through the study of the above factors, it is determined that the solid–liquid ratio is 1:12. The defatted powder is added into the PBS (containing 0.5 mol/L NaCl) whose concentration is 0.01 mol/L. This is stirred at high speed for 10 min, the pH is regulated to 7.9, and the pulp is left in a bath oscillator at the extraction temperature of 50°C for 70 min for reaction. Afterward, it is filtered with gauze, and the filtrate is centrifuged for 10 min at a speed of 4000 r/min, and the supernatant is the total protein leaching solution. The fractional separation flow of the arachin and conarachin components is as shown in Fig. 4.23.

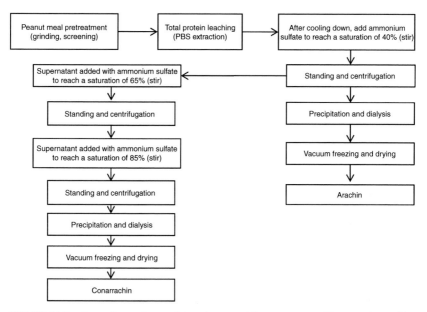

**FIGURE 4.23** Chart of preparing arachin and conarrachin components with ammonium sulfate precipitation method.

**a.** Preparation of Peanut Arachin

Fetch and put the aforesaid peanut meal total protein leaching solution in an ice-bathed tank for rapid cooling to about 4°C, slowly add the well-ground solid ammonium sulfate powder (finished within 5–10 min) to make the saturation of the ammonium sulfate reach 40%, leave it to stand for 3 h, centrifuge it for 20 min at 10,000 × g, and collect the precipitates. Put the precipitates into the dialysis bag (preboiled with ethanol). Use double-distilled water to dialyze for 24 h (4°C, constantly stirred), changing the water four or five times during this period. The samples are then frozen and dried to obtain arachin.

**b.** Preparation of Conarachin

Add solid ammonium sulfate into the solution, which has reached 40% ammonium sulfate saturation and had the arachin removed, to reach a saturation of 65%. Leave the solution to stand for 3 h, centrifuge it for 20 min at 10,000 × g, and remove the precipitates. After the removal, continue to add ammonium sulfate into the supernatant to reach a saturation of 85%, leave it to stand for 3 h, centrifuge it for 20 min at 40,000 × g, and remove the precipitates. Put the precipitates into the dialysis bag (preboiled with ethanol). Use double-distilled water to dialyze for 24 h (4°C, constantly stirred),

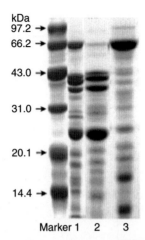

**FIGURE 4.24** The electrophoretograms of arachin and conarachin prepared with improved ammonium sulfate precipitation method. 1. Peanut protein isolate; 2. Arachin; 3. Conarachin.

changing the water four or five times during this period. The samples are then frozen and dried to obtain conarachin. The electrophoretograms of arachin and conarachin which are prepared with ammonium sulfate precipitation method are, respectively, as shown in Fig. 4.24.

## 4.5.2 Fractional Separation With Low-Temperature Cold Precipitation Method

1. Arachin
   Compound the defatted peanut powder with PBS solution of certain pH value and ionic strength. Stir for 1 h at normal temperature and centrifuge for 30 min at 8000 r/min at normal temperature. Leave the supernatant to stand for certain period of time at 2°C, and then centrifuge it for 30 min at 8000 r/min. The precipitates, after being frozen and dried in vacuum, become arachin. The technological process is as shown in Fig. 4.25.

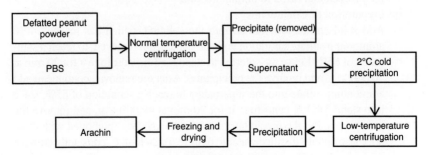

**FIGURE 4.25** Flow chart of preparing arachin component with cold precipitation method.

Through optical density analysis, the arachin obtained with the above process has a purity of 61.07± 0.15%, and therefore further exploration is required to improve the extraction process for higher purity.

During the extraction process, the protein content of the supernatant from the first extraction of peanut protein powder is represented by M1; the protein content of the centrifuged supernatant from the second cold precipitation after overnight standing is represented by M2. By calculating the protein (conarachin) content in the supernatant coming from cold precipitation, the precipitates obtained from low-temperature configuration are arachin. Therefore,

$$\text{Peanut protein extraction rate}(\%) = [(M^1 - M^2)/$$
$$\text{Total Content of Protein in Peanuts}] \times 100\%$$

$$(4.2)$$

**a.** Number of Leaching Times of Peanut Total Protein

Under the conditions of the solid–liquid ratio being 4:10, ionic strength being 0.2 mol/L, and the pH value of leaching solution being 7.5, the solution was leached once, twice, three times, four times, and five times. The optimal number of leaching times was determined according to the total protein extraction rate. The experimental results are shown in Fig. 4.26.

As seen in Fig. 4.26, with the increase of the number of leaching times, the peanut protein extraction rate is on the rise constantly. However, the extraction rate no longer changes significantly after the second time of leaching, and the peanut protein extraction rate only increases by

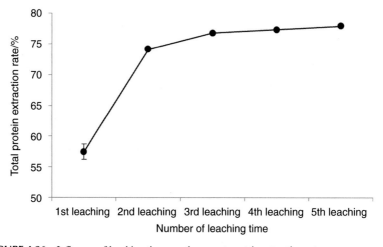

**FIGURE 4.26** Influence of leaching times on the peanut protein extraction rate.

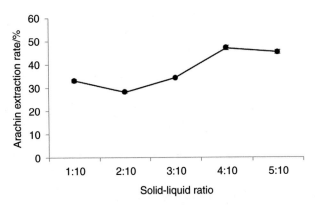

**FIGURE 4.27** Influence of solid-liquid ration on peanut protein extraction rate.

2.62% at the third leaching. The production cost rises with the increase of the number of leaching times, as does the industrial waste generated. Taking into consideration the factors of energy conservation and emission reduction, the optimal number of leaching times is determined to be twice.

**b.** Arachin Single-Factor Experiment

- Solid–Liquid Ratio

  Under the conditions of ionic strength being 0.2 mol/L, the pH value of leaching solution being 7.5, and the cold precipitation time being 16 h, the solution was leached once, The optimal solid–liquid ratio was determined according to the arachin extraction rate. The experimental results are shown in Fig. 4.27.

  As shown in Fig. 4.27, with the change of the solid–liquid ratio, the arachin extraction rate assumes a gradual upward trend. However, when the solid–liquid ratio reaches 4:10 and 5:10, the arachin extraction rate no longer changes significantly. Therefore, the optimal solid–liquid ratio is determined to be 4:10.

- Ionic Strength

  Under the conditions of the solid–liquid ratio being 4:10, the pH value of leaching solution being 7.5, and the cold precipitation time being 16 h, the solution was leached once. The optimal solid–liquid ratio was determined according to the arachin extraction rate. The experimental results are shown in Fig. 4.28. It can be seen that the arachin extraction rate increases gradually with the increase of the ionic strength. However, after the ionic strength becomes greater than 0.3 mol/L, the arachin extraction rate no longer changes significantly and begins to level off. Therefore, the optimal ionic strength is determined to be 0.3 mol/L.

- Under the conditions of the solid–liquid ratio being 4:10, the ionic strength being 0.2 mol/L, and the cold precipitation time being 16 h,

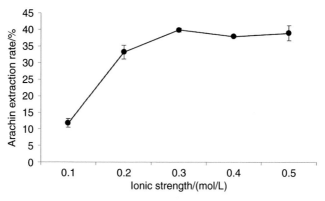

**FIGURE 4.28**  Influence of ionic strength on peanut protein extraction rate.

the solution was leached once. The optimal pH value of the solution was determined according to the arachin extraction rate, as shown in Fig. 4.29. It is obvious that with the increase of the pH value of the solution, the arachin extraction rate rises gradually. It is because the protein extracted with the cold precipitation method has low denaturation, and the pH value of the solution should be kept at a low level so as to suppress the protein denaturation. In addition, there are also reports on the separation of peanut protein component with the cold precipitation method (Neucere et al., 1969), which state that the extraction pH value should be maintained within 8.0. Therefore, the optimal pH value is determined to be 7.5.

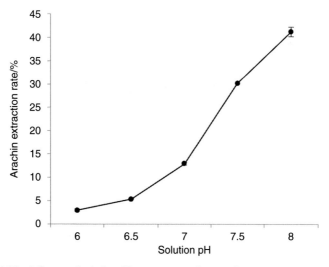

**FIGURE 4.29**  Influence of solution pH on peanut protein extraction rate.

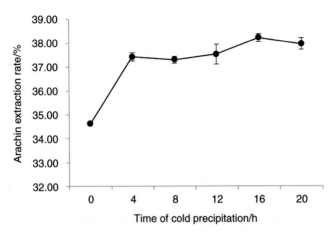

**FIGURE 4.30** Influence of cold precipitation time on peanut protein extraction rate.

- Cold Precipitation Time

  Under the conditions of the solid–liquid ratio being 4:10, the ionic strength being 0.2 mol/L, and the solution pH being 16 h, the solution was leached once. The optimal cold precipitation time was determined according to the arachin extraction rate, as shown in Fig. 4.30. It is obvious that with the prolonging of the cold precipitation time of the solution, the arachin extraction rate is on the rise. Within 0–4 h, the arachin extraction rate shows a sharp increase with the prolonging of the cold precipitation time. However, after 4 h of cold precipitation, the arachin extraction rate slows down. Further prolonging of the precipitation time has negligible effect on the extraction rate, possibly because the cold precipitation of arachin has reached a dynamic equilibrium. Therefore, the appropriate time for cold precipitation is 4 h.

c. Arachin Cold Precipitation and Extraction Conditions

- Orthogonal Optimization Experiment

  On the basis of single-factor experiments, an $L_9(4^3)$ four-factor and four-level orthogonal experiment was designed, and Duncan's new multiple range method was adopted to carry out significance analysis for the influence of various factors. The optimal combination is: solid–liquid ratio of 5:10, solution pH of 7.9, ionic strength of 0.4 mol/L, and time of cold precipitation of 4 h.

- Experiment Optimization and Verification

  The optimal leaching process is used to extract arachin, and the supertanant obtained from centrifugation is dried to become arachin, whose purity is $76.40 \pm 0.53\%$ (Fig. 4.31).

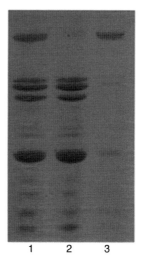

**FIGURE 4.31**  Electrophoresis spectrums of arachin and conarachin prepared with cold precipitation method. 1. Peanut protein isolate; 2. Arachin; 3. Conarachin.

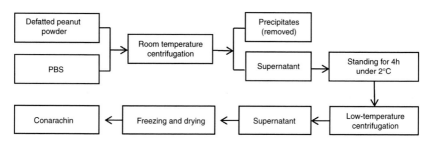

**FIGURE 4.32**  Flow chart of preparing conarachin component with cold precipitation method.

**2.** Conarachin

The defatted peanut powder is compounded with PBS of certain pH and ionic strength. Stir it at a normal temperature for 1 h and centrifuge it for 30 min at 8000 r/min at room temperature. Then at a temperature of 2°C, centrifuge it for 30 min at 8000 r/min. The supernatant is directly frozen in vacuum or dried after acid precipitation to obtain conarachin (refer to Fig. 4.32 for the technological process), whose purity is $58.35 \pm 0.35\%$ (Fig. 4.31).

## 4.6  PEANUT PROTEIN FILM PREPARATION TECHNOLOGY

In recent years, protein films displaying biological degradation and environmental properties have drawn more and more attention. The peanut protein molecules, which contain a large amount of hydrogen bonds, hydrophobic bonds, disulfide bonds, Van der Waals force, ionic bonds, and coordinate bonds, have high film formation performance (Feng and Liu, 2004). In addition, the peanut protein film can prevent the evaporation of water from the inside of food, retard the oxidization between oxygen in the air and food, prevent the breeding of microorganisms, and prolong the shelf life of food. Compared with synthetic packaging materials, protein film is biodegradable, thus avoiding any form of environmental pollution, and it is therefore a form of environmentally friendly packaging material with promising market prospects. Therefore, the research on the development of products related to peanut protein film has very important significance in developing new ways of peanut protein utilization, providing added value to the peanut industry, raising the farmers' incomes in the production area, and reducing environmental pollution and the environmental cost of enterprise production.

### 4.6.1   Production Principle of Peanut Protein Film

The natural protein relies on the hydrogen bonds, ionic bonds, hydrophobic interaction, dipolar interaction, and disulfide bonds in the molecules to maintain its stable structure. When the protein is dissolved in solution, different treatment methods are adopted to dissociate the subunits, denature the molecules, expose the internal hydrophobic groups, and reinforce the intermolecular interaction. In the meantime, the molecular interaction is enhanced, and some disulfide bonds in the molecules break and reform new disulfide bonds to establish a three-dimensional network structure. The hydrophobic groups in the protein molecules on the air–water interface extend into the air while the hydrophilic groups remain in water so that the directionally aligned protein layer is formed at the water level, and protein film with certain mechanical strength and barrier property can be obtained under proper conditions.

### 4.6.2   Peanut Protein Film Production Process

The protein film forming process generally comprises: Compound protein solution of certain concentration → Add certain amount of various modifiers → Adjust pH → Water-bath shaking under constant temperature → Vacuum degassing → Coat and dry under certain conditions → Store the film under certain conditional after removal. Therefore, the factors that affect the protein film performance include pH, temperature, some modifiers, and modification methods. There are generally three kinds of protein film modification methods: physical methods, chemical methods, and enzyme methods. The physical methods include microwave, ultrasonic, ultraviolet radiation, and ultrahigh pressure, etc.; the chemical methods include plasticizing agent, reducing agent, crosslinking agent, and surface active agent, etc.; and the enzyme methods mainly include TG enzyme and catalase, etc. The studies on modification mainly focus on the improvement of the mechanical properties of protein film and water vapor permeability, which are the main indexes of protein film modification studies. Each study has its own emphasis, which results in great variation in the selection of the remaining indexes. Due to the hydrophilicity of peanut protein film and its vulnerability to the influence of environmental temperature and moisture, the peanut protein film has yet to be put into practical use, and its preparation process still lingers at the laboratory level. On the basis of the previous research, Liu et al., (2010b) optimized the peanut protein film forming process by means of the response surface method. They developed the modification process to improve the tensile strength of peanuts with the method of UV radiation-TGase compound modification by means of studies on physical-chemical and physical-enyzme compound modification processes.

1. Peanut Protein Film Preparation Process

   Compound peanut protein solution of 2, 4, 6, 8, and 10% ($w_B$), and add plasticizers (glycerol, sorbitol, and PEG) of 20, 25, 30, 35, and 40% ($w_B$, calculated based on protein mass) to adjust the pH to be respectively 7, 8, 9, 10, and 11; shake the solution at 50, 60, 70, 80, and 90°C for 0, 10, 20, 30, and 40 min respectively, and then degas the solution in vacuum for 5 min at 0.1 MPa. Coat the solution on a plastic disk and dry it in a drier at 65°C (Zhiping, Feng, and Huang, 2007). The film, after being peeled off the disk, is balanced for 2 days at a relative humidity (RH) of 50%. Afterward, the tensile strength (TS) and elongation at break (EB), transmittance, whiteness, and water vapor permeability (WVP) of the peanut protein film are measured (Bamdad et al., 2006).

   The process conditions for the peanut protein film preparation are determined through single-factor experiments. Based on the TS index, four factors are selected for response surface optimization experiments, namely, peanut protein concentration, glycerol concentration, pH, and heating temperature, and the heating time is determined to be 20 min.

   The preparation conditions are finally determined to be: peanut protein concentration 6% ($w_B$), addition of glycerol 25% ($w_B$, calculated based on protein mass), pH 9, heating temperature 70°C, and heating time 20 min. Under these conditions, the TS of the prepared protein film is (14.87 ± 3.00) MPa, the EB is 12.06 ± 2.56%, and the transmittance is 69.94 ± 2.42%, the sensory quality is good (the brightness ($L$), redness ($a$), and yellowness ($b$) are respectively 92.43 ± 0.60, 0.78 ± 0.16, and 5.54 ± 0.70), and the WVP is (2.75 ± 0.050) $\times 10^{-11}$ g·mm/(h·kPa·m$^2$).

2. Peanut Protein Film Modification and Preparation Process

   The peanut protein film that is obtained with the process optimization has shortcomings, which limit its application in the food industry. The research team analyzed the influence of microwave, ultrasonic wave, high pressure, UV radiation, reducing agent, cross-linking agent, surfactant, and TGase on the performance of the peanut protein film, and carried out studies on the physical-chemical and physical-enzyme compound modification methods.

a. **Microwave**

Microwave is a kind of electromagnetic wave whose frequency is 0.3–3000 GHz (the corresponding wavelength is 0.1–100 cm). Its interaction with biological tissues is mainly manifested in thermal effect and nonthermal effect. The microwave can penetrate biological tissue to enable the bipolar molecules and the polar side chains of the protein to oscillate at an extremely high speed so as to cause molecular electromagnetic oscillation, increasing molecular movement, and heat generation. The microwave can also act on the hydrogen bonds, hydrophobic bond force, and Van der Waals force and redistributed them so that the conformation and activity of the protein can be changed, the protein type edible film can be re-cross-linked, and the network structure of the edible film can be changed to improve the film performance.

With the improvement of the microwave power, the TS of the protein first rises and then declines (Fig. 4.33a). At low microwave power, the thermal effect of the microwave cannot be displayed within a short period of 70 s, and the mechanical strength of the protein film is not improved. When the microwave power reaches 480 W, the TS of the peanut protein

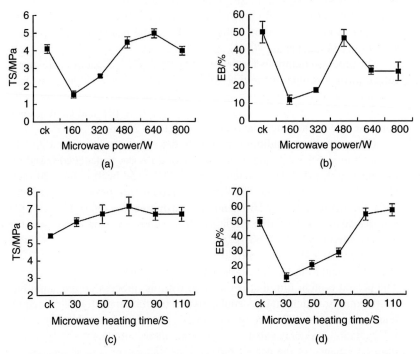

**FIGURE 4.33** Influence of microwave on the performance of peanut protein film. (a) Influence of microwave power on TS; (b) Influence of microwave power on EB; (c) Influence of microwave heating time on TS; (d) Influence of microwave heating time on EB.

film exceeds that of the control film, and it reaches its highest level when the microwave power is 640 W. At 640 W, the TS of protein film is 4.94 ± 0.24 MPa, which is an increase of 21.08% compared with the TS, 4.08 ± 0.25 MPa of the control film. It indicates that the thermal effect of the microwave can replace the water-bath heating link in the peanut protein film preparation process. Therefore, the optimal microwave heating power is determined to be 640 W. The EB of the peanut protein film prepared by means of microwave heating is inferior to that prepared by the means of water-bath at 70°C (Fig. 4.33b). If the microwave heating is only carried out for the peanut protein film-forming solution, the EB of the film is relatively low. For the peanut protein film prepared with microwave heating, the EB of protein film first rises and then declines with the increase of the microwave power. When the microwave is 480 W, the EB reaches the highest level (46.66% ± 4.83%), decreasing by 6.70% compared with the EB (50.01% ± 5.91%) of the control film. It indicates that the microwave heating modification fails to improve the fluidity of the molecules inside the protein film. With the prolonging of the time of microwave heating, the TS of the protein film first rises and then declines (Fig. 4.33c), and reaches its highest level (7.14 ± 0.50 MPa) at 70 s of microwave heating, increasing by 30.05% compared with the TS (5.49 ± 0.10 MPa) of the control film. At 70 and 90 s of microwave heating, no significant difference can be detected in terms of the TS of the protein film, and the optimal microwave heating time is determined to be 90 s taking into consideration the TS of the protein film at the same time. At this point, the TS of protein film is 6.70 ± 0.35 MPa, increasing by 22.04% compared with the control film. With the prolonging of the microwave heating, the EB of the peanut protein film assumes an upward trend (Fig. 4.33d) and reaches its highest level (57.40 ± 4.50%), at 110 s of microwave heating, improving by 15.87% compared with the EB (49.54% ± 2.96%) of the control film. The EB of the peanut protein film is similar to that of the control film at 90 s of microwave heating. At this time, the EB of the protein film is 54.48 ± 4.74%, increasing by 9.97% compared with the EB of the control film. Since the mechanical strength of the peanut protein film is vulnerable to environmental influence (Tian et al., 2005; Meng et al., 2005), the preparation of the control protein film is added to the studies on modification experiment.

**b.** Ultrasonic Wave

The effects of ultrasonic waves on the medium include cavitation, mechanical effect, and super-mixing effect, etc. These effects cause the exposure of many reactive groups contained inside the molecules and increase the sulphydryl content on the surface to form new disulfide bonds so that the intermolecular interaction is enhanced. In the meantime, part of the amino acids are rearranged and combined to form molecular covalent cross-linking so that the gelling property is improved and the special network structure is enhanced. The ultrasonic wave acts on the film-forming solution in two ways: the ultrasonic wave generated by the ultrasonic cell crusher directly acts on the film-forming solution, while the ultrasonic cleaning machine acts on the film-forming solution by means of transmission. The experimental results indicate that with the improvement of the ultrasonic power of the ultrasonic cell crusher, the TS of the peanut protein film assumes a downward trend and reaches $1.24 \pm 0.13$ MPa when the ultrasonic power is 400 W, decreasing by 40.38% compared with the TS ($2.08 \pm 0.16$ MPa) of the control film. The peanut protein film-forming solution treated by the ultrasonic cell crusher does not improve the TS of the film. This is possibly because that the ultrasonic effect directly acts on the film-forming solution, and causes serious destruction to the protein structure as well as the breakage of disulfide bonds, thus reducing the mechanical strength of the peanut protein film.

The peanut protein film-forming solution treated by the ultrasonic cleaning machine improves the TS of the peanut protein film, and with the improvement of the ultrasonic frequency, the TS of the protein film assumes a downward trend (Fig. 4.34a). When the ultrasonic frequency is 28 kHz, the TS of the protein film is $5.13 \pm 0.44$ MPa, increasing by 56.40% compared with the TS ($3.28 \pm 0.23$ MPa) of the control film. The ultrasonic treatment will change the conformation of the film-forming solution and cause the protein to expose more hydrophobic groups so as to promote the formation of disulfide bonds and hydrophobic interactions (Liu et al., 2004). With the increase of the ultrasonic frequency, the EB of the peanut protein film assumes an upward trend (Fig. 4.34b) and reaches its highest level ($80.12 \pm 4.82\%$) when the ultrasonic frequency reaches 100 kHz, increasing by 54.61% compared with the EB ($51.82\% \pm 0.70\%$) of the control film. The ultrasonic treatment promotes the formation of the disulfide bonds of the protein, and therefore the TS increases. However, with the increase of the ultrasonic frequency, the TS of the protein film decreases and the film fluidity is enhanced. Therefore, the EB of the protein film increases. With the increase of the ultrasonic power, the TS of the peanut protein film increases significantly (Fig. 4.34c). When the ultrasonic power reaches 270 W, the TS reaches its highest level ($4.66 \pm 0.33$ MPa), increasing by 42.07% compared with the TS ($3.28 \pm 0.30$ MPa) of the control film. Within the range of 210–300 W, there is no significant difference in terms of the TS of the prepared peanut protein film. However, the EB of the protein film is high at 300 W, and therefore it is determined that the ultrasonic power should be set at 300 W. With the increase of the ultrasonic power, the EB first rises and then declines (Fig 4.34d) and reaches the highest level, $57.74 \pm 5.87\%$, at 330 W, increasing by 15.48% compared with the EB ($50.00\% \pm 5.91\%$) of the control film. At 300 W, with the prolonging of the ultrasonic treatment, the TS of the protein film first rises and then

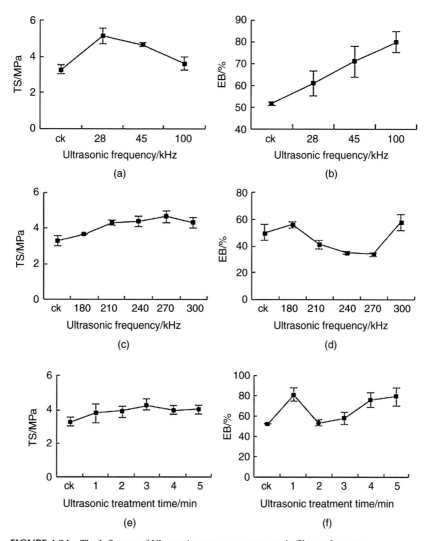

**FIGURE 4.34**    The Influence of Ultrasonic wave on peanut protein film performance.

declines (Fig. 4.34e) and reaches its highest level (4.30 ± 0.33 MPa) at 3 min of ultrasonic treatment, increasing by 31.10% compared with the TS (43.28 ± 0.23 MPa) of the control film. This is possibly because during the film formation process, it is hard for large amounts of broken chemical bonds to form a compact net structure, and therefore the TS decreases. The ultrasonic treatment time is set to be 3 min. With the prolonging of the ultrasonic treatment, the EB of the peanut protein film first declines and then rises (Fig. 4.34f) and reaches its highest level

(80.62 ± 6.67%) at 1 min of ultrasonic treatment, increasing by 55.58% compared with the EB (51.82 ± 0.70%) of the control film. Under the excessive ultrasonic treatment, it is hard for the broken chemical bonds to recombine and the fluidity of the protein molecules increases, and therefore the EB of the protein film increases.

c. UV Radiation

Differently to the other physical treatment methods, UV radiation acts directly on the protein film. The UV radiation can also influence the conformation and composition of protein through enabling amino acid oxidation, covalent bond cleavage, formation of protein free radicals, as well as resynthesis and polymerization. The solid protein is subject to cross-linking or molecular degradation (Liu et al., 2004) depending on the variation in protein nature and radiation dose. There are relatively few studies on the application of UV radiation in protein film modification but the application of UV radiation technology can save a lot of financial and material resources for studies on edible film, and it has promising market prospects. The experimental results indicate that, with the prolonging of radiation time, the TS of the protein film first rises and then declines slightly (Fig. 4.35a). Under the effect of 4 h UV radiation with a 40 W UV lamp, the TS of the protein film can reach its highest level (2.91 ± 0.29 MPa), increasing by 24.89% compared with the TS (2.33 ± 0.13 MPa) of the control film. It indicates that the cross-linking inside the protein film has been completed under the effect of 4 h UV radiation with a 40 W UV lamp. The 40 W UV lamp that is used to radiate the peanut protein film does not contribute to the EB improvement, and the longer the radiation time, the more significantly the EB of the protein film decreases (Fig. 4.35b), and the EB of the protein film reaches its lowest point (95.70 ± 6.00%) under the effect of 12 h UV radiation, decreasing by 20.63% compared with the EB (120.57 ± 13.41%) of the control film. This is possibly because the increase of the cross-link bonds in the protein film results in the decrease of the molecular fluidity inside the film.

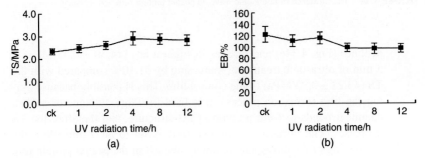

**FIGURE 4.35**    Influence of UV radiation time on peanut protein film.

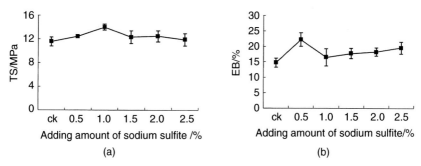

**FIGURE 4.36**    Influence of the adding amount of sodium sulfite on peanut film.

**d.** Reducing Agent

The reducing agent breaks the disulfide bonds in the protein molecules, unfolds the protein molecules, exposes the internal hydrophobic groups, and increases the sulfydryls (SH) at the same time. Therefore, the large amount of sulfydryl and hydrophobic groups reform the intermolecular disulfide bonds and hydrophobic interaction in the following coating, drying, and film formation processes, during which the network structure of the film is enhanced and the tensile strength is improved. The reducing agent has an effect opposite to that of the plasticizer. The reducing agents used for protein film modification mainly include sodium sulfite, glucose, and homocysteine, etc. The experiment investigates the modifying effects of three reducing agents, including sodium sulfite, glucose, and homocysteine, among which the sodium sulfite has the most significant effect. With the addition of more sodium sulfite, the TS of the protein film first rises and then declines (Fig. 4.36a) and reaches its highest level (13.95 ± 0.50 MPa) when the adding amount reaches 1.0% ($w_B$, calculated based on protein mass). The TS increases by 20.78% compared with the TS (11.55 ± 0.74) of the control film. This is because the addition of sodium sulfite breaks the disulfide bonds of peanut protein and causes the film-forming solution to have more sulfydryls, which reform disulfide bonds that contribute to the formation of the protein film network structure at the drying stage, and therefore the TS of the protein film increases. The adding amount of the sodium sulfite is set to be 1.0% ($w_B$, calculated based on protein mass). The addition of sodium sulfite helps to improve the EB of the peanut protein film (Fig. 4.36b). When the adding amount of the sodium sulfite exceeds 1.0% ($w_B$, calculated based on protein mass), the EB of the film improves with the increase of the adding amount, but not in a significant manner. When the adding amount of the sodium sulfite reaches 0.5% ($w_B$, calculated based on protein mass), the EB of the protein film is high (22.13 ± 2.17%), increasing by 50.03% compared with the EB (14.75 ± 1.47%) of the control film. For the protein, the addition of the small molecular substances increases

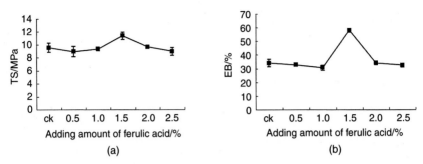

**FIGURE 4.37**   Influence of ferulic acid on the performance of peanut protein film.

the fluidity of the molecules inside the film, and therefore the EB of the film increases.

**e. Cross-Linking Agent**

Cross-linking agent refers to the substance that plays a bridging role among the linear molecules to enable multiple linear molecules to bond and cross-link to form network structure, and it can promote or adjust the formation of covalent bonds or ionic bonds among polymer molecules. The commonly used cross-linking agents include divalent calcium ions and ferulic acid. The experiment investigates the modifying effects of three cross-linking agents, namely, ferulic acid, calcium chloride, and epoxy chloropropane, among which ferulic acid has the most significant modifying effect. With the increase of the adding amount of ferulic acid, the TS of the protein film first rises and then declines (Fig. 4.37a). The TS of the peanut protein film increases only when the adding amount of ferulic acid ranges from 1.5% to 2.0% ($w_B$, calculated based on protein mass), and it indicates that the proper amount of ferulic acid can improve the mechanical property of the peanut protein film. When the adding amount of ferulic acid reaches 1.5% ($w_B$, calculated based on protein mass), the TS reaches its highest level (11.35 ± 0.59 MPa), increasing by 18.72% compared with the TS (9.56 ± 0.70 MPa) of the control film. Ferulic acid reacts with SP I to cause the red shift of the absorption peak of the product, and it indicates that the ferulic acid cross-links with the protein (Yi et al., 2003). The adding amount of ferulic acid is set to be 1.5% ($w_B$, calculated based on protein mass). When the adding amount of ferulic acid is 1.5% ($w_B$, calculated based on protein mass), the EB of the peanut protein film is 57.96 ± 1.31%, increasing by 69.52% compared with the EB (34.19 ± 2.76%) of the control film.

**f. Surfactant**

The surfactant refers to the substance that has fixed hydrophilic and lipophilic groups, and can be directionally aligned on the surface of solution and significantly decreases the surface tension. The main surfactants include sodium dodecyl sulfate (SDS), stearic acid, and lecithinfatty

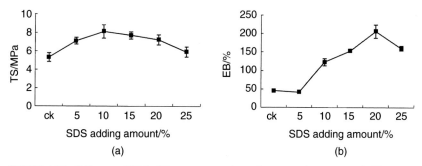

**FIGURE 4.38**  Influence of SDS adding amount on the performance of peanut protein film.

glyceride. Kester and Fennema (1989) studied the influence of various lipids on WVP. It was found that bees wax performs best in obstructing water vapor, followed by stearyl alcohol, acetyl monoglyceride, glyceryl tristearate, and stearic acid. Moreover, the barrier property of the bees wax for water vapor can compete with that of the polyethylene terephthalate and polyvinyl chloride. The experiment investigated the modifying effect of two surfactants, namely, SDS and oleic acid, of which the SDS has the most significant modifying effect. The experimental results indicate that with the increase of the adding amount of SDS, the TS of the protein film first rises and then declines (Fig. 4.38a). When the adding amount of SDS is 10% ($w_B$, calculated based on protein mass), the TS of the film reaches its highest level (8.10 ± 0.71 MPa), increasing by 52.26% compared with the TS (5.32 ± 0.48 MPa) of the control film. As a kind of anionic surfactant, SDS has relatively strong protein dissociation and denaturation capability. It can expand the protein molecules in the film-forming solution and expose more side chains for more cross-linking to increase the TS of the protein film. The adding amount of SDS is set to be 10% ($w_B$, calculated based on protein mass). With the increase of the adding amount of SDS, the EB of the protein film increases significantly (Fig. 4.38b). The EB of the protein film reaches its highest level when the adding amount of SDS reaches 20% (calculated based on the protein mass), and at this time, the EB of the protein film is 205.47 ± 17.74%, increasing by 350.39% compared with the EB (45.62 ± 1.98%) of the control film. The combination of the hydrophobic part in SDS molecules and the hydrophobic amino acid residues in the protein prevents the intermolecular hydrophobic interaction of the protein. The unfolding of the polypeptide chain makes the protein molecules assume a linear structure, and the EB of the film increases (Zhang et al., 2008a).

**g.** TG Enzyme

The commonly used enzyme cross-linking agents include transglutaminase (TG), lipoxygenase, lysyl oxidase, polyphenol oxidase, and

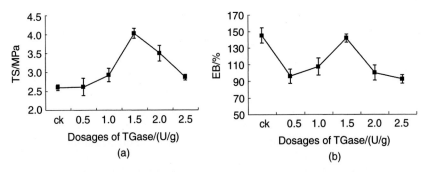

**FIGURE 4.39**    Influence of dosage of TGase on the performance of peanut protein film.

peroxidase. TG enzyme is a kind of enzyme that catalyzes the formation of intermolecular or intramolecular ε- (γ-glutamyl) lysine covalent bonds, and it can catalyze the connection of the γ-amide groups on the glutamine residues and ε-amino groups on the lysine residues of the protein, such as casein, lactoglobulin, and soybean protein, etc. In this way, the ε- (γ-glutamyl) lysine covalent bonds can be formed to change the protein structure and functional properties, and give unique structure and taste to the food protein.

The experimental results indicate that the addition of TGase can significantly improve the TS of the film. When the dosage of TGase reaches 1.5 U/g (protein), the TS of the protein film reaches its highest level (4.02 ± 0.13 MPa), increasing by 54.62% (Fig. 4.39) compared with the TS (2.60 ± 0.07 MPa) of the control film. The TGase can catalyze the connection of the γ-amide groups on the glutamine residues and ε-amino groups on the lysine residues of the protein to enable them to form ε-(γ-glutamyl) lysine covalent bonds so as to increase the intermolecular cross-linking in the protein film and improve the TS of the peanut protein film. The dosage of TGase is set to be 1.5 U/g (protein). The addition of TGase significantly reduces the EB of the protein film. However, with the increase of the dosage of TGase, the EB of the film first rises and then declines (Fig. 4.39b). The EB of the protein film reaches its highest level, 141.64 ± 4.89%, when the dosage of the enzyme reaches 1.5 U/g (protein), decreasing by 2.52% compared with the EB, 145.30 ± 9.35%, of the control film. Since the molecular cross-linking in the protein film is increased by TGase, the molecular fluidity is reduced, and consequently, the EB of the protein film is reduced.

**h.** Compound Modification

Compound modification refers to the modification process that incorporates two or more physical, chemical, or enzymic modification methods. The current research mainly focuses on hybrid modification by combining two chemical modification methods. Hybrid modification is the trend

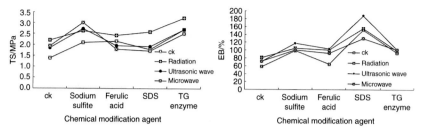

**FIGURE 4.40** Influence of various modification methods on the peanut protein film.

for modification for protein performance optimization. In the above experiments, three physical modification methods, three chemical modification methods, and one enzymic modification method were determined by means of the single-factor experiments for the physical, chemical, and enzymic modification methods. On this basis, physical-chemical and physical-enzyme compound modification studies were carried out. Where, the microwave process parameters are 640 W, 90 s; ultrasonic process parameters are 28 kHz, 300 W, and 3 min; UV radiation process parameters are 40 W and 4 h; the dosage of sodium sulfite is 1.0% ($w_B$, calculated based on protein mass); the dosage of ferulic acid is 1.5% ($w_B$, calculated based on protein mass); the dosage of SDS is 10% ($w_B$, calculated based on protein mass); and the dosage of TGase is 1.5 U/g (protein). The influence of various compound modification methods on the TS of the peanut protein film is as shown in Fig. 4.40a.

   The experimental results indicate that various physical, chemical, and enzymic modification methods and hybrid modification methods can all significantly improve the TS of the peanut protein film. The radiation-TGase modification renders the highest TS for the peanut protein film, which is 3.22 ± 0.07 MPa, increasing by 133% compared with the TS, 1.38 ± 0.09 MPa, of the control film; significantly increasing by 46.36%, compared with the TS, 2.20 ± 0.13 MPa, of the radiated film; significantly increasing by 19.70% compared with the TS, 2.69 ± 0.13 MPa, of the TGase film. Therefore, the hybrid modification process is adopted where the peanut protein film added with 1.5 U/g (protein) TGase is subject to radiation for 4 h by means of the 40 W UV lamp. Various compound modification methods in combination with SDS can render relatively high EB for the peanut protein film (Fig. 4.40b), where the peanut protein film prepared with the ultrasonic wave-SDS method renders the highest EB of the peanut protein film, which is 187.9 ± 14.86%, increasing by 222.85% compared with the EB, 58.20 ± 4.62%, of the control film. No significant difference of EB is detected among the peanut protein films (with relatively high TS) prepared with TGase combined modification methods, where the EB of the film prepared with the radiation-TGase method is 98.61 ± 6.31%, increasing by 69.43% compared with that of the control film.

Finally, it was determined that the UV radiation-TGase compound modification method is used as the modification method to improve the TS of the peanut protein film. The process parameters of the compound modification process parameters are: add TGase (1.5 U/g protein) in the peanut protein to prepare peanut protein film and adopt 40 W UV lamp to radiate the peanut protein film for 4 h. Under this condition, the TS of the peanut protein film is 3.22 ± 0.070 MPa, increasing by 133.33% compared with the TS, 1.38 ± 0.09 MPa, of the control film; significantly increasing by 46.36%, compared with the TS, 2.20 ± 0.13 MPa, of the radiated film; significantly increasing by 19.70% compared with the TS, 2.69 ± 0.13 MPa, of the TGase film. In the meantime, the EB of the peanut protein film is 98.61 ± 6.31%, increasing by 69.43% compared with the EB, 58.20 ± 4.62%, of the control film. The transmittance is 50.32 ± 0.33%, the sensory quality is good (*L*, *a*, and *b* are respectively 87.20 ± 0.80, 1.78 ± 0.22, and 10.88 ± 0.32), and the WVP is 2.67 ± 0.021 × $10^{-11}$g·mm/(h·kPa·m$^2$).

The SEM analysis experiment investigated the microstructure of the peanut protein powder and the peanut protein films prepared with the physical method, chemical method, enzyme method, and composite modification method. The results indicate that, after the peanut protein is made into peanut protein film, the spherical structure is destroyed. Through UV radiation, the protein articles of the peanut protein film that has been treated with TGase compound modification become smaller and more structurally compact. Therefore, the mechanical strength of the peanut protein film is greatly improved. No significant difference is detected in terms of the microstructure between the peanut protein films prepared with different modification methods and the control film. The protein molecules in the protein film are opened to assume an irregularly globular and lamellar state but the protein is still a complete structure, indicating that the various modification processes that are adopted to the prepare peanut protein film will not destroy the molecular structure of protein molecules. Although the molecular structure of the protein molecules is not destroyed, the performance of the peanut protein film is changed to varying degrees, and this suggests that the intermolecular interaction of the protein is closely related to its film-forming performance. The moderate denaturation of the protein is the precondition of the film formation, and the condition of the network structure will influence the film performance. The film performance might be improved by enhancing the intermolecular interaction of the protein molecules so as to enable them to form a more compact and uniform network structure (Want et al., 2010). The SEM results indicate that the control film that has not been subject to modification has relatively loose structure, but the peanut protein particles are relatively large. The microstructure of the protein film modified by UV radiation is also loose, but the peanut protein particles become smaller and the more pores can be observed. The protein particles of the TGase-modified protein film have similar size with those of the control film; however, the former have more compact structure and interprotein cross-linking (Jiang et al., 2007). The protein particles of the peanut protein film that has been subject to UV radiation-TGase compound modification become smaller and more structurally compact, and therefore the mechanical strength of the peanut protein film is greatly enhanced (Liu, 2011).

### 4.6.3   Influence of Saccharide Grafting Modification on the Performance of Peanut Protein Film

Most of the recent studies on the improvement of protein film through chemical modification focus on the addition of modifying agents, such as cross-linking and reducing agents, into the film-forming solution to influence the interaction between the protein structure and protein so as to improve the mechanical property and water resistance of the protein film. However, the improved protein film can still not be used in food packaging. Studies have shown that the protein modified by saccharide grafting can overcome some shortcomings of the protein film to certain degrees, such as poor water resistance and great brittleness (Su et al., 2010). This is because the low-molecular saccharide can be used as plasticizer and weaken the interaction among the molecular chains so as to improve the flexibility of the protein film (Wihodo and Moraru, 2013).

In addition, the amino-carbonyl condensation between the protein and the carbohydrates extends the protein molecules to certain degrees and further exposes their internal hydrophobic groups and sulfydryls. Under these effects, the protein molecules cross-link during the film formation so that the intermolecular interaction is reinforced and the special network structure of the protein becomes compact to improve the mechanical property and water resistance of the protein film. Studies have shown that some substances, such as xylose, galactose, glucose, sucrose, starch, carboxymethyl cel, and sodium alginate, have the effects of increasing strength, improving ductility and water resistance (Su et al., 2010; Zhang et al., 2011; Ghanbarzadeh et al., 2007; Lee et al., 2007; Galus et al., 2012). The research team carried out a comparative study on the influence of the different low-molecular carbohydrate grafting modifications on the mechanical property and water resistance of peanut protein film, and the carbohydrates with good improvement effects were selected and used for further studying the process of improving the performance of protein film modified with the carbohydrate grafting.

1. Screening of Carbohydrate Type

   10% (pB) peanut protein isolate (PPI) solution is compounded with 10% (pB) various kinds of carbohydrate solutions (Xylose, mannose, lactose, galactose, glucose, and sucrose). Mix different amounts of PPI solution and carbohydrate solution to make PPI/carbohydrate 5, 10, 20, 50, and 100, and add water to make the ultimate concentration of PPI 5%. Group the PPI added with Xylose, mannose, galactose, lactose, glucose, and sucrose as X, M, L, Ga, G, and S group according to the carbohydrate type, and the group is further divided into X-5, X-10, X-20, X-50, X-100, M-5, and M-10, etc.

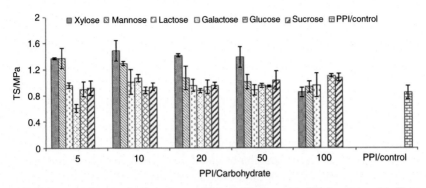

**FIGURE 4.41** Influence of carbohydrate grafting modification on the TS of peanut protein film.

After being mixed evenly, each group of solutions was respectively subject to water bath heating at 70°C for 2 h, and then rapidly cooled down with cold water to room temperature and added with a glycerin equivalent to 25% of the protein content. After being mixed evenly, the solution was poured on a plate and dried for 1 h at 65°C. Then, the plate was placed in a drier containing saturated sodium bromide (58% relative humidity), and the film was peeled off for determination after the plate was left to stand in room temperature for 48 h. The PPI prepared under the same condition but without any added carbohydrate was used as control film.

**a.** Influence of Carbohydrate Grafting Modification on the TS of Peanut Protein Film

The carbohydrate grafting modification can improve the TS of the peanut protein film to a certain extent (Fig. 4.41), where the xylose has the most significant influence (TS is 1.01–1.77 of the nonmodified PPI control film), followed by mannose (TS is 1.10–1.63 of the nonmodified PPI control film). However, lactose does not have significant influence on the TS of the protein film ($P > 0.05$) possibly because different carbohydrates manifest different activity and speed when having Maillard reaction with protein. The xylose is pentose, and compared with the hexose, such as galactose and glucose, it has relatively short carbon chains and is therefore subject to smaller stereo-hindrance effect of the carbon frame, making the Maillard reaction easier to happen. The lactose is a reducing disaccharide, but it has low reaction due to its relatively large molecular mass. Sucrose is a nonreducing sugar, and therefore it does not participate in the Maillard reaction.

The TS reinforcement effect of the protein film exerted by the xylose, mannose, and galactose increases with the increase of the concentration. The TS of the X-50 protein film is 1.38 MPa, significantly higher than the 0.84 MPa ($P < 0.05$ A) of the control film. With the increase of the xylose concentration, the TS of the X-20 and X-10 protein film is, respectively, 1.41 and 1.48 MPa, and there is no significant difference

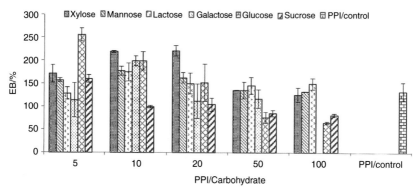

**FIGURE 4.42** Influence of carbohydrate grafting modification on the ductility of peanut protein film.

between the two ($P > 0.05$). However, with the further increase of the carbohydrate content, when the PPI/carbohydrate is lower than 10, the TS of X-5 and Ga-5 protein film declines. This is because at the low level of dosage, the addition of the carbohydrate reinforces the intermolecular interaction to make the network structure denser, but the further addition of carbohydrate increases the intermolecular voids of the system and causes the decrease of the TS of the protein film (Osés et al., 2009).

**b.** Influence of Carbohydrate Grafting Modification on the Ductility of Peanut Protein Film

Glucose has the greatest influence on the ductility of peanut protein film (Fig. 4.42), and the EB is as high as 255.22%, followed by xylose, and the EB of the protein film of X-10 and X-20 is respectively 1.66 and 1.67 times of the PPI control film. In general, the influence of various small-molecular carbohydrates on the ductility of the protein film increases with the increase of the carbohydrate concentration. This is because the low-molecular carbohydrates, when being inserted into the protein molecules, can weaken the intermolecular force and increase the mobility and lubrication of the molecular chains, thereby reinforcing the flexibility and ductility of the film (Jangchud and Chinnan, 1999). However, when the PPI/carbohydrate is lower than 10, the EB of the X-5, M-5, L-, and Ga-5 protein film will decline. This is because the over-high carbohydrate content can greatly weaken the protein film structure and exert negative influence on the ductility of the film (Ghanbarzadeh et al., 2006).

In summary, the peanut protein film treated with xylose grafting modification has the best effect, and the X-10 protein film has the best comprehensive performance: the TS is 1.48 MPa, and the EB is 218.92%.

In addition, it was discovered by analyzing the solubility of protein film, the addition of five types of carbohydrates could not significantly improve the water resistance of the peanut protein film except for xylose. The solubility of X-10 protein film is at least 60.80%, and the film still remains complete after being immersed for 24 h. However, the solubility of 60% is still a relatively high value in the actual applications, and therefore further studies and improvement should be carried out for the xylose grafting modification process for peanut protein.

2. Study on Xylose Grafting Modification Process

On the basis of the carbohydrate screening results, and under the conditions that the xylose was selected as the best carbohydrate for grafting modification and the peanut protein isolate/xylose was set at 10, the influence of different modification conditions on the performance of the peanut protein film was studied, including the modification pH (3, 5, 7, 9, 11), modification temperature (30, 50, 70, and 90°C), and modification time (30, 60, 90, 120, and 180 min).

On the basis of the single-factor experiments, the peanut protein modification process was preliminarily determined. By setting the pH at 9.0 m, temperature at 90°C, and time of 90 min as the zero level, the three-factor and five-level quadratic orthogonal rotational combination experiment was designed to analyze the influence of various factors on the performance xylose modified peanut protein film.

Through model-based calculation, the modification conditions were finally determined to be pH 9.5, temperature 91.5°C, and time 95 min. Under these modification conditions, the TS of the protein film is 10.37 MPa and the EB is 96.47%, consistent with the predicted value (the TS and EB are respectively predicated to be 10.11 MPa and 95.63%). The solubility of the modified protein film decreases from 96.94% before the modification to 35.94%. Therefore, the xylose grafting modification can not only improve the mechanical properties of but can also significantly improve the water resistance of the peanut protein film.

## 4.7 MODIFICATION TECHNOLOGY OF PEANUT PROTEIN

Modification of peanut protein means to strengthen and improve the structural and functional properties of protein through changing one or several physical and chemical properties of protein and to improve the utilization of nutrition though inhibiting enzymatic activity and removing toxic and harmful substances. The modification of protein is actually the modification of radical groups to obtain a unique function by changing the functional groups and molecular structure (Zheng and Zhong, 2009). The modification of peanut protein mainly includes physical modification, chemical modification, and enzymatic modification (Zhou, 2005b).

Physical modification is improving the functional characteristics and nutritive value of protein through physical methods including heating, freezing, and mechanical action, etc. Physical modification only normally changes the higher structure of protein with advantages including lower processing cost, lower time consumption, nontoxicity, lower damages to the nutritive value of protein, and disadvantages including a narrow modification range (Jiao et al., 2010). Chemical modification is improving the structural and functional characters of protein through chemical methods including chemical reagents, taking advantage of the chemical activity of protein side chain to change electrostatic charge, and hydrophobic grouping of protein structure (Zhao et al., 2011). Enzymatic modification mainly includes enzymatic polymerization modification and enzymatic degradation modification. Enzymatic polymerization modification uses TGase to cross-link protein and increase its gelling property and elasticity and create favorable functional character (Wu et al., 2009a, 2009b). Enzymatic degradation modification uses protease hydrolysis to hydrolyze protein and thus improve its properties, such as solubility, etc.

To gain peanut protein products with better functional characters and stimulate the development of peanut protein deep processing and utilization and food processing industry, we extract peanut protein isolate, PPC, and peanut protein components (arachin and conarachin) by the above-mentioned methods and use physical modification, chemical modification, and enzymatic modification to improve their functional characters, including gelling property, solubility, etc.

## 4.7.1 Improvement of PPI Gelling Property

Peanut protein isolate (PPI) possesses many functional characters including emulsibility, water binding capacity, gelling property, solubility, foamability, film-forming property, etc. It can be used in many food processing sectors, including for meat products, aquatic products, bakery products, dairy products, beverages, ice cream, and candy, etc. The gelling property, as one of the main functional characters, can increase the viscosity, plasticity, and elasticity of protein and help to form gel, which can be a carrier of water, flavor agent, saccharides, and other complexes. Using ultrahigh pressure technique, our research team modified the PPI. To provide theoretical foundation for the application of PPI in sausage, we first test the change rule of PPI gelling property under different ultrahigh treatment conditions, and second select the optimal ultrahigh pressure treatment process through response surface design.

Prepare protein sample in certain m/V and put it in polypropylene vacuum bag, expel the bubble (to prevent rupture of the bag during the high pressure treatment process) and heat-seal the bag under vacuum. Check the sealing condition of high pressure equipment and put the vacuum package into the high pressure equipment. The pressure (pressure surge ± 10 MPa) and holdup time should be set. Take out the PPI after disposing and then freeze dry for use. Add deionized water in the freeze-drying PPI to prepare solution with 14% mass concentration. Heat PPI at 95°C for 1 h and then take out for rapid cooling at 4°C for 24 h. Then use TA-TX2i texture analyzer (probe tip diameter 12 mm) to measure the hardness, elasticity, and cohesion of gel. Apply Pinterits and Arntfield's method (2008) with a little modification: running mode, TPA; pretest speed, 2.0 mm/s; test speed, 0.8 mm/s; pushing distance, 50%; posttest speed, 0.8 mm/s; data collection speed, 200/s. The gel hardness/g is the peak force (force 2) in the first compression process; elasticity is the time difference between the second compression and the end of the compression/the time difference between the first compression and the end of the compression (time diff 4:5/time diff 1:2); cohesion is the area under the positive peak of the second compression/the area under the positive peak of the second compression.

1. Pressure and Time

See Figs. 4.43 and 4.44 for the influence on the hardness, elasticity, and cohesion of PPI thermal gel under different pressure and treating time. It is shown in Fig. 4.43 that the hardness of thermal gel will first rise and then decline with the increase of pressure ($P < 0.05$). When the pressure is 100 MPa, the hardness of PPI thermal gel will reach the highest value, namely 172.52 g. When the pressure is between 100 and 500 MPa the hardness will sharply decrease. The elasticity of PPI thermal gel is relatively high, more than 0.7, with no significant change when the pressure is between 50 and 120 MPa. When the pressure is over 120 MPa, the elasticity will decrease. The cohesion, between 0.33 and 0.38, of PPI thermal gel will slightly rise first and then slightly decline with the increase of pressure ($P > 0.05$) and there is no

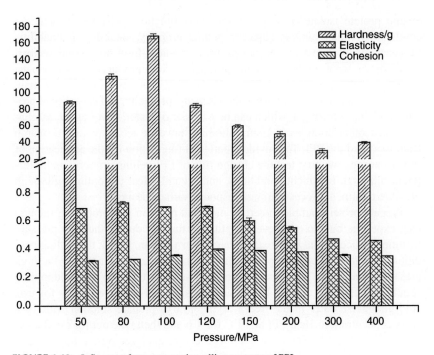

**FIGURE 4.43** Influence of pressure on the gelling property of PPI.

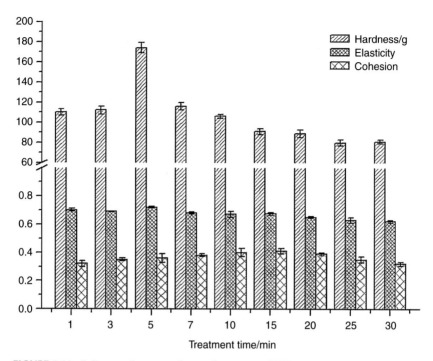

**FIGURE 4.44**  Influence of treatment time on the property of PPI.

significant change. The natural structure of most protein is stabilized by the interaction between noncovalent bonds, whose balance will be damaged in the high pressure treatment, and thus cause the extension of protein primary structure and changes of noncovalent binding, including hydrogen binding, hydrophobic binding, and ionic binding. As a result, the property of the protein is changed (Angsupanich et al., 1999). This is the reason why ultrahigh pressure can change the PPI thermal gelling property. Other research shows that the modification of protein will be more efficient with the increase of pressure. The gel network structure will be more compact and fine, and the hardness will also increase. However, the protein aggregation occurs when the pressure reaches a certain level and thus the gelling property will decrease (Jin et al., 1999). In the research, when the pressure is greater than 100 MPa, the protein aggregation may increase with the increase of pressure and thus cause the rapid decrease of hardness of thermal gel. It is shown in Fig. 4.43 that there is great difference of influence on thermal gel hardness under different treating time ($P < 0.05$). The hardness will reach the highest value at 5 min of treatment time, and it will rapidly decrease after 5 min. The elasticity and cohesion of PPI thermal gel will slightly increase first and then slightly decrease with the increase of time. The elasticity will reach the highest value at 5 min, namely 0.704. The cohesion will reach the highest value at 15 min, namely 0.374. Ultrahigh pressure treatment will accelerate the agglomeration of the protein subunit and when it reaches a certain level, the agglomeration effect is not obvious with the increase of time. On the contrary, the network structure of gel will be destroyed and the hardness of gel will decrease because of aggregation (Guadalupe, 2005).

2. Protein Concentration and pH

Figs. 4.45 and 4.46 show the influence on the property of PPI thermal gel under different protein concentrations and pH. Fig. 4.45 shows that the hardness of thermal gel will first rise and then decline with the increase of protein concentration. When the protein concentration is 5%, the hardness of PPI thermal gel will reach the highest value and then gradually decrease with the gradual increase of protein concentration. It is also shown in Fig. 4.45 that the elasticity and cohesion of PPI thermal gel will first rise and then decline with the increase of protein concentration. When the protein concentration is 10%, the elasticity of PPI thermal gel will reach the highest value, namely 0.796. When the protein concentration is 15%, the cohesion of PPI thermal gel will reach the highest value, namely 0.402. When the protein concentration increases to 20%, the cohesion of PPI thermal gel will decrease to 0.372. At a certain pressure, the gel hardness of protein solution with higher concentration will exceed the gel hardness of protein solution with lower concentration because of the decrease of noncovalent liquid. If the concentration is too high, it is difficult to form the compact and ordered network

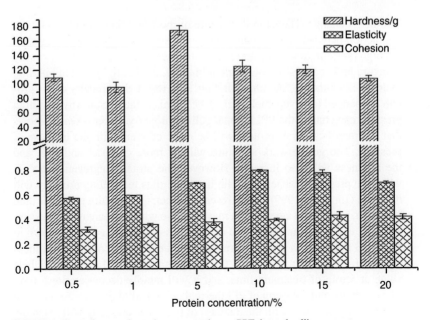

**FIGURE 4.45**   Influence of protein concentration on PPT thermal gelling property.

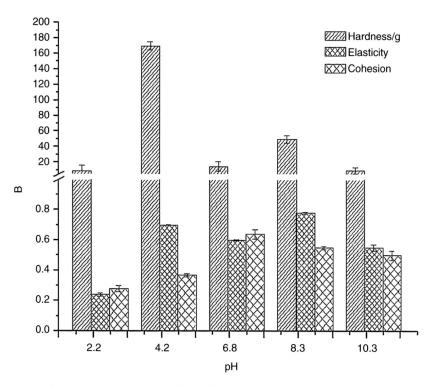

**FIGURE 4.46**  Influence of pH on PPI thermal gelling performance.

structure of gel and thus the hardness will decrease (Briscoe et al., 2002; Gosal and Ross-Murphy, 2005).

It is shown in Fig. 4.46 that pH has significant influence on the hardness and elasticity of gel. When the pH is 4.2, the hardness of gel is 172.52 g. When the pH is 6.8, the hardness of gel will decrease to 10.56 g. When the pH is 8.3, the hardness of gel will increase to 46.33 g. When the pH is 10.3, the hardness of gel will decrease to 7.67 g. When the pH is 4.2, the gel has good elasticity. When the pH is 6.8, the elasticity of gel will decrease to 0.60. When the pH is 8.3, the elasticity of gel will reach the highest value, namely 0.78. When the pH is 10.3, the elasticity of gel will decrease to 0.59. The variation trend of cohesion is totally different from the variation trend of hardness and elasticity. The cohesion of PPI will first rise and then decline with the increase of pH. When the pH is neutral, the cohesion will reach the highest value, namely 0.667. Comprehensively analyzing the influence on the hardness, elasticity, and cohesion of gel under different pH, we find that it is best for the formation of thermal gel when the PPI is near the isoelectric point. The reason is that when the pH is near the isoelectric point, the attraction is bigger than the repulsion and the spatial conformation is relatively loose and thus the network structural strength of the gel is relatively low. The experimental results show that when the pH is 4.2, the PPI thermal gel obtains preferable hardness and elasticity. In this case, the subsequent process optimization will be conducted under the condition of pH 4.2.

As seen from the above single-factor results, ultrahigh pressure does not have significant influence on the thermal gel elasticity and cohesion of PPC, and their variations are respectively 0.23–0.78 and 0.26–0.67. However, hardness is subject to a great influence of ultrahigh pressure, with the change ranging from 18 to 172.52 g.

3. Ultrahigh Pressure Condition
   Through the mathematic analysis of the regression model, it can be found that after ultrahigh pressure treatment, the process parameters of thermal gel of PPI are: pressure 115 MPa, time 5 min, and PPI concentration 3.11%. Under these process conditions, the gel hardness of PPI is 176.96 g. To further test the reliability of the response surface analysis method, the above optimal conditions are adopted for ultrahigh pressure treatment, and the actually measured PPI gel hardness is 174.37 g. Compared with the theoretical value, there is a small relative error, being only 1.46%. Therefore, the ultrahigh pressure parameters obtained by means of response surface analysis and optimization are both accurate and reliable, and they can be used in actual operation.

## 4.7.2   Improvement of the Functional Properties of PPC

The natural peanut protein has functional properties such as solubility, emulsifying, foaming, and gelling properties. However, it is still required to modify the functional properties of the protein so as to meet different requirements in product processing. For example, the requirements of the plant protein drink processing can be met by improving the solubility; and by improving the gelling property, it can be better applied in the meat products. Therefore, Wu (2009) adopted physical, chemical, and enzyme methods to modify the PPC (PPC) prepared with alcohol precipitation process, and Ma (2009) modified PPC with the physical method and protease so as to improve the gelling property and the solubility of PPC, respectively. They also carried out studies on factors that influence the modification.

1. Improvement of the Gelling Property of PPC
   a. Physical Modification
      - Thermal Treatment
        As shown in Fig. 4.47, after being treated at a temperature of 40–100°C, the content of dissociated sulfhydryls of the protein reaches 10.66 μmol/g at 50°C. With the temperature rising, the content of –SH declines slightly and reaches 9.98 μmol/g at 80°C, and after that, the content of –SH declines rapidly. However, the changing

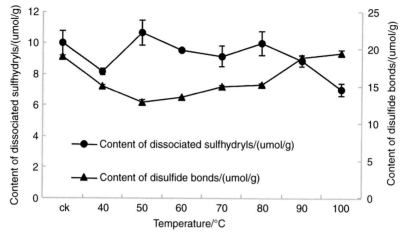

**FIGURE 4.47**    Influence of thermal treatment temperature on the content of –SH/-S-S- in protein.

trend of the content of disulfide bonds in protein is just the opposite of that of the content of sulfhydryls. At 50°C, the content of disulfide bonds is only 12.84 μmol/g. With the temperature rising, the content of disulfide bonds increases slowly and reaches 15.25 μmol/g at 80°C, and after that, the content increases rapidly. The variance analysis indicates that at a temperature of 50–80°C, there is no significant difference between the dissociated sulfhydryls and the disulfide bonds ($P > 0.05$) in terms of the changing of content. Seen from Table 4.2, the surface hydrophobicity index (So) of protein increases

**TABLE 4.2** Influence of Thermal Treatment on Hydrophobicity Index So on Protein Surface

| Thermal Treatment Temperature (°C) | Surface Hydrophobicity Index (So) | Thermal Treatment Time (min) | Surface Hydrophobicity Index (So) |
|---|---|---|---|
| 40 | 7.77 | 2 | 8.13 |
| 50 | 7.55 | 4 | 8.82 |
| 60 | 8.06 | 6 | 9.52 |
| 70 | 8.53 | 8 | 9.08 |
| 80 | 9.01 | 10 | 8.53 |
| 90 | 9.08 | 12 | 8.23 |
| 100 | 7.41 | 14 | 8.19 |
| ck | 7.46 | ck | 7.46 |

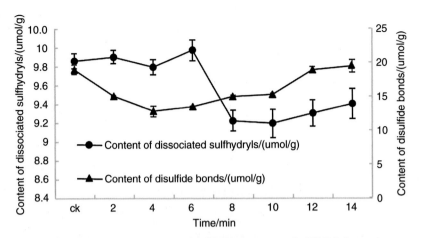

**FIGURE 4.48**  Influence of thermal treatment time on the content of –SH/-S-S- in protein.

gradually with the rise of the thermal treatment temperature and reaches its highest level (9.08) at 90°C. Afterward, with the rise of the temperature, the So value declines rapidly. The variance analysis indicates that the surface hydrophobicity index (9.01) at 80°C does not change significantly compared with 90°C. Therefore, the optimal thermal treatment temperature is set to be 80°C on the basis of evaluating the changing of the dissociated sulfhydryls, the content of disulfide bonds, and the surface hydrophobicity index at different thermal treatment temperatures.

As shown in Fig. 4.48 and Table 4.2, when the treatment time lasts for 0–6 min, the content of dissociated sulfhydryls of the protein gradually increases while the content of disulfide bonds gradually decreases, and So value increases. At the 6 min of thermal treatment, ideal values are obtained, which are 9.97, 13.48, and 9.52 μmol/g respectively. With the prolonging of the treatment time, the content of the dissociated sulfhydryls of the protein decreases sharply while the content of disulfide bonds increases. This indicates that with the prolonging of the treatment time, the dissociated sulfhydryls in protein undergo an exchange reaction with the disulfide bonds to form intermolecular and intramolecular sulfhydryl bondings, which cause the hydrophobic groups on the surface of the protein to be rewrapped and the surface hydrophobicity index to decrease. This also confirms the conclusion of Friedman et al. (1982), and the thermal denaturation of plant protein influences its structure in two ways: First, the globular structure of the globulin unfolds to form linear macromolecules where the hydrophobic groups originally wrapped in the coiling structure are exposed; second, with the increase of the treatment

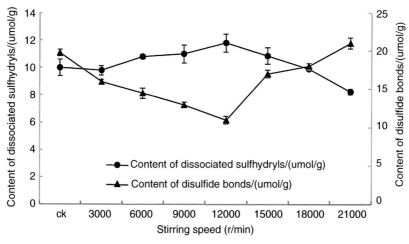

**FIGURE 4.49**   Influence of high-speed stirring speed on the content of -SH/-S-S- in protein.

temperature or time, the hydrophobic groups interact or –SH is oxidized to generate –S–S– to form compact network structure, which cause the originally exposed groups to be rewrapped.

- High-Speed Stirring

  As seen from Fig. 4.49 and Table 4.3, the speed changing of high-speed stirring has great influence on the gelling property of protein. The content of dissociated sulfhydryls and surface hydrophobicity index So of protein first rises and then declines with the increase of the stirring speed. However, the change in content of disulfide bonds is the opposite of that of the dissociated sulfhydryls. Specifically,

**TABLE 4.3** Influence of High-Speed Stirring on Surface Hydrophobicity Index So of Protein

| High-Speed Stirring Intensity (r/min) | Surface Hydrophobicity Index (So) | High-Speed Stirring Time (s) | Surface Hydrophobicity Index (So) |
|---|---|---|---|
| 3000 | 7.94 | 12 | 7.28 |
| 6000 | 7.88 | 24 | 7.68 |
| 9000 | 8.17 | 36 | 7.89 |
| 12,000 | 8.43 | 48 | 8.27 |
| 15,000 | 8.65 | 60 | 8.89 |
| 18,000 | 7.86 | 72 | 8.87 |
| 21,000 | 7.24 | ck | 7.46 |

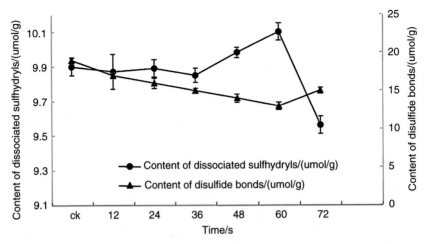

**FIGURE 4.50**    Influence of high-speed stirring time on the content of -SH/-S-S- in protein.

when the rotating speed is 3,000–12,000 r/min, the content of dissociated sulfhydryls in protein increases gradually and reaches its highest level at 12,000 r/min, reaching 11.57 μmol/g, while the content of disulfide bonds reaches its lowest level, only 11.83 μmol/g. At this time, the So value of protein is 8.425. When the stirring speed is 12,000–21,000 r/min, the content of dissociated sulfhydryls decreases rapidly while the content of disulfide bonds increases, and the So value rises slightly at 15,000 r/min and then declines rapidly. The variance analysis indicates that the So value at the stirring speed of 12,000 r/min does not change significantly compared with that at 15,000 r/min ($P > 0.05$). Therefore, according to the experimental results, 12,000 r/min is selected as the optimal speed for high-speed stirring.

Fig. 4.50 and Table 4.3 show that when the protein is treated for 12–72 s at the rotating speed of 12,000 r/min, the content of dissociated sulfhydryls increases slightly with the prolonging of the stirring time and reaches its highest level at the stirring time of 60 s, being 9.99 μmol/g, while the content of disulfide bonds is at its lowest level, being only 11.83 μmol/g. At this time, the So value of protein is 8.89. With the prolonging of stirring time, the gelling property of protein decreases instead. This is mainly because during the high-speed stirring process, the high-speed shearing effect disintegrates the molecular aggregates of protein, that is, it breaks the weak intermolecular forces of protein, such as hydrogen bonds and Van der Waals force, etc. During this process, the hydrophobic groups previously wrapped inside the molecules are exposed more frequently.

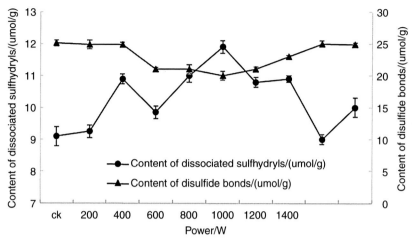

**FIGURE 4.51**    Influence of ultrasonic treatment power on the content of -SH/-S-S- in protein.

- Ultrasonic Treatment

    Fig. 4.51 and Table 4.4 show the ultrasonic treatment has significant influence on the gelling property of PPC. When the ultrasonic power increases from 600 to 1000 W, the content of dissociated sulfhydryls increases to 11.91 umol/g ($P < 0.01$); while the content of disulfide bonds decreases to 19.47 umol/g. At this time, the So value of protein is 12.89. After that, the increase of ultrasonic power rapidly decreases

**TABLE 4.4** Influence of Ultrasonic Treatment on Surface Hydrophobicity Index So of Protein

| Ultrasonic Treatment Power (Hz) | Surface Hydrophobicity Index (So) | Ultrasonic Treatment Time (min) | Surface Hydrophobicity Index (So) |
| --- | --- | --- | --- |
| 200 | 10.13 | 1 | 10.12 |
| 400 | 10.74 | 2 | 10.54 |
| 600 | 11.26 | 4 | 10.38 |
| 800 | 11.59 | 6 | 11.72 |
| 1000 | 12.89 | 8 | 11.08 |
| 1200 | 11.68 | ck | 7.46 |
| 1400 | 10.09 | | |
| 1600 | 10.60 | | |
| 1800 | 10.49 | | |

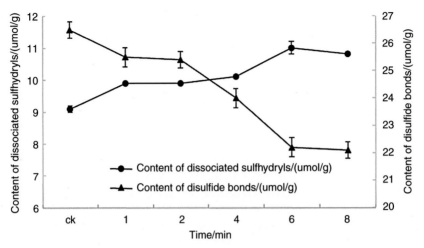

**FIGURE 4.52** Influence of ultrasonic treatment time on the content of -SH/-S-S- in protein.

the gelling property of the protein, and therefore it was determined to set 1000 W as the optimal ultrasonic power for protein treatment. In the meantime, while keeping the ultrasonic power stable at 1000 W, the influence of different ultrasonic treatment times (1–8 min) on the gelling property of protein was investigated (Fig. 4.52), and it was concluded that the length of the ultrasonic treatment time can significantly influence the gelling property of protein. When the time of ultrasonic treatment is 1–6 min, the content of dissociated sulfhydryls of protein is always on the rise, while the content of disulfide bonds is always on the decrease. At 6 min of ultrasonic treatment, the content of dissociated sulfhydryls and disulfide bonds is 11.91 and 22.25 umol/g, respectively. When the treatment time is prolonged to 8 min, the content of dissociated sulfhydryls and disulfide bonds does not change significantly compared with that at 6 min. It can be seen that the prolonging of ultrasonic treatment time does no good to the improvement of the gelling property of protein.

An ultrasonic wave is a kind of elastic mechanical wave that can travel through physical media, whose frequency is $2 \times 10^4$–$2 \times 10^9$ Hz. When traveling through the media, the ultrasonic wave will generate a thermal effect, mechanical effect, or cavitation effect, and cause changes to some properties of the medium (Mason and Paniwnyk, 1996). Studies have shown, for the ultrasonic waves of certain frequency and emitting surface, the increase of power results in the increase of acoustic intensity, which will subsequently lead to the increase of acoustic pressure amplitude and liquid pressure. In this case, the time required by the collapse of cavitation bubbles

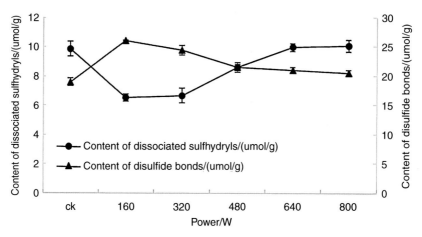

**FIGURE 4.53**    Influence of microwave treatment power on the content of -SH/-S-S- in protein.

will be shortened, that is, the cavitation events generated by the ultrasonic wave will increase in number within unit time (Huang, 2004), and this contributes to the solvent dissolution in a liquid system. However, the acoustic pressure should not be an unlimited increase because the increase of acoustic intensity results in the increase of acoustic pressure amplitude, which consequently causes the failure of the cavitation bubbles to fall apart within the acoustic wave inflation phase. In addition, the large amount of cavitation bubbles generated by overhigh acoustic intensity reduces the energy transmission through the reflected acoustic waves. This effectively explains the variation of surface hydrophobicity index of the protein during ultrasonic treatment (Singh and Mac, 2001).

- Microwave Treatment

As shown in Fig. 4.53 and Table 4.5, with the increase of the microwave power, the gelling property of the PPC changes significantly. Under the treatment condition of 160–800 W, the content of dissociated sulfhydryls of protein is always on the rise while the content of disulfide bonds is on the decrease. When the treatment power is 800 W, the content of dissociated sulfhydryls of protein is 10.04 umol/g, the content of disulfide bonds is 20.49 μmol/g, and the So value of protein at this time is 8.15. Therefore, the microwave power 800 W is selected as the optimal condition of microwave treatment. At this power, the influence of treatment time (20–60 s) on the gelling property of protein is investigated. Seen from Fig. 4.54 and Table 4.5, at 30 s of treatment time, the protein has the highest content of dissociated sulfhydryls and the lowest content of disulfide bonds. At this point, the So value of protein is 7.98. Although this

**TABLE 4.5** Influence of Microwave Treatment on Surface Hydrophobicity Index So of Protein

| Microwave Treatment Power (W) | Surface Hydrophobicity Index (So) | Microwave Treatment Time (s) | Surface Hydrophobicity Index (So) |
|---|---|---|---|
| 160 | 7.41 | 20 | 7.58 |
| 320 | 7.27 | 30 | 7.98 |
| 480 | 7.93 | 40 | 8.09 |
| 640 | 7.93 | 50 | 7.46 |
| 800 | 8.15 | 60 | 7.61 |
| ck | 7.46 | ck | 7.46 |

is not the highest value (the So value is highest at 40 s of treatment time, being 8.09), the variance analysis indicates that there is no significant difference between the So value of protein at 30 and 40 s of treatment time. However, with the prolonging of treatment time, the protein solution overflows due to boiling, which does not improve the gelling property of protein, and this practice is both time-consuming and energy-consuming. Generally speaking, the selection of microwave power should be in line with the principle of efficiently improving the dissolution rate while not causing material denaturation. The microwave power should not be overlarge and the microwave radiation time should not last too long so as to avoid system overheating

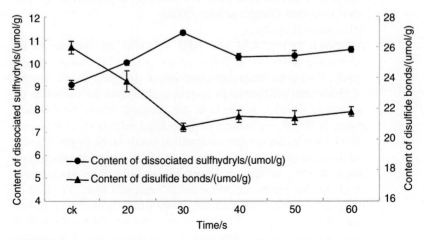

**FIGURE 4.54** Influence of microwave treatment time on the content of -SH/-S-S- in protein.

or inaccurate temperature control caused by boiling, or any other risk (Liompart et al., 2007). Therefore, this study selects the conditions of microwave treatment to be power 800 W and time 30 s.

Microwave energy is a kind of nonionizing radiation energy which drives the molecular movement through ion migration and dipole rotation. When the energy acts on molecules, it promotes the molecular transformation or polarizes the molecules, that is, it can generate instantaneous polarization under the effect of the microwave field, and make polar transformation at the speed of 2450 million times/s (Zhou and Liu, 2006). Under this condition, the vibration and rupture of bonds, as well as the interparticle friction and collisions, generate a large quantity of heat energy at high speed, causing the rupture of intermolecular and intramolecular noncovalent bonds and $-S-S-$, etc. Therefore, part of the insoluble protein molecules previously wrapped in the molecules are released and diffused in the solvent, and that is why the microwave treatment can increase the content of dissociated sulfhydryls of protein and increase the surface hydrophobicity index.

**b.** Chemical Modification

Aminlari et al. (1977) found that an effective method to improve the protein dispersibility index (PDI) is to add a certain amount of reducer, such as $NaHSO_3$ and Cysteine, in the protein. Since $NaHSO_3$ and Cysteine (Cys) are both reducers specific to disulfide bonds, some scholars explain that it is because the use of reducers breaks the disulfide bonds in protein, that is, it is disulfide bond cross-linking that causes the decrease of protein solubility.

- Na2SO3 Treatment

As shown in Fig. 4.55 and Table 4.6, when adding relatively small amounts of $Na_2SO_3$ (0–0.625 mmol/L), the content of dissociated sulfhydryls in protein increases rapidly. When the dosage reaches 0. 625 mmol/L, the content of dissociated sulfhydryls in protein is 15.89 µmol/g, increasing by 74.82% compared with the original PPC. At this time, the So value of protein is 13.48, increasing by 80.65% compared with the original PPC. By continuously increasing the dosage of $Na_2SO_3$, no significant change can be detected in the content of dissociated sulfhydryls and sulfide bonds and So value ($P > 0.05$). Therefore, it is determined that the optimal dosage of $Na_2SO_3$ is 0.625 mmol/L. Under this condition, the influence of treatment time on gelling property of protein was investigated. Seen from Fig. 4.56 and Table 4.6, when the treatment time is 15–9 min, the content of dissociated sulfhydryls in protein and So value are on slow rise. When the treatment time reaches 90 min, the content of dissociated sulfhydryls in protein is 13.01 µmol/g and the So value is 12.81, increasing by 43.16% and 71.72%, respectively, compared with those

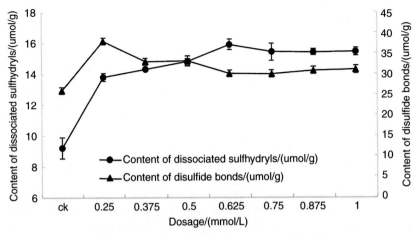

**FIGURE 4.55** Influence of dosage of $Na_2SO_3$ on the content of -SH/-S-S- in protein.

of the previous PPC. The longer the treatment time, the less significant the changes of the content of dissociated sulfhydryls in protein and So value ($P > 0.05$). Therefore, the $Na_2SO_3$ treatment conditions were determined to be dosage 0.625 mmol/L, treatment time 90 s.

- Cys Treatment

Refer to Fig. 4.57 and Table 4.7 for the influence of dosage of Cys on gelling property of protein. With the increase of the dosage of Cys, the content of dissociated sulfhydryls in the protein increases. When the dosage reaches 0.75 mmol/L, the content of dissociated

**TABLE 4.6** Influence of $Na_2SO_3$ Treatment Time on Surface Hydrophobicity Index So of Protein

| $Na_2SO_3$ Dosage (mmol/L) | Surface Hydrophobicity Index (So) | $Na_2SO_3$ Treatment Time (min) | Surface Hydrophobicity Index (So) |
|---|---|---|---|
| 0.250 | 8.42 | 15 | 10.86 |
| 0.375 | 9.43 | 30 | 11.97 |
| 0.500 | 118.00 | 45 | 12.06 |
| 0.625 | 13.48 | 60 | 12.49 |
| 0.750 | 12.90 | 90 | 12.81 |
| 0.875 | 12.92 | 120 | 11.61 |
| 1.000 | 13.00 | 150 | 11.83 |
| ck | 7.46 | 180 | 11.86 |

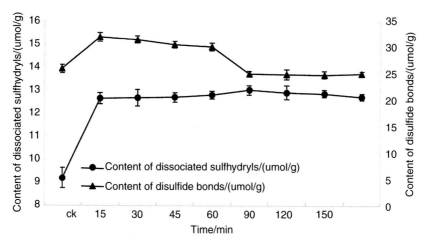

**FIGURE 4.56**   Influence of $Na_2SO_3$ treatment time on the content of -SH/-S-S- in protein.

sulfhydryls in protein increases, being 15.91 µmol/g, increasing by 75.06% compared with the original PPC. At this time, the content of dissociated sulfhydryls is basically the same as that when the dosage of $Na_2SO_3$ is 0.625 mmol/L; however, the dosage increases by 20% compared with that of the $Na_2SO_3$. In the following period, the increase of the dosage of Cys does not have significant influence on the gelling property of protein, and therefore the dosage of Cys was

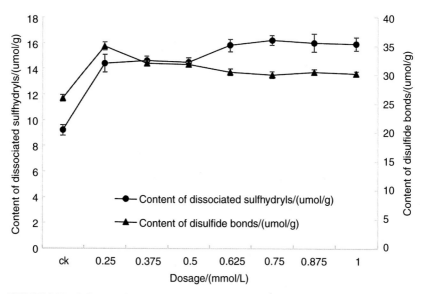

**FIGURE 4.57**   Influence of dosage of Cys on the content of -SH/-S-S- in protein.

**TABLE 4.7** Influence of Cys Treatment on Surface Hydrophobicity Index So of Protein

| Cys Dosage (mmol/L) | Surface Hydrophobicity Index (So) | Cystreatment Time (min) | Surface Hydrophobicity Index (So) |
|---|---|---|---|
| 0.250 | 8.55 | 15 | 8.03 |
| 0.375 | 8.69 | 30 | 8.40 |
| 0.500 | 8.65 | 45 | 8.85 |
| 0.625 | 9.39 | 60 | 8.93 |
| 0.750 | 12.33 | 90 | 10.24 |
| 0.875 | 11.02 | 120 | 11.09 |
| 1.000 | 10.20 | 150 | 10.22 |
| ck | 7.46 | 180 | 10.22 |

determined to be 0.75 mmol/L. Under this condition, the influence of different treatment times (0–180 min) on the gelling property of protein was investigated. As shown in Fig. 4.58 and Table 4.7, the gelling property of protein differs greatly at 0 and 15 min ($P < 0.01$). During this period, the content of dissociated sulfhydryls increases rapidly, increasing by 40.72% compared with the original PPC. At the same time, the content of disulfide bonds also increases by 24.72%. This is possibly because the Cys treatment of protein is carried out under oscillating or stirring status, and the oscillation or stirring increases

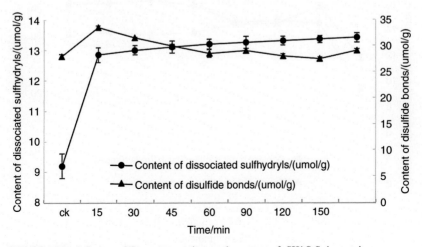

**FIGURE 4.58** Influence of Cys treatment time on the content of -SH/-S-S- in protein.

the opportunities of intermolecular collision and friction in the protein solution. After the sulfhydryls become dissociated from the protein, it is easy to form disulfide bond cross-linking, which will result in the increase of disulfide bonds. However, if the increasing speed of the content of dissociated sulfhydryls is considerably higher than the speed of disulfide bond cross-linking, the content of dissociated sulfhydryls will still be determined to be on the rise while the content of disulfide bonds declines relatively. This speculation is confirmed in the following experiment: with the prolonging of the treatment time, the content of the dissociated sulfhydryls rises slowly while the content of disulfide bonds declines slowly. The two reach a dynamic balance at 120 min of the treatment. At 150 min, the protein solution has the highest content of dissociated sulfhydryls and lowest content of disulfide bonds. With the continuous prolonging of treatment time, no significant change is detected with regard to the content of dissociated sulfhydryls, disulfide bonds, and So value ($P > 0.05$).

Reducers are added into the protein solution. On one hand, it can increase the –SH content of protein and improve the surface hydrophobicity of protein and increase the protein solubility; also, it can maintain the –SH content in protein and make the –SH function at gelling stage. On the other hand, during the protein processing period, it is easy for the dissociated sulfhydryls generated due to oscillation or stirring to form disulfide bonds. Therefore, when the protein is treated with reducers, it is very important to control the oscillating or stirring speed. In summary, the addition of the proper amount of reducers can help to improve the gelling property of protein.

**c.** Transglutaminase-B (TG-B) Enzymic Modification
- TG-B Enzymic Modification Single-factor Experiment
  **(1)** Substrate concentration

  Within 5–25% of substrate concentration, the –TG-B enzyme system of peanut protein is subject to apparent substrate inhibition, that is, with the increase of the substrate concentration, the gel hardness increases to a certain value and then decreases (Fig. 4.59). When the substrate concentration is 20%, the gel hardness is 9.87 g, which is 0.41 g lower than that at the 15% substrate concentration and 0.51 g higher than that at the 25% substrate concentration. It is obvious that when the substrate concentration increases from 20% to 25%, the gel hardness decreases greatly. Theoretically speaking, the over-large substrate concentration results in over-low effective moisture in the system, and therefore reduces the diffusion and movement of substrate and enzyme and suppresses the cross-linking effect. When the substrate concentration is over-low, the probability of collision between the enzyme and substrate will be increased and the cross-linking effect will

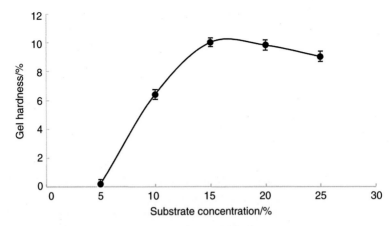

**FIGURE 4.59**    Influence of substrate concentrate on gel hardness.

be suppressed. As seen from the experimental results, the 5% substrate concentration has reduced the probability of collision between the enzyme and substrate. However, the 20–25% substrate concentration reduces the diffusion and movement of the substrate and enzyme system. In actual production, an unoptimal protein concentrate, whether too high or too low, will increase the energy consumption and reduce equipment utilization rate. Therefore, it is practically meaningless to set the substrate concentration to be too high or too low. The substrate concentration is finally set to be 15% through experimental results.

**(2)** Enzyme dosage

With 1 U/g as the basic dosage, Wu et al. (2009) studied the influence of enzyme dosage on gel hardness of protein and the results are as shown in Fig. 4.60. It can be seen that within 1–3 U/g, the gel hardness increases rapidly with the increase of enzyme dosage. When the enzyme dosage reaches 3 U/g, the gel hardness reaches 10.35 g, an increase by 7.79 g and 1.94 g respectively of that at 1 and 2 U/g. However, the gel hardness remains basically unchanged within the dosage range of 3–5 U/g. Therefore, the optimal enzyme dosage was determined to be 3 U/g.

**(3)** Reaction temperature

The enzyme reaction temperature is closely related to the stability of enzyme molecules because the enzyme molecules have specific spatial structures which are very vulnerable to secondary bond dissociation when the reaction temperature exceeds certain limits, and this will cause the loss of enzyme or partial loss of enzyme activity. However, if the reaction temperature is too low, the molecular movement intensity in the system will be greatly

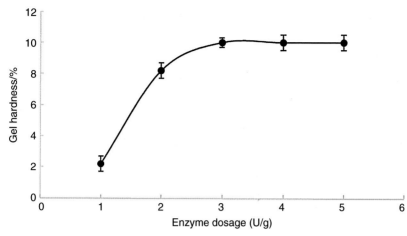

**FIGURE 4.60**  Influence of enzyme dosage on gel hardness.

reduced to decrease the collision probability between enzyme and substrate. The research team of the author studied the variation of the gel hardness of protein from 20 to 60°C, and the results are as shown in Fig. 4.61. During the process, as the temperature rises from 20 to 40°C, the gel hardness of the system increases gradually and reaches the highest value at 40°C, being 10.56 g. The gel hardness of the system begins to decline when the temperature exceeds 40°C. Therefore, the optimal TG-B enzyme reaction temperature was determined to be 40°C.

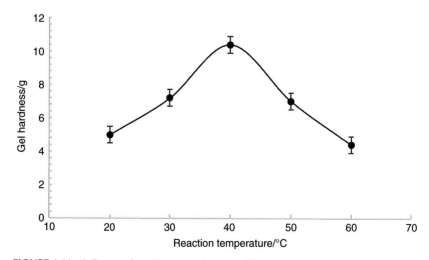

**FIGURE 4.61**  Influence of reaction temperature on gel hardness.

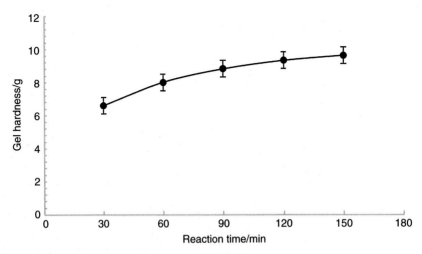

**FIGURE 4.62** Influence of reaction time on gel hardness.

(4) Reaction Time

Analysis (Fig. 4.62) is carried out for the changing trend of gel hardness of the system within different enzyme reaction times (30–150 min). During the initial 120 min of enzyme reaction, the gel hardness is always on the rise and reaches 9.89 g at 120 min of hydrolysis. When the hydrolysis lasts for 150 min, although the gel hardness continues to rise, the rising range is only 3.24%. Therefore, when the reaction time is 120 min, the gel hardness of the system has approached the maximal value and no longer exhibits significant increase with the prolonging of reaction time. Therefore, the reaction time is initially determined to be 120 min.

(5) pH

The influence of pH on enzyme-catalyzed reactions is mainly in the form of influence on the enzyme activity. Wu et al. (2009) analyzed the variation of protein gel hardness of the system within pH4–8. The results are as shown in Fig. 4.63. With the increase of pH, the gel hardness of protein increases rapidly; when the system pH reaches 7.0, the gel hardness reaches 9.26 g; when the system pH reaches 8.0, the gel hardness assumes a downward trend compared with pH 7.0. Therefore, pH 7.0 is preliminarily selected as the optimal pH for TG-B enzyme.

- Orthogonal Experiment of TG-B Enzyme Modified PPC

The optimal process combination for enzyme modified peanut protein is: enzyme dosage 3.0 U/g, substrate concentration 15%, reaction temperature 40°C. The reaction has little influence on the whole process, and the variance analysis for single-factor results indicates that the optimal gel time is 140 min, and pH7. Under these conditions,

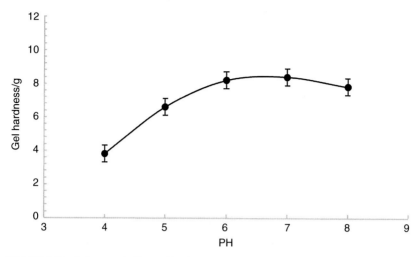

**FIGURE 4.63** Influence of pH on gel hardness.

the gel hardness is 11.85 g. However, since the hardness no longer changes significantly after the reaction time exceeds 60 min, the reaction time was determined to be 60 min according to the actual production.

The strong gel-type PPC has greater hardness 11.85 ± 0.56 g than that of gel-type soybean protein concentrate. The two have similar water-absorbing and oil-absorbing capacity, which are respectively 176.43 ± 1.33% and 205.12 ± 1.43%. However, compared with the gel-type soybean protein isolate, the water-absorbing and oil-absorbing capacities of strong gel-type peanut protein concentrate are slightly poorer. In view of the absence of a standard for the functionality of gel-type protein products both in China and abroad, the gelling property of the product was investigated based on enterprise criteria, that is, the gel hardness of gel-type soybean protein concentrate ≥10 g, water-holding capacity ≥150%, and oil-holding capacity ≥200%. Through analysis, it was shown that the quality of the PPC product is superior to the gel property of similar soybean products on market.

**2.** Improvement of PPC Solubility

**a.** Physical Modification

- Process of Solubilization and Modification by Means of High-Speed Stirring

**(1)** Single-Factor Experiment

The influence of stirring speed on the NSI of solubilized and modified PPC is shown in Fig. 4.64. The results indicate that with the increase of the stirring speed, the NSI of the protein also

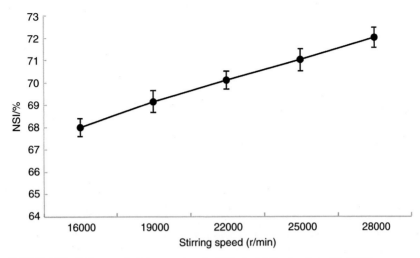

**FIGURE 4.64**   Influence of stirring speed on the NSI of solubilized and modified PPC.

rises. This is because then the stirring enables the redispersion of protein structure aggregated due to the denaturation induced by alcohol extraction, and the solubility also rises. The increase of stirring speed can reinforce the material friction and collision and enhance the protein particle distribution effect. However, although the experimental equipment can reach the designed stirring speed of 28,000 r/min, its applicable working speed is not expected to exceed 22,000 r/min, over which, the equipment load will become too high and the heat generated by the motor will sharply increase, and therefore it will be hard to maintain for a long-term operation exceeding 60 s. In summary, it is practical to set 22,000 r/min as the optimal stirring speed (single factor) for the solubilization and modification by means of high-speed stirring.

The influence of stirring time on the NSI of PPC solubilized and modified by means of high-speed stirring is shown in Fig. 4.65. The results indicate that with the stirring time prolonging from 12 to 66 s, the NSI of protein also rises. This is because the proper prolonging of stirring time means that it can be evenly dispersed, and therefore the protein solubility will rise as a result. However, when the stirring time reaches 84 s, the NSI declines instead. This is because the excessive stirring will result in excessive protein denaturation as evidenced by the changed color of the protein solution. The protein solution changes from the previously milky color to a dark color, indicating that the denaturation has exceeded the required level. Therefore, it is determined that

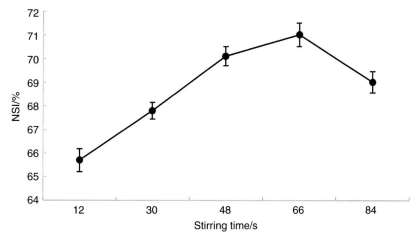

**FIGURE 4.65** Influence of stirring time on NSI of solubilized and modified PPC.

the optimal stirring time (single-factor) for the solubilization and modification by means of high-speed stirring is 66 s.

Solid–liquid

The influence of the solid–liquid ratio on the NSI of PPC solubilized and modified by means of high-speed stirring is shown in Fig. 4.66. The results indicate that when the solid–liquid ratio changes from 1:6 to 1:7, the NSI of protein increases slightly. With the increase of the solid–liquid ratio from 1:7, the NSI declines rapidly. When the solid–liquid ratio becomes 1:10, the NSI becomes basically the same as the original alcohol-extracted protein because the stirring solubilization can result in material

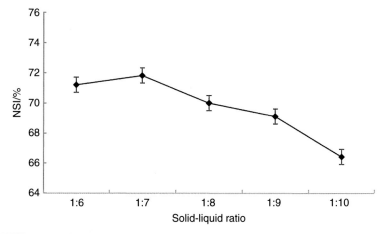

**FIGURE 4.66** Influence of solid-liquid ratio on the NSI of solubilized and modified PPC.

collision and friction. However, the solution with over-high concentration will retard the movement of the solid particles, and the collision and friction forces are inadequate to disintegrate the particle structure. In contrast, the solution with over-low concentration will greatly reduce the probability of interparticle collision and friction, and the stirring can only make them flow around instead of breaking apart. For the sample used in this experiment, the proper solid–liquid ratio for the solubilization and modification by means of high-speed stirring is 1:7.

**(2)** Process Conditions

The experiment was carried out under the conditions of stirring speed of 22,000 r/min, optimal stirring time of 66 s, and optimal solid–liquid ratio of 1:7 obtained by means of single-factor experiments at this rotating speed. The NSI of PPC solubilized and modified by means of high-speed stirring reaches 72.39 ± 0.42%, which is a significant increase compared with the NSI (65.30%) of the alcohol-extracted PPC, close to the 72.89% of the raw peanut protein powder. This indicates that after high-speed stirring treatment, the aggregation of PPC formed due to the denaturation induced by alcohol extraction is broken apart, and the side effect of alcohol extraction for the protein solubility is basically eliminated.

- Solubilization and Modification Process by Means of Microwave Treatment

**(1)** Single-Factor Experiment

The influence of microwave power on the NSI of solubilized and modified PPC is shown in Fig. 4.67. With the increase of the

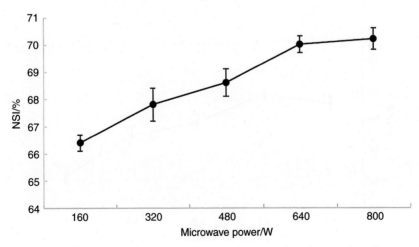

**FIGURE 4.67**   Influence of microwave power on the NSI of solubilized and modified PPC.

microwave power, the NSI of protein rises but the rise slows after 640 W. The main role played by microwave treatment for protein modification is thermal denaturation. The higher microwave power is beneficial to the increase of the solution temperature so as to break apart the aggregated protein structure caused by denaturation induced by the alcohol extraction. Although higher microwave power might result in a better solubilization effect, it is very easy for the boiling protein solution at 800 W to overflow from the container in actual operation, and the accident may disrupt the operation. In addition, when microwave modification is applied to actual industrial operation, it is impossible for the solution to keep boiling for a long time. Therefore, it is practical to set 640 W as the optimal power (single-factor) for solubilization and modification by means of microwave treatment.

The influence of microwave treatment time on the NSI of solubilized and modified PPC is as shown in Fig. 4.68. The results indicate that when the microwave treatment time increased from 10 to 40 s, the NSI of protein also rose; the significant rise could be observed from 20 to 30 s. However, when the microwave treatment time exceeds 40 s, the NSI could no longer change. In actual operation, when the microwave treatment time exceeds 30 s, it is very easy for the boiling protein solution at 800 W high power to overflow from the container in actual operation, and the accident may disrupt the operation. Therefore, it is practical to set 30 s as the optimal microwave treatment time (single-factor) for the solubilization and modification.

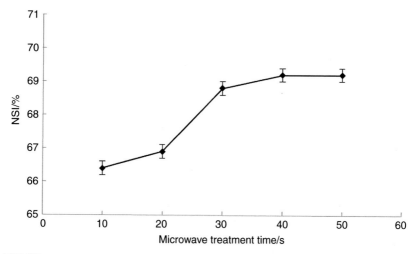

**FIGURE 4.68**  Influence of microwave treatment time on NSI of solubilized and modified PPC.

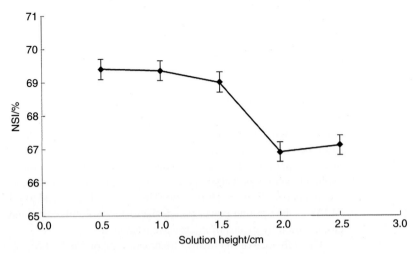

**FIGURE 4.69** Influence of solution height on NSI of solubilized and modified PPC.

The influence of solution height on the NSI of solubilized and modified PPC is shown in Fig. 4.69. The results indicate that when the solution height increases from 0.5 to 1.0 cm, the NSI of protein remains unchanged. With the continuous increase of solution height, the NSI begins to decrease and no longer changes after exceeding 2.0 cm. This is mainly because that the microwave penetration is limited, and the over-thick solid–liquid layer does not have good heating effect, and therefore the large amount of protein that is not at the surface of the solution cannot be denatured by the microwave treatment. However, in actual operation, the solution rapidly reaches the boiling point during microwave treatment even when the solution height is lower than 1.5 cm, which will disrupt the experiment. Therefore, through comprehensive consideration, it was determined to set 1.5 cm as the optimal solution height (single factor) for solubilization and modification by means of microwave treatment.

The influence of the solid–liquid ratio on NSI of PPC solubilized and modified by means of microwave treatment is shown in Fig. 4.70. The results show that during the process the solid–liquid ratio changes from 1:7 to 1:11, the influence on the NSI of the PPC is insignificant. When the actual process is considered, the low solid–liquid ratio will help to improve the treatment efficiency. Therefore, it was determined to set 1:7 as the optimal solid–liquid ratio (single factor) for the solubilization and modification by means of microwave treatment.

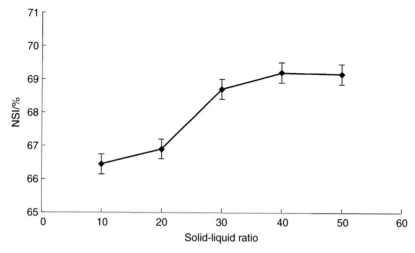

**FIGURE 4.70** Influence of solid-liquid ratio on NSI of solubilized and modified PPC.

**(2)** Process Conditions

According to the above experimental results and analysis, the microwave power is set to be 640 W, the microwave treatment time is set to be 30 s, the solution height is set to be 1.5 cm, and the solid–liquid ratio is set to be 1:7 based from the comprehensive experimental results of the solubilization effect. A verification experiment was carried out under these conditions. The NSI of PPC solubilized and modified by means of microwave treatment reached 69.83 ± 0.80%, higher than the NSI (65.30%) of alcohol extracted PPC. However, it still lags behind the NSI (72.89%) of raw protein powder. This indicates that, through microwave treatment, the solubility of PPC lost due to the denaturation induced by alcohol extraction is recovered to a certain extent.

- Solubilization and Modification Process by Means of Thermal Treatment

**(1)** Single-factor Experiment

The influence of heating temperature on the NSI of PPC solubilized and modified by means of water bath heating treatment is shown in Fig. 4.71. When the temperature of the water bath increased from 70 to 75°C, the NSI of protein also increased slightly. This indicates that the intermolecular covalent bonds of the protein break and the originally compact structure becomes loose, and therefore the solubility increases. After the temperature exceeds 75°C, the further increase of the temperature of the water bath results in the decrease of NSI, and the NSI decreases rapidly after 85°C. This indicates that the protein in water solution

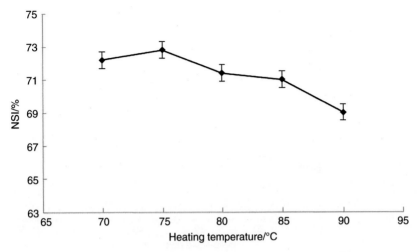

**FIGURE 4.71**   Influence of heating temperature on NSI of solubilized and modified PPC.

begins to denaturize and aggregate above 75°C and the solubility disappears gradually. Therefore, it is determined that the optimal heating temperature (single factor) for solubilization and modification by means of thermal treatment is 75°C.

The influence of heating temperature on the NSI of PPC solubilized and modified by means of water bath heating treatment is as shown in Fig. 4.72. When the heating time is 1 min, the NSI of protein becomes basically the same with that of the original

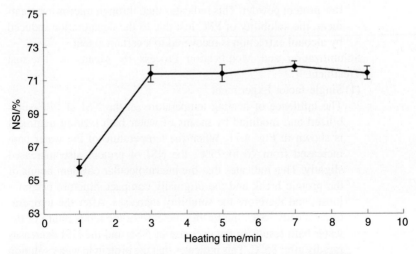

**FIGURE 4.72**   Influence of heating time on NSI of solubilized and modified PPC.

alcohol-extracted protein, and this indicates the heat of water bath does not enter the solution yet. When the heating time increased from 1 to 3 min, the NSI increased sharply, with the highest value appearing after 7 min of thermal treatment. At this point, the internal structure of the protein is disintegrated to the status relatively suitable for solubility. Afterward, with the further prolonging of heating time, the NSI begins to decline, and at this time, the protein not suitable for solubility begins to aggregate. Therefore, it is determined that the optimal heating time (single factor) for solubilization and modification is 7 min.

The influence of solid–liquid ratio on the NSI of PPC solubilized and modified by means of water bath heating treatment is shown in Fig. 4.73. When the solid–liquid ration changes from 1:7 to 1:9, the NSI of protein increases accordingly.

However, with the further change of solid–liquid ratio, the NSI begins to decrease. The NSI decreases rapidly after the solid–liquid ratio exceeds 1:10. This indicates that the solid–liquid ratio, whether it is too low or too high, is not beneficial to the thermal treatment modification of protein. The solvent, if inadequate, might increase the steric hindrance for molecular expansion of protein and prevent the structural disintegration. The solvent, if excessive, might speed up the heat transfer and prevent the aggregation of dissolved protein. Therefore, the optimal solid–liquid ratio for solubilization and modification by means of thermal treatment is 1:9.

**(2)** Simple Orthogonal Experiment

According to the results of single-factor experiment, the three factors involved in the simple orthogonal experiment of water

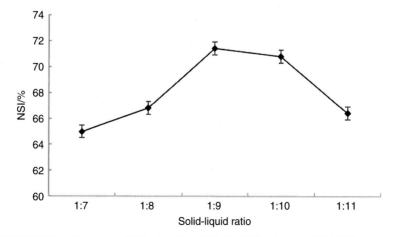

**FIGURE 4.73** Influence of solid-liquid ratio on NSI of solubilized and modified PPC.

bath thermal treatment include: heating temperature (A), heating time (B), and solid–liquid ratio (C). Duncan's new multiple range method is adopted to analyze the results of the orthogonal experiment. The heating temperature has the greatest influence during the solubilization process, followed by solid–liquid ratio, and both of them have significant influence on the NSI. However, the heating time does not have statistical significance. The analysis results indicate that the optimal combination of level of factors are heating temperature 75°C, heating time 7 min, and solid–liquid ratio 1:9. Under these conditions the optimal solubilization and modification effects can be obtained. A verification experiment was carried out with these process parameters, the NSI of PPC solubilized and modified by means of thermal treatment reaches 72.94 ± 0.61%, which is a significant increase compared with the NSI (65.30%) of alcohol-extracted PPC, reaching the level of 72.89% similar to that of the peanut protein powder. This indicates that after thermal treatment, the solubility of PPC lost due to the alcohol extraction is fully restored, and the adverse influence caused by alcohol extraction is offset. In the actual production, the heating time can be shortened so as to improve production efficiency and reduce production cost.

- Solubilization and Modification Process by Means of Ultrasonic Wave Treatment

  **(1)** Single-factor Experiment

  The influence of ultrasonic power on NSI of PPC solubilized and modified by means of ultrasonic treatment is shown in Fig. 4.74. The experimental results indicate that when the ultrasonic power

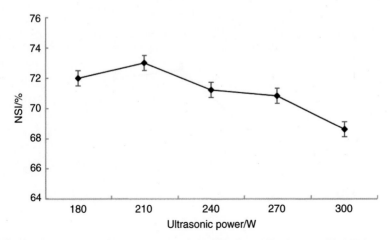

**FIGURE 4.74**  Influence of ultrasonic power on the NSI of solubilized and modified PPC.

increases from 180 to 210 W, the NSI of protein rises slightly. This indicates that the increase of ultrasonic power can increase the energy density of the solution so as to generate more intense cavitation effect, and the protein solubility also increases accordingly. However, after exceeding 200 W, the further increase of ultrasonic power will result in the significant decrease of the NSI, and this is possibly because a too high energy density generates too much local heat and makes the protein conformation more unsuitable for solubility. Therefore, the optimal power (single factor) for solubilization and modification by means of ultrasonic treatment is 210 W.

The influence of water bath temperature on NSI of PPC solubilized and modified by means of ultrasonic treatment is shown in Fig. 4.75. The experimental results show that when the water bath temperature of ultrasonic treatment increases from 60 to 80°C, the NSI of protein also rises, and the rising speed begins to slow down at above 70°C. This is possibly because the relatively high temperature will help to disintegrate the protein structure so as to subject more protein molecules to the ultrasonic action and enable the protein to denature in a manner more suitable for solubility. However, the temperature that can cause the overdenaturation of the protein is higher than 80°C, which is higher than the temperature (75°C) for thermal treatment, and this is possibly related to the ultrasonic effect. For the ultrasonic cleaner used in this experiment, the highest water bath temperature is set to be 80°C, and the cleaner will stop automatically for self-protection when the water bath temperature reaches 81°C.

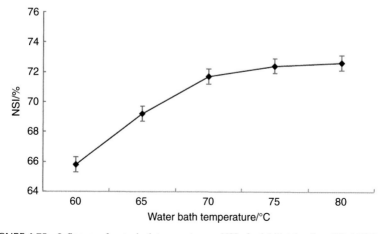

**FIGURE 4.75** Influence of water bath temperature on NSI of solubilized and modified PPC.

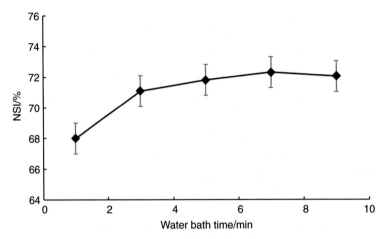

**FIGURE 4.76**  Influence of water bath time on NSI of solubilized and modified PPC.

Furthermore, during the operation of the cleaner, the water bath temperature will rise constantly with the operation time, rising by 1–2°C after 5 min of operation. A too high setting of the initial temperature will cause the machine to shut down immediately for self-protection. Therefore, taking the comprehensive conditions into consideration, the optimal water bath temperature (single factor) for the solubilization and modification by means of ultrasonic treatment is 75°C, and this factor is not involved in the design of the simple orthogonal experiment.

The influence of ultrasonic treatment time on the NSI of PPC solubilized and modified by means of ultrasonic treatment is shown in Fig. 4.76. The experimental results indicate that when the ultrasonic treatment increases from 1 to 7 min, the NSI of the protein increases accordingly. It increases at a relatively higher speed before 3 min, and then the speed slows down. This is bewcause it takes time for the ultrasonic treatment to improve the protein hydration. However, after the ultrasonic treatment time exceeds 7 min, the NSI decreases slightly. This is possibly because the overlong ultrasonic treatment can overdenature the protein, and it also might be that the overlong water bath heating makes the protein generate aggregation not suitable for the solubility improvement. Therefore, the optimal time for solubilization and modification by means of ultrasonic treatment is 7 min.

The influence of solid–liquid ratio on the NSI of PPC solubilized and modified by means of ultrasonic treatment is shown in Fig. 4.77. The experimental results indicate that when the solid–liquid ratio changes from 1:7 to 1:10, the NSI of protein also

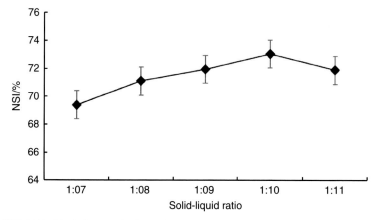

**FIGURE 4.77**  The influence of solid-liquid ratio to NSI of the modified PPC.

increases. When the solid–liquid ratio exceeds 1:10, the further raising of the ratio will reduce the NSI. This indicates that the solid–liquid ratio, whether too high or too low, will not help to improve the protein modification by means of ultrasonic treatment. The solution concentration, if excessive, might increase the steric hindrance for molecular expansion of protein and prevent the structural disintegration. The concentration, if inadequate, might weaken the cavitation effect and prevent the increase of solubility. Therefore, the optimal solid–liquid ratio for solubilization and modification by means of ultrasonic treatment is 1:10.

**(2)** Orthogonal experiment

According to the results of the single-factor experiments, the three factors involved in the simple orthogonal experiment of ultrasonic treatment include: heating temperature (A), heating time (B), and solid–liquid ratio (C). Duncan's new multiple range method is adopted to analyze the results of the orthogonal experiment. The solid–liquid ratio has the greatest influence during the solubilization process, followed by ultrasonic power, and both of them have significant influence on the NSI. However, the ultrasonic treatment time does not have statistical significance. The analysis results indicate that the optimal combination of level of factors are: ultrasonic power 210 W, ultrasonic treatment time 7 min, and solid–liquid ratio 1:10, under which conditions the optimal solubilization and modification effects can be obtained. When a verification experiment was carried out with these process parameters, the NSI of PPC, solubilized and modified by means of thermal treatment, reached 73.72 ± 0.57%, which was a significant increase compared with the NSI (65.30%) of alcohol-extracted

PPC, and it reached the level of 72.89%, similar to that of the peanut protein powder. This indicates that after ultrasonic treatment, the solubility of PPC lost due to the alcohol extraction is not only fully restored but also increases relative to that of the raw protein powder. In the actual production, the ultrasonic treatment time can be shortened so as to improve production efficiency and reduce production cost.

- Compound Process of Physical Solubilization and Modification Method

The NSI of protein increases to varying degrees by adopting four physical methods to modify the alcohol-extracted PPC. With selected optimal process parameters, the solubilization effects of the four physical means are as follows in a descending order: Ultrasonic treatment > thermal treatment > high-speed stirring > microwave treatment. Since different solubilization and modification methods have different acting mechanisms, it is probable that the protein which has been modified with one method can be further modified with another method. The experiment investigates the changes of the solubility through the compound use of the four methods to select the optimal compounding of physical methods. All the process parameters used by the compound experiment are the optimal parameters through optimization, as shown in Table 4.8.

All the methods have most suitable solid–liquid ratios except microwave treatment which does not have any influence on the solubilization and modification effects within the solid–liquid ratio range of 1:7 to 1:11. All of these most suitable solid–liquid ratios are different. If a compound modification process includes microwave treatment, the solid–liquid ratio will be the most suitable solid–liquid ratio of the other treatment method. If a compound modification process does not include microwave treatment and the solid–liquid ratio of the first process is lower than the succeeding one, then solvent will be added between the two processes to meet the solid–liquid requirement of the succeeding one. If the solid–liquid ratio of the first process is higher than the succeeding one, then the solid–liquid ratio of the succeeding one will adopt the optimal solid–liquid ratio. Therefore, the high-speed stirring treatment and microwave treatment subsequent to the ultrasonic treatment will adopt the solid–liquid ratio of 1:10 while the high-speed stirring treatment subsequent to thermal treatment will adopt the solid–liquid ratio of 1:9. The experimental results of the compound solubilization and modification with four different physical methods are as shown in Table 4.9. The font-weight results indicate that the NSI of the PPC subject to compound treatment is higher than any of

**TABLE 4.8** Optimal Parameters for PPC Solubilized and Modified With Different Physical Methods

| Modification Method | NSI/% After Modification | Suitable Solid–Liquid Ratio | Other Parameters of Optimal Process |
|---|---|---|---|
| High-speed stirring | 72.39 ± 0.42 | 1:7 | Stirring speed 22,000 r/min, stirring time 66 s |
| Microwave treatment | 69.83 ± 0.80 | 1:(7–11) | Microwave power 640 W, microwave time 30 s, solution height 1.5 cm |
| Thermal treatment | 72.94 ± 0.61 | 1:9 | Heating temperature 75°C, heating time 7 min |
| Ultrasonic treatment | 73.72 ± 0.57 | 1:10 | Ultrasonic power 210 W, ultrasonic time 7 min, water bath temperature 75°C |
| High-speed stirring | 72.39 ± 0.42 | 1:7 | Stirring speed 22,000 r/min, stirring time 66 s |
| Microwave treatment | 69.83 ± 0.80 | 1:(7~11) | Microwave power 640 W, microwave time 30 s, solution height 1.5 cm |
| Thermal treatment | 72.94 ± 0.61 | 1:9 | Heating temperature 75°C, heating time 7 min |
| Ultrasonic treatment | 73.72 ± 0.57 | 1:10 | Ultrasonic power 210 W, ultrasonic time 7 min, water bath temperature 75°C |

the individual process used alone, that is, the compound process has partial superposing effect for improving the solubility.

The experimental results indicate that high-speed stirring is the most suitable method to be used in combination with other physical modification methods. Whether used before or after the microwave treatment, thermal treatment, and ultrasonic treatment, all these methods can further improve the solubility of protein on the original basis so as to achieve good effects that might not have been achieved without compound treatment. This is mainly because the solubilization mechanism via high-speed stirring is different from the other methods, and with high-speed stirring, the protein particles can be crushed through material collision and friction caused by mechanical force. However, the microwave treatment and thermal treatment mainly take effect by destroying the intramolecular

**TABLE 4.9** Experimental Results of PPC Solubilization and Modification by Means of Compound Physical Method

| | Subsequent Process | | | |
|---|---|---|---|---|
| First Process | High-Speed Stirring | Microwave Treatment | Thermal Treatment | Ultrasonic Treatment |
| High-speed stirring | | 73.34 ± 0.74 | 76.20 ± 0.67 | 78.18 ± 0.54 |
| Microwave treatment | 74.82 ± 0.77 | | 72.81 ± 1.03 | 72.73 ± 0.83 |
| Thermal treatment | 73.83 ± 0.50 | 71.35 ± 0.90 | | 72.44 ± 0.65 |
| Ultrasonic treatment | 75.06 ± 0.36 | 74.25 ± 0.61 | 74.78 ± 0.46 | |

and intermolecular covalent bonds by means of thermal energy, and the water bath and ultrasonic treatment also enhance solubilization through thermal denaturation. Although long-time high-speed stirring will also raise the temperature of protein solution, the optimal stirring time selected for this study is only 66 s, which barely has any heating effect. Therefore, the mechanical force and the heating effect of other treatment methods can be well combined to avoid the problems of excessive thermal or excessive mechanical denaturation.

The sequence of operation also influences the effect of compound modification that involves high-speed stirring, and better effects can be obtained by arranging the high-speed stirring before other operations. This is because first, the solid–liquid ratio is no longer the most suitable ratio of 1:7 if the high-speed stirring is arranged after other operations, and this will affect the solubilization effects; second, it is possible that the high-speed stirring can disintegrate the protein structures that have aggregated due to alcohol extraction, and the molecules can be heated in a more uniform manner as a result. However, when the high-speed stirring is used in combination with microwave treatment, it is better to arrange the microwave treatment to be prior to other operations. This is because the microwave treatment is not affected by the solid–liquid ratio, and the high-speed stirring, even arranged after the microwave treatment, is carried out under the optimal ratio of 1:7. In addition, different from traditional heating methods, the microwave heating has high penetration and rapid heating speed, which can evenly heat the protein without the need of stirring for disintegration. In this case, there is no advantage by arranging the high-speed stirring ahead of other operations. It is speculated that the change of electrical properties of protein, as a result of the microwave treatment, leads to more effective impacting and crushing in the centrifugal force field generated by the stirring so that better superposition effect of solubilization can be achieved by arranging the microwave treatment to be prior to other operations.

Certain superposition effect was also achieved by combining ultrasonic treatment (arranged prior to other operations) with other solubilization mechanisms featuring thermal denaturation. However, the effect was not so significant, only making the NSI of protein increase by 1.06% maximally on the basis of the original ultrasonic treatment. This is because the ultrasonic treatment was carried out at the water bath temperature of 75°C, and the operation is substantially equivalent to the compounding of ultrasonic modification and thermal modification. The further application of other thermal modification methods will definitely restrict the increase of the solubility. By arranging the ultrasonic treatment to be prior to other operations, superposition effect can be obtained possibly because the cavitation effect generated by the ultrasonic waves can deform and crush the protein particles, which are then effectively heated in a more even manner. By putting the ultrasonic treatment after other operations, the highly denatured part of the protein from the originally unevenly heated protein will be excessively denatured, which will decrease the solubility instead.

The compounding of microwave treatment and thermal treatment does not have any superposition effect because the thermal treatment is totally equivalent to thermal modification and solubilization and the microwave treatment also solubilizes by this means. The difference between the two is the latter takes effect in a more rapid and even manner. All the processes used in the compounding experiment have been optimized, that is, the denaturation degree of the protein after being subject to microwave treatment and heat treatment is suitable for solubilization, and the continuous heating will inevitably result in the decrease of the solubility.

In summary, by adopting the method that combines high-speed stirring and thermal treatment, the PPC will be subject to the effect of mechanical energy, thermal energy, and ultrasonic energy, and the disintegration of the structures aggregated during the alcohol extraction offsets the adverse effects of alcohol extraction on solubility. In addition, some of the intramolecular and intermolecular covalent bonds which are insoluble are broken apart because of the compound denaturation, which further enhances the solubility but avoids the excessive denaturation that leads to the decrease of solubility. By arranging the high-speed stirring before the ultrasonic treatment,

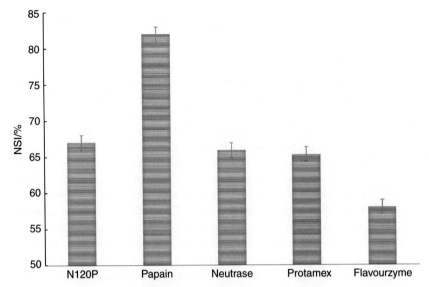

**FIGURE 4.78** Influence of different proteases on the NSI of solubilized and modified PPC.

various denaturation effects can be better synergized to obtain better effects of physical solubilization and modification.

**b.** Modification by Restricted Enzyme

- Protease

**(1)** Protease

The influences of different proteases on NSI of PPC solubilized and modified by restricted enzyme are shown in Fig. 4.78. The experimental results indicate that the solubilization and modification effect of papain is significantly better than the four other kinds of protease. N120P, neutrase, and protanmex have similar solubilization and modification effects for protein while flavourzyme has a relatively poor effect. The NSI of the restricted enzymatic hydrolytes of the four kinds of enzyme is all lower than that obtained after compound physical solubilization. After enzymolysis, the NSI of flavourzyme is even lower than that of the raw alcohol-extracted PPC. This is possibly because first, the high temperature of 95°C causes excessive denaturation of the protein during enzyme deactivation, and the slight increase of the protein solubility caused by the restricted enzymolysis is far from enough to offset the decrease caused by thermal denaturation; second, the process parameters selected according to the factory instructions are not suitable for the solubilization and modification of PPC. When solubilizing with restricted enzyme, the relative insufficiency of enzymolysis might cause the protein under the hydrolysis level to aggregate, which will reduce the solubility. However, the relatively excessive enzymolysis might expose more hydrophobic amino acid residues, and the increase of the surface hydrophobicity of the protein will also reduce the solubility.

Simple sensory evaluation was carried out for the enzymatic hydrolytes of the above five proteases. The results indicate that bitterness is found in the enzymatic hydrolytes of the five enzymes, except papain, and the enzymatic hydrolytes of N120P have apparent bitterness. Since the high-soluble PPC products are mainly used as functional food additives, especially as additives for protein drinks, it is obvious that the bitterness might cause damage to the food quality of the final products. Therefore, taking into comprehensive consideration the solubilization, modification, and sensory evaluation results, papain is finally selected as the protease most suitable for solubilization and modification of PPC, and the relevant enzymatic hydrolysis parameters are further optimized.

**(2)** Single-Factor Experiment

The influence of solid–liquid ratio on the NSI of PPC solubilized and modified by means of restricted enzyme is shown in Fig. 4.79. The experimental results indicate that when the solid–liquid ratio changes from 1:10 to 1:12, the NSI of the restricted enzymatic hydrolytes basically remain unchanged. Thereafter, the solid–liquid ratio changes and the NSI began to decline. Since the

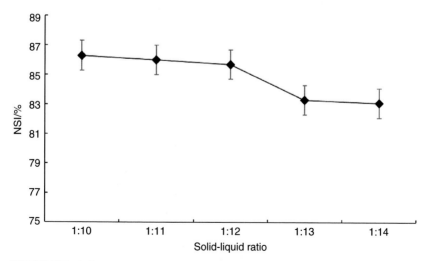

**FIGURE 4.79**  Influence of solid-liquid ratio on The NSI of solubilized and modified PPC.

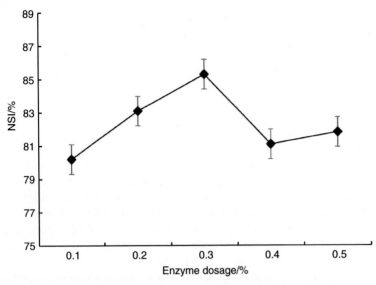

**FIGURE 4.80** Influence of enzyme dosage on the NSI of solubilized and modified PPC.

method for restricted enzymatic hydrolysis shows that there is no change of solid-liquid ratio for ultrasonic treatment, the optimal solid–liquid ratio (single factor) for solubilization and modification by means of restricted enzyme is considered to be 1:10.

The influence of the enzyme dosage on the NSI of PPC solubilized and modified by means of restricted enzyme is shown in Fig. 4.80. The experimental results show that when the enzyme dosage (E:S) increases from 0.1% to 0.3%, the NSI rises accordingly. This indicates that the substrate is not saturated with the enzyme below the 0.3% of enzyme dosage. The increase of enzyme dosage can enable the full participation of substrate in the reaction, and avoid the protein aggregation at low levels of hydrolysis caused by the relative enzymolysis insufficiency. However, after the enzyme dosage exceeds 0.3%, the NSI begins to decrease, and this is possibly because the relatively excessive restricted enzymolysis causes the exposure of more hydrophobic amino acid residues, and therefore the surface hydrophobicity of the protein increases to decrease the solubility. Therefore, it is determined that the optimal enzyme dosage (single factor) for the solubilization and modification by means of restricted enzymolysis is 0.3%.

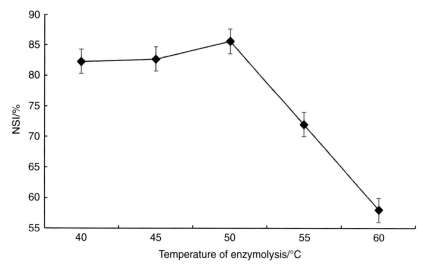

**FIGURE 4.81**    Influence of the temperature of enzymolysis on the NSI of solubilized and modified PPC.

The influence of the temperature of enzymolysis on the NSI of PPC solubilized and modified by means of restricted enzyme is shown in Fig. 4.81. The experimental results indicate that when the temperature of enzymolysis increases from 40 to 50°C, the NSI of the enzymatic hydrolytes slightly increases, and this indicates that relatively higher temperature can help the protease to fully take effect. However, above 50°C, the further increase of temperature will make the NSI of the enzymatic hydrolytes decrease rapidly. This is because the temperature exceeds the tolerance range of the protease, which is deactivated as a result. Therefore, the optimal temperature of enzymolysis (single factor) for the solubilization and modification by means of restricted enzyme is set at 50°C.

The influence of the time of enzymolysis on the NSI of PPC solubilized and modified by restricted enzyme is shown in Fig. 4.82. The experimental results show that when the time of enzymolysis increases from 10 to 30 min, the NSI of the enzymatic hydrolytes increases significantly. This indicates that it takes time for the protease to take effect so as to obtain a better solubilization effect. However, after 30 min, the NSI begins to decrease. This is similar to the effect of overdosage of enzyme, where the overdosage causes the relative excess of restricted enzymolysis, and the solubility of the hydrolytes decreases. Therefore, the optimal time of enzymolysis (single factor) for solubilization and modification by means of restricted enzyme is set at 30 min.

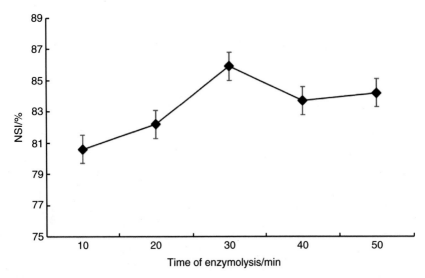

**FIGURE 4.82** Influence of the time of enzymolysis on the NSI of solubilized and modified PPC.

(3) Response surface optimization experiment

Through single-factor experiment, the three factors involved in the solubilization and modification by means of restricted enzyme include enzyme dosage (A), time of enzymolysis (B), and temperature of enzymolysis (C), while the solid–liquid ratio is directly determined to be 1:10.

Through model-based calculation, the NSI of the PPC obtained under optimal process conditions is 86.58% and the verification value is 86.34 ± 0.69%, which is consistent with the predicted value of the model. In comparison, the NSI increases significantly compared with that (78.18%) of the PPC solubilized and modified by means of compound physical methods. All of the following experiments of postenzymolysis solubilization and modification are carried out with the optimal process parameters for the solubilization and modification by means of restricted enzyme.

- Postenzymolysis Solubilization and Modification Process
(1) Posttreatment Method

The influence of different postenzymolysis treatment methods on the NSI of solubilized and modified PPC is shown in Fig. 4.83. The experimental results indicate that the solubilization effect of the high-pressure homogenization treatment is obviously better than the surface activators or reducers. In actual production, high-pressure homogenization is widely applied in the food industry. The addition of the two surface activators, Tween 80 and soybean

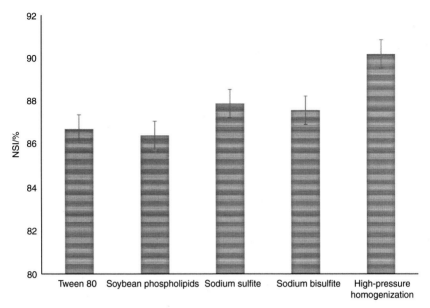

**FIGURE 4.83** Influence of different posy-enzymolysis treatment methods on the NSI of solubilized and modified PPC.

phospholipids, barely makes any contribution to the protein solubility, and this is because the surface activator has poor hydrophilicity and can therefore solubilize the PPC more effectively. The two reducers, sodium sulfite and sodium bisulfite, can increase the protein solubility to a certain degree because the reducers can effectively suppress the formation of intramolecular and intermolecular disulfide bonds. However, considering that the security of some chemical modification methods, such as the addition of reducer, still comes under question in the food processing industry, this experiment only carried out a preliminary attempt to find out the effect of solubilization by means of reducer, and the specific process parameters, such as the dosage, are not optimized. Therefore, taking into comprehensive consideration the solubilization effect and the security, it was determined to select the high-pressure homogenization treatment as the posttreatment method for solubilizing and modifying PPC by means of restricted enzyme, and the process parameters will be further optimized.

**(2)** Process of Solubilization and Modification by Means of High-pressure Homogenization

   **a.** Single-Factor Experiment

   The influence of homogenization pressure on the NSI of PPC solubilization and modification is as shown in Fig. 4.84.

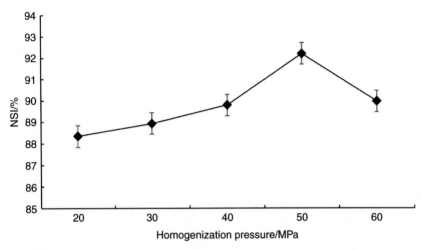

**FIGURE 4.84** Influence of the homogenization pressure on the NSI of solubilized and modified PPC.

When the homogenization pressure increases from 20 to 50 MPa, the NSI of the protein also rises. This indicates that relatively higher homogenization pressure can help to generate more intense shear force and impact force, and therefore can crush the protein more effectively to improve its solubility. However, after exceeding 50 MPa, the further increase of protein NSI causes the decrease of NSI. This is possibly because the thermal energy generated by a too high pressure enables the solution temperature to rise sharply, and the protein is denatured to be insoluble. Therefore, it is determined that the homogenization pressure (single factor) for solubilization and modification by means of high-pressure treatment is 50 MPa.

The influence of solid–liquid ratio on the NSI of the PPC solubilized and modified by means of homogenization pressure is shown in Fig. 4.85. The experimental results show that when the solid–liquid ratio increases from 1:10 to 1:13, the NSI of protein increases significantly. However, when the solid–liquid ratio continues to change up to 1:14, the NSI begins to decrease again. This is because that a too high solution concentration increases the motion resistance of the protein so as to weaken the acting effect of the shearing force, impacting force, and cavitation blasting force. At the same time, a too low solution concentration will also influence the intermolecular interaction, cause the shearing force to be dispersed to a certain extent, and weaken the solubilization and modification effect. Therefore, the optimal solid–liquid ratio

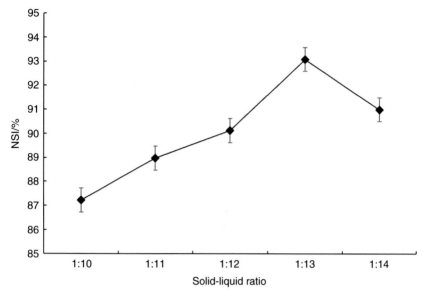

**FIGURE 4.85**  Influence of the solid-liquid ratio on the NSI of solubilized and modified PPC.

for solubilization and modification by means of high-pressure homogenization was determined to be 1:13.

The influence of homogenization time on the NSI of PPC solubilization and modification is shown in Fig. 4.86. The experimental results show that when the homogenization time

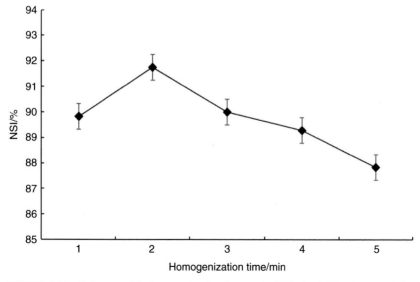

**FIGURE 4.86**  Influence of the homogenization time on the NSI of solubilized and modified PPC.

increases from 1 to 2 min, the NSI of protein increases significantly. However, the NSI begins to decrease rapidly after 2 min. This indicates that the proper prolonging of homogenization time will enable the shearing force and impacting force to work on the materials more effectively so that the protein solubility can be increased. However, the homogenization time, if too long, will significantly increase the solution temperature. As a result, the protein will have excessive thermal denaturation and its solubility will be reduced. Therefore, the optimal time for solubilization and modification by means of high-pressure homogenization treatment was determined to be 2 min.

**b.** Orthogonal Experiment

According to the single-factor experimental results, three factors involved in the simple orthogonal experiment of solubilization treatment by means high-pressure homogenization were selected, namely, homogenization pressure (A), solid–liquid ratio (B), and homogenization time (C). Duncan's new multiple range method was used to analyze the results of orthogonal experiment.

It can be seen that the homogenization pressure plays the largest role during the solubilization process, followed by homogenization time and solid–liquid ratio. All three have a significant influence on the NSI. The analysis results indicate that under the optimal conditions of homogenization pressure 50 MPa, solid–liquid ratio 1:13, and homogenization time 2 min, the best solubilization and modification effects can be obtained. When a verification experiment was carried under these process parameters, the NSI of the PPC solubilized and modified by means of high-pressure homogenization reached 96.57 ± 0.40%, which is a significant increase compared with the NSI (86.34%) of the PPC solubilized and modified by means of restricted enzyme.

With the alcohol-extracted PPC as the raw material, the influence of different physical treatment methods on the enzymolysis treatment effects was investigated to select the protease that is suitable for solubilization and modification by means of the restricted enzyme of PPC. Afterward, the response surface method was adopted to optimize the process, and then different postenzymolysis treatment methods were investigated and the posttreatment process of high-pressure homogenization was optimized.

By comparing the treatment methods adopted before the restricted enzymolysis, we figured out that the ultrasonic treatment was the most suitable physical treatment method to be used before the solubilization and modification by means of restricted enzyme. In this study, the compound method of high-speed stirring and ultrasonic treatment was adopted to simulate actual industrial production. The solubilization and modification effects as well as the sensory evaluation results were taken into comprehensive consideration in selecting the enzyme used for solubilization, and papain was ultimately selected. Of the three factors, enzyme dosage (X1), time of enzymolysis (X2), and temperature of enzymolysis (X3), the dominant factor that influences the NSI of the enzymatic hydrolytes is enzyme dosage, followed by the time of enzymolysis. The optimal process of solubilization by means of restricted enzymolysis is enzyme dosage (E:S) 0.2913%, the time of enzymolysis 18.95 min, and the temperature of enzymolysis 46.70°C. By taking the solubilization effect and security into comprehensive consideration, the high-pressure homogenization treatment was determined to be the optimal postenzymolysis treatment method, and the factors that influence the effect of solubilization by means of high-pressure homogenization are respectively (arranged in a descending order): Homogenization pressure > homogenization time > solid–liquid ratio. The optimal process parameters for high-pressure homogenization were determined to be homogenization pressure 10 MPa, solid–liquid ratio 1:13, and homogenization time 2 min.

## 4.7.3  Improvement of Gelling Property of Peanut Protein Component

1. Physical Modification
   a. Improvement of Gelling property of Arachin
      - Single-Factor Experiment
        (1) Arachin Concentration
            Fig. 4.87 shows the gelling property of thermal gel formed by arachin under different protein concentrations. The arachin cannot form gel at 6% concentration. When the concentration increases from 8 to 18%, its hardness, gelling property, and water-holding capacity all increase continuously. This is because with the increase of the protein concentration, the number of molecules within unit volume increases and the intermolecular interaction also increases. The elasticity is basically stabilized after the concentration reaches 14% and the cohesion first decreases and then increases.

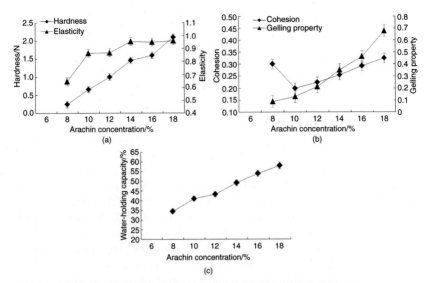

**FIGURE 4.87** Influence of solid-liquid ratio on the NSI of solubilized and modified PPC.

### (2) Phosphate Concentration

Fig. 4.88 shows the variation of the gelling property of arachin under different PBS concentrations. It can be seen that when the phosphate concentration is 0.01 mol/L, the hardness, elasticity, and gelling property are all at their highest points but the

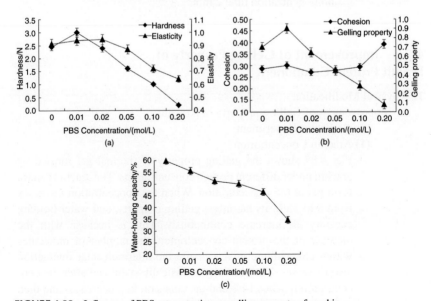

**FIGURE 4.88** Influence of PBS concentration on gelling property of arachin.

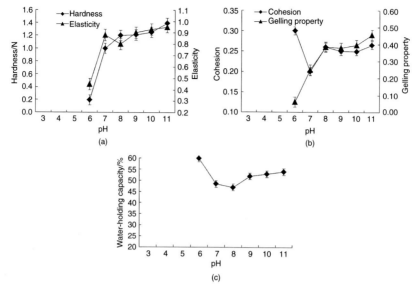

**FIGURE 4.89** Influence of pH on gelling property of arachin.

cohesion does not change significantly. The water-holding capacity declines with the increase of phosphate concentration, indicating that there is strong interaction between low-concentration phosphate solution and protein to form more free water through combination, and it has good water-holding capacity.

**(3)** pH

pH can change the ionization of functional groups of proteins and influence the thickness of the double electric layer, and influence the interprotein or protein–solvent interaction. Fig. 4.89 reflects the variation of the gelling property of arachin with the changing of pH. It can be seen that when pH < 6, the peanut arachin cannot form gel, and the changing trends of hardness, elasticity, and gelling property are basically the same, that is, when the pH changes from acidic to alkaline state, each property increases sharply; however, when pH > 7, the change levels off. However, the cohesion and water-holding capacity are just the opposite, reaching their lowest value at a neutral state.

**(4)** Heating Temperature

Thermal treatment is a common method that enables protein to gel. Fig. 4.90 shows the variation of the gelling property of arachin with the change of heating temperature. The hardness, elasticity, gelling property, and water-holding capacity of arachin gel all rise with the increase of temperature. However, the hardness and gelation index change sharply between 90 and 95°C, while the

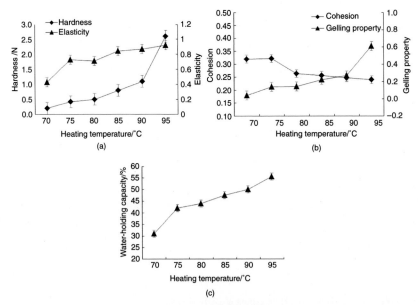

**FIGURE 4.90** Influence of heating temperature on gelling property of arachin.

elasticity and water-holding capacity change sharply from 70 and 75°C. The results are consistent with the theory that the arachin gelation requires protein denaturation. The underlying mechanism is that during the thermal denaturation of protein molecules, the annular structure in the gel network unfolds to form stretched nonpolar amino acid peptides, exposing the hydrophobic groups which are originally wrapped into the coiled structures, so that interprotein interaction occurs (Kella and Poola, 1985). In the meantime, with the intensification of the thermal-induced movement of the protein molecules, the probability of intermolecular collision increases greatly, which is more conducive to intermolecular cross-linking through hydrophobic interaction, and therefore the gel hardness of the protein is increased (Molina, 2004).

(5) Heating Time

Fig. 4.91 shows the variation of the gelling property of arachin at different heating times. It can be seen that the gel hardness and elasticity of the gel increase with the prolonging of heating time at 95°C. However, the influence of the heating time on the gel hardness and elasticity is only restricted to the initial 30 min. This indicates that during the heating process, the protein must be fully denatured and unfolded to form the ordered three-dimensional gel network structure.

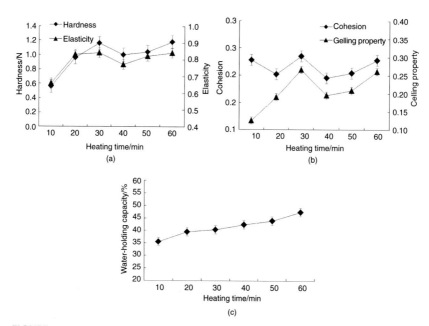

**FIGURE 4.91** Influence of heating time on gelling property of arachin.

- Quadratic Rotational Orthogonal Experiment

  Response surface methodology (RSM) is a kind of effective method to optimize process conditions, and it can determine the influence of various factors and their interactions on the indexes under investigation (response value) during technological operation, it can be used to fit a complete quadratic polynomial model through central composite experiment, and it can present more excellent experiment design and result expression. Du (2012) adopted the RSM to optimize the main process parameters for the arachin-formed gel.

  **(1)** Results of Orthogonal Rotational Composite Experiment

  With the gel hardness as the index, a Box-Behnken response surface optimization experiment was designed. On the basis of single-factor experiments, the three-factor and three-level quadratic rotational orthogonal experiment was designed, where the arachin concentration was set to be 14%, the PBS concentration was set to be 0.01mol/L, and the heating temperature in °C (X1), the heating time in min (X2), and the pH (X3) were all set to be zero. The arachin hardness (N) was deemed as the response value (Y). Statistical analysis software Design-Expert 7.0 was used to carry out regression analysis and test significance for the experimental data.

**(2)** Variance of Regression Model

Design Expert 7.0 software was used to carry out multiple regression fitting for the experimental results, and see below for the quadratic polynomial regression model equation [Eq. (4.3)] of arachin gel hardness Y to the variables, heating temperature, heating time, and pH.

$$Y = 1.303568 + 0.908916 X_1 + 0.125097 X_2 + 0.632747 X_3$$
$$+ 0.080238 X_1 X_2 + 0.353845 X_1 X_3 + 0.001393 X_2 X_3 \quad (4.3)$$
$$+ 0.434018 X_1^2 + 0.055309 X_2^2 \times 0.4205 X_3^2$$

As seen from the significance analysis for the regression equation coefficient, the influence of the monomial X1 (heating temperature) and X3 (pH) on the arachin gel hardness has reached an extremely significant level ($P < 0.01$), while the influence of X2 (heating time) on the arachin gel hardness does not reach a significant level. The influence of various factors on the arachin gel hardness is as follows (in a descending order): heating temperature > pH > heating time.

**(3)** Optimal Conditions

The statistical analysis software Design-Expert 7.0 was used to predict the optimal conditions for arachin gel hardness, and the experimental points that can ensure the optimal arachin gel hardness are: heating temperature 94.4°C, heating time 39.6 min, and pH 8.0. Under these conditions, the theoretical value of the arachin gel hardness is 321 N. Minute adjustment is made for the preparation conditions: heating temperature 94°C, heating time 40 min, and pH 8.0. In the verification experiment, the arachin gel hardness was determined to be 317 ± 0.18 N, which approaches the theoretical value. The results indicate that the measured values are consistent with the theoretical predictions for the parameters of the arachin gel preparation process that is optimized by means of the response surface method, indicating that the model is accurately established and the process parameters are reliable.

**b.** Improvement of Conarachin Gelling Property

- Single-Factor Experiment

**(1)** Conarachin Concentration

The influence of concentration variation on the gelling property of conarachin is shown in Fig. 4.92. Similar to the arachin, when the conarachin concentration is below 6%, gel cannot be formed. When the concentration increased from 8 to 18%, both its hardness and gelling property also increased constantly. This is because with the increase of the protein concentration,

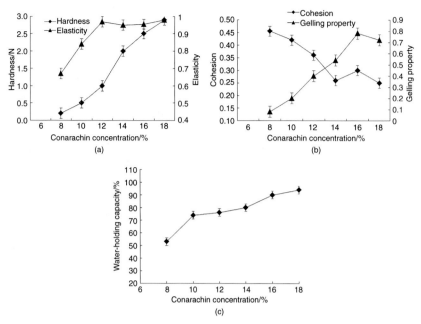

**FIGURE 4.92** Influence of conarachin concentration on gelling property of conarachin.

the number of molecules within the same volume increases and the intermolecular force is reinforced. The elasticity is basically stabilized after reaching the level of 12%, where the cohesion begins to decline and the gelling property first rises and then declines, and best gelling property is obtained at 16% concentration. Therefore, the optimal conarachin concentration is selected to be 16%.

**(2)** Calcium Chloride Concentration

The calcium ions have a complicated influence on protein gelling, and the salting-in effect and salting-out effect are involved. As shown in Fig. 4.93, the protein gel formed by adding calcium ions has greater surface tenacity and hardness. With an increase in calcium ions, the gel hardness also increases accordingly. When the calcium ion concentration reaches 0.2 mol/L, the formed gel presents great hardness as well as good elasticity and gelling property, which is possibly because of the salting-in effect. When the calcium ion concentration exceeds 0.2 mol/L, the gel hardness, elasticity, and gelling property decrease accordingly, this is possibly because of the salting-out effect. It is necessary to carry out further study on the internal mechanism. The water-holding capacity of the gel increases constantly with the increase in the calcium ion concentration, and then the changing trend levels off.

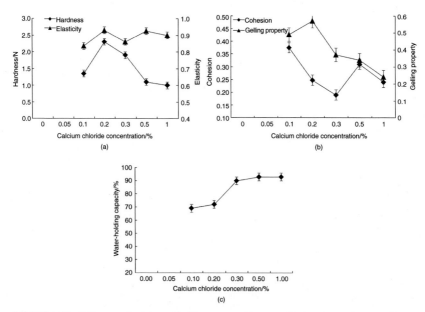

**FIGURE 4.93**    Influence of calcium chloride concentration on gelling property of conarachin.

This is possibly because the presence of calcium ions loosens the internal structures of the formed gel, where more free water is combined.

**(3)** pH

As shown in Fig. 4.94, pH is an important factor that influences conarachin gel formation and its textural characteristics, and there is a large difference between the gelling property of conarachin under alkaline conditions and acid conditions. The change of pH can influence the ionization and net charge values of protein molecules so that the attraction and repulsion of the protein molecules, as well as their capacity to combine with water molecules, will be changed, and also, the acting force to form and maintain the gel is also influenced. From Fig. 4.94, when the pH is within acid range (3–6), the gel hardness of the conarachin first increases and then decreases, and the hardness is highest when the pH value is 5. Within the alkaline range (7–11), the gel hardness first increases and then decreases with the increase of pH. When the pH is 8, it has the greatest hardness. The elasticity and water-holding capacity of gel changes with pH but not in a significant manner. When the pH is lower than 3 or higher than 11, the conarachin cannot be turned into gel, possibly because under acid conditions, the conarachin is already partially dissociated before

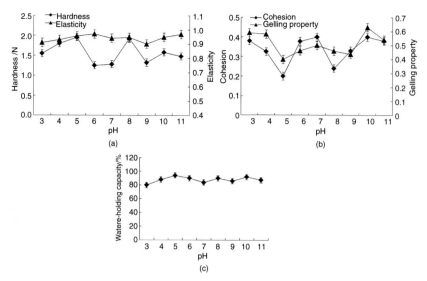

**FIGURE 4.94**    Influence of pH on gelling property of conarachin.

the heating, and floccules can be seen in the heated protein solution. However, under alkaline condition, the too high pH might cause the protein to denature, and therefore the protein aggregate to have water precipitation. Therefore, in the subsequent optimization experiments, the experiments were respectively carried out under acid and alkaline conditions to study the conarachin gelling property under different solution environments so as to provide certain theoretical reference for its application in the food industry.

**(4)** Heating Temperature

As shown in Fig. 4.95, the gel hardness, elasticity, and gelling property of the conarachin, as well as the water-holding capacity of the gel, increases with the rise of the heating temperature. This is because with the increase of temperature, the conarachin undergoes thermal denaturation and the active groups in the molecules are fully exposed. In addition, the molecules are crosslinked to form a three-dimensional network structure. The higher the heating temperature, the greater the conarachin exposure, the more the intramolecular active groups are exposed, and the more complete is the three-dimensional network structure formed. The denaturation level of the conarachin is determined by means of differential scanning calorimetry (DSC) during the thermal treatment process, and the temperature ($T_d$) of denaturation is 87.49°C. Therefore, the conarachin has been fully denatured after being heated above 90°C, and the hardness, elasticity, and

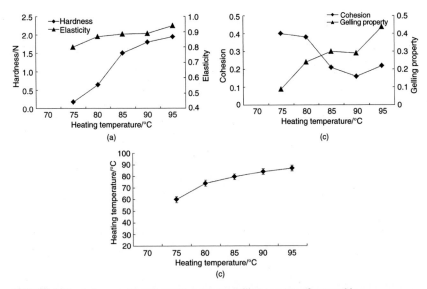

**FIGURE 4.95** Influence of heating temperature on gelling property of conarachin.

water-holding capacity of the protein gel do not change significantly compared with those at the heating temperature of 95°C.

**(5)** Heating Time

Fig. 4.96 shows the changes of the gelling property of conarachin subject to different heating times. It can be seen that with the prolonging of the heating time, the hardness is generally on the rise. This indicates that with the intensification of thermal denaturation, the gel network structure formed by the conarachin becomes more compact with greater hardness. The slight decrease in the middle is possibly because the thermal motion of the molecules is intensified to a level higher than the denaturation level. However, the elasticity first decreases, then increases, and decreases again. The decrease at the initial stage might be caused by the intensification of the thermal motion of molecules, while the increase in the middle is because the fine three-dimensional network of gel can entrap more water molecules, cause protein hydration, form new disulfide bonds, and increase elasticity as a result. However, with the continuous increase of the heating time, the value of elasticity decreases instead. This is possibly because the long-term heating increases the chance of intermolecular interaction of the protein, and some active groups might undergo a chemical reaction, such as aggregation of thiol oxidation, which is detrimental to the protein gelling property. It can also be seen that the heating time does not have a significant influence on the

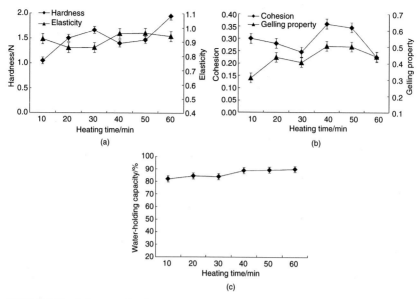

**FIGURE 4.96**   Influence of heating time on gelling property of conarachin.

water-holding capacity, and the cohesion and gelling property are the highest at 40 min of heating time. Therefore, the optimal heating time is ultimately selected to be 40 min.

- Orthogonal Optimization Test

pH is a critical factor that influences the formation of conarachin gel, as well as its textural characteristics. The hardness changing trend under acid conditions is very different from that under alkaline conditions. Due to the fact that the conarachin hardness changing trend differs between acid and alkaline conditions, orthogonal experiments were respectively carried out under the two conditions. Hardness ($N$) was used as an investigation index. Duncan's new multiple range method was used to analyze the results of orthogonal optimization.

**(1)** Acid Condition

According to the above single-factor experimental results, it was presumed that the conarachin concentration was 16%, the heating time was 40 min, and three main factors and their valuation range were also determined. DPS software was adopted for L9 (43) orthogonal experiment design. Duncan's new multiple range method was adopted for comparative analysis on the significance of the influence of various factors. The various factors influencing the conarachin gel hardness include (in descending order): concentration of calcium chloride > pH > heating temperature.

The optimal combination is: heating temperature 95°C, concentration of calcium chloride 0.3 mol/L, and pH of 5.

**(2)** Alkaline Condition

According to the above single-factor experimental results, it was presumed that the conarachin concentration was 16%, the heating time was 40 min, and three main factors and their valuation range were also determined. DPS software was adopted for L9 (43) orthogonal experiment design. Duncan's new multiple range method was adopted for comparative analysis on the significance of the influence of various factors. The various factors influencing the conarachin gel hardness include (in descending order): pH > concentration of calcium chloride >heating temperature. The optimal combination is: pH of 8, concentration of calcium chloride 0.3 mol/L, and heating temperature 95°C.

**2.** Enzymatic Modification

    **a.** Improvement of Conarachin Gelling Property

      - Single-Factor Experiment

        **(1)** Enzyme Dosage

        Under the conditions of pH being 7.0, cross-linking time being 1.5 h, and cross-linking temperature being 40°C, the influence of enzyme dosage on the arachin gel hardness was investigated, as shown in Fig. 4.97. It can be seen that when no TG enzyme is added, the gel hardness of the arachin is below 10 g. Within the dosage range of 0–12 U/g, as the enzyme dosage increases, the arachin gel hardness increases sharply and reaches the highest value at 12 U/g. Above 12 U/g, when the enzyme dosage

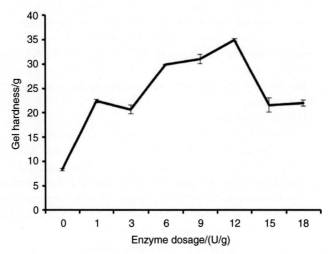

**FIGURE 4.97** Influence of enzyme dosage on arachin gel hardness.

is continuously increased to 15 and 18 U/g, the gel hardness assumes a slow downward trend. It can be seen that the addition of TG enzyme can significantly help to improve the gel strength of the arachin, especially at the level of 12 U/g of enzyme dosage. When the enzyme dosage is 15 and 18 U/g, the gel hardness decreases slightly instead. On one hand, it might be because the covalent bonds that are required to maintain the stability of protein gel network have been saturated to a certain extent. On the other hand, it is possible that the large enzyme dosage does not contribute to the formation of intermolecular G–L bonds in the protein. These results are consistent with the study results of Tian and Liang (2005a, 2005b). Xu (2003) made use of TG enzyme to cross-link different protein, and they found that with the increase of the TG enzyme, the gel hardness also increases accordingly. However, they are not always positively related. The enzyme dosage differs for different proteins to form the highest gel hardness. Therefore, the enzyme dosage is 12 U/g.

**(2)** pH

Under the conditions of enzyme dosage being 12 U/g, cross-linking time being 1.5 h, and cross-linking temperature being 40°C, the influence of pH on arachin gel hardness was investigated, as shown in Fig. 4.98. It can be seen that with the change of pH, the gel hardness first rises and then decreases. When the pH is 10.0, the gel hardness reaches its highest point, and the

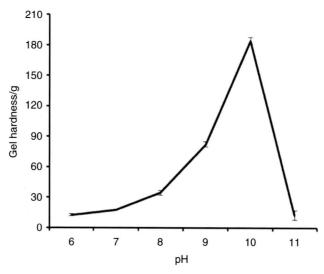

**FIGURE 4.98**  Influence of pH on arachin gel hardness.

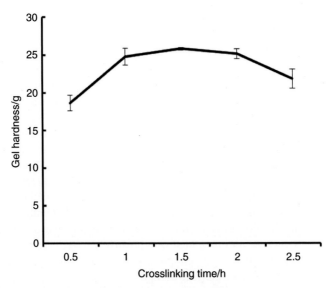

**FIGURE 4.99**   Influence of crosslinking time on arachin gel hardness.

gel hardness decreases sharply when the pH exceeds 10.0. This is mainly because the pH of the system influences not only the enzyme activity and stability but also the solubility of arachin, as well as the network structure of the gel (Yu et al., 2010a; Yuan et al., 2010). Therefore, the optimal pH is determined to be 10.0, and under this condition, the arachin gel has high hardness, reaching above 180 g.

(3) Cross-Linking Time

Under the conditions of enzyme dosage being 12 U/g, pH being 7.0, and cross-linking temperature being 40°C, the influence of cross-linking time on arachin gel hardness was investigated, as shown in Fig. 4.99. It can be seen that with the increase of the reaction time, the gel hardness increases significantly. The gel hardness reaches its highest level at 1.5 h, and then it levels off before declining slightly. This is possibly because the covalent bonds that are required to maintain the stability of the protein gel network have been saturated to a certain extent, and therefore the overlong cross-linking will not increase the arachin gel hardness. Xu (2003) cross-linked different proteins with TG enzyme. They found that certain periods of enzymatic reaction contributed to the improvement of the hardness. However, with the prolonging of the reaction time, the gel hardness decreased gradually. The conclusions are consistent with the experimental results. Therefore, the optimal cross-linking time is 1.5 h.

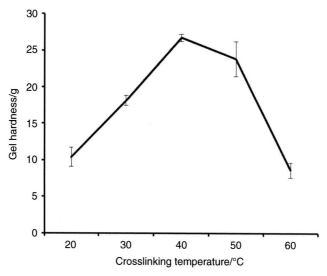

**FIGURE 4.100**    Influence of crosslinking temperature on arachin gel hardness.

**(4)** Cross-Linking Temperature

Under the conditions of enzyme dosage being 12 U/g, pH being 7.0, and cross-linking time being 1.5 h, the influence of arachin gel hardness was investigated, as shown in Fig. 4.100. The influence of temperature on arachin cross-linked by TG enzyme is mainly evidenced in the enzymatic reaction. Fig. 4.100 shows that, with the increase of the cross-linking temperature, the arachin gel hardness first rises and then declines. The gel hardness reaches its highest level at 40°C. The increase of temperature speeds up the reaction time and also influences the activity of TG enzyme. However, high temperature can deactivate TG enzyme and consequently weakens the cross-linking ability of enzyme to protein. Therefore, the optimal temperature is 40°C.

- Orthogonal Rotational Combination Design Experiment

On the basis of single-factor experiments, the arachin concentration was set to be 14%, the PBS concentration was set to be 0.01 mol/L, and the heating temperature in °C (X1), heating time in min (X2), and pH (X3) were all set to be at zero, and the arachin hardness (g) was set as the response value (Y). On this basis, the three-factor and five-level quadratic orthogonal rotational combination experiment was designed to obtain the optimal theoretical process parameters for arachin cross-linked by TG enzyme: the enzyme dosage 12 U/g, pH 10.0, cross-linking time 1.5 h, and cross-linking temperature 40°C. Under these conditions, the arachin hardness (g) was theoretically predicted to be 199.98 g. To verify the reliability of the results obtained by means of the response surface method, three repeated verification experiments were carried out under the above conditions, and then the results were compared with the previous results. The arachin gel hardness was determined to be 195.91 ± 11.56 g. It can be seen that the actual value is consistent with the theoretical value.

**b.** Improvement of Conarachin Gelling Property

- Single-Factor Experiment

**(1)** Enzyme Dosage

Under the conditions of pH being 7.0, cross-linking time being 1.5 h, and cross-linking temperature being 40°C, the influence of enzyme dosage on the conarachin gel hardness was investigated, as shown in Fig. 4.101. It can be seen that before the TG enzyme is added, the conarachin gel hardness is about 10 g; after the TG enzyme is added, the gel hardness increases sharply. Within the range of the enzyme dosage 0–15 U/g, the conarachin gel hardness increases to about 105 g as the enzyme dosage increases, and it can be seen that the addition of TG enzyme can greatly improve the conarachin gel hardness. However, with the continuous increase of enzyme dosage, the gel hardness of protein declines significantly. When the enzyme dosage reaches 18 U/g, the gel hardness is about 100 g. After modification by means of TG enzyme, the protein gel hardness increases, possibly because the cross-linking of TG enzyme enables the formation of intermolecular G–L bonds so that a spatial network structure is formed. However, the increase of enzyme dosage will decrease

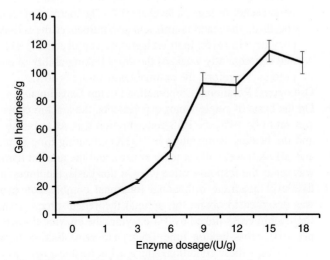

**FIGURE 4.101**    Influence of enzyme dosage on conarachin gel hardness.

the gel hardness mainly because on one hand, the covalent bonds needed to maintain the stability of protein gel network have been saturated to a certain extent, and it was reported that (Tian and Liang, 2005a) the increase of enzyme dosage might lead to the acting sites to cross-link on the molecular surface of protein and consequently reduce the chance of cross-linking other protein molecules around. On the other hand, it is possible that the large enzyme dosage is not conducive to the intermolecular cross-linking of the protein. Therefore, the optimal enzyme dosage was determined to be 15 U/g.

**(2)** pH

Under the conditions of enzyme dosage being 15 U/g, cross-linking time being 1.5 h, and cross-linking temperature being 40°C, the influence of pH on the conarachin gel hardness was investigated, as shown in Fig. 4.102. It can be seen that, with the change of pH, the gel hardness rises at first and then declines. Within the pH range from 6.0 to 7.0, the gel hardness increases sharply to reach its highest point. When exceeding 7.0, the gel hardness declines sharply, possibly because on the one hand, each enzyme has its optimal working condition, and the pH in the system influences the enzymatic activity and stability; on the other hand, the change of pH might influence the solubility of conarachin, which prevents the effective formation of the gel network structure, as has been reported by Yu et al. (2010a). Therefore, the optimal pH was determined to be 7.0.

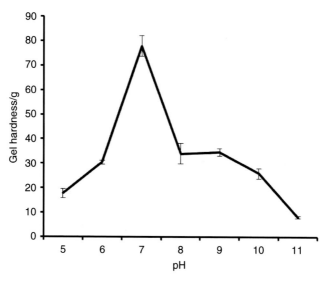

**FIGURE 4.102**    Influence of pH on conarachin gel hardness.

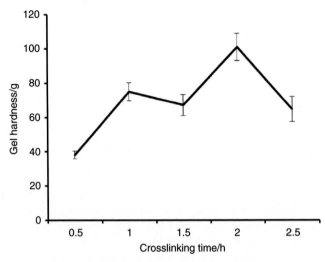

**FIGURE 4.103**    Influence of ciosslinking time on conarachin gel hardness.

**(3)** Cross-Linking Time

Under the conditions of enzyme dosage being 15 U/g, pH being 7.0, and cross-linking temperature being 40°C, the influence of cross-linking time on the conarachin gel hardness was investigated, as shown in Fig. 4.103. It can be seen that between 0 and 2 h, the conarachin gel hardness increases significantly with the increase of the reaction time. However, the gel hardness declines slightly after 2 h. This indicates that within a certain period of time, the reaction of the TG enzyme is conducive to the increase of the gel hardness possibly because that the cross-linking of the TG enzyme enables the formation of network structures in the protein. However, with the prolonging of reaction time, the gel hardness decreases gradually possibly because the covalent bonds needed to maintain the stability of the protein gel network have been saturated to a certain extent, and the cross-linking for too long a time might cause the rupture of the bonds and reduce the conarachin gel hardness. Therefore, the optimal cross-linking time was determined to be 2 h.

**(4)** Cross-Linking Temperature

Under the conditions of enzyme dosage being 15 U/g, pH being 7.0, and cross-linking time being 1.5 h, the influence of the cross-linking time on the conarachin gel hardness was investigated, as shown in Fig. 4.104. The influence of the temperature on the conarachin cross-linked by the TG enzyme is mainly evidenced by the enzymatic reaction. As shown in Fig. 4.104, with the increase of the cross-linking temperature, the conarachin gel hardness first

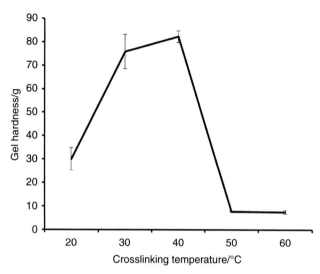

**FIGURE 4.104**  Influence of crosslinking temperature on conarachin gel hardness.

rises and then declines. Between 20 and 40°C, the conarachin gel hardness increases significantly with the increase of cross-linking temperature. However, between 40 and 60°C, the conarachin gel hardness decreases sharply. With the increase of temperature, the protein gel hardness first rises and then declines possibly because the increase in temperature speeds up the reaction speed and influences the activity of TG enzyme. At 40°C, the activity of TG enzyme reaches its highest level to display optimal enzymatic activity. However, after 40°C, the high temperature deactivates the TG enzyme, and the cross-linking capacity of the enzyme for protein is weakened. Therefore, the optimal temperature is 40°C.

- Orthogonal Rotational Combination Design Experiment
  On the basis of single-factor experiments, the enzyme dosage (X1), the pH (X2), the cross-linking time (X3), and the cross-linking temperature (X4) were selected to carry out the orthogonal rotational combination design experiment, and the optimal theoretical process parameters for the conarachin cross-linked by TG enzyme were obtained: the enzyme dosage is 15 U/g, the pH is 7.0, the cross-linking time is 2.8 h, and the cross-linking temperature is 40°C. Under these conditions, the protein gel hardness of conarachin was theoretically predicted to be 105.33 g. To verify the reliability of the response surface method, the verification experiment was carried out three times, and the results were compared with the previous results to determine that the conarachin gel hardness was 100.67 ± 2.56 g. Therefore, it can be concluded that the actual value is consistent with the theoretical value.

## 4.8 PEANUT PROTEIN PRODUCTION LINE EQUIPMENT AND RELEVANT EQUIPMENT

### 4.8.1 Technological Process of Production Line

1. Peanut Tissue Protein

   After the peanut meal is ground to 40–60 mesh with a hammer-type grinder, the meal and additive are transported to be mixed in the mixer. After being evenly mixed, the powder is conveyed to the double-screw extruder for extrusion. The extruded granular particles are then transported by the pneumatic conveyor and discharger for degassing and drying. After the materials are dewatered and cooled by the drying and cooling machine, the measuring and packing machine is used to process them into peanut tissue protein products (Fig. 4.105).

2. Peanut Protein Powder

   The cold pressing peanut meal is first entered into the complete set of subcritical extraction equipment or the complete set of supercritical extraction equipment for efficient extraction.

   The specific operational procedures of the complete set of subcritical extraction equipment are as follows: the peanut meal is first conveyed with a conveyor to an extractor with a pressure-vessel structure. After the feeder is closed, most of the air in the extractor is pumped out. Then, the extraction solvents, such as liquefied propane and butane, are pumped in for several sets of countercurrent extraction. When the process requirements for optimal extraction are met, the outlet valve of the extractor is opened, after pumping out all the liquid solvent in the extractor so that the solvent absorbed in the peanut meal is vaporized, separated from the materials, and then discharged from the extractor. The whole process is called the desolventizing process. The temperature of the peanut meal will decrease with the evaporation of the solvent during this process. Therefore, heat should be replenished for the materials during the desolventizing process to compensate for the heat lost during the solvent evaporation. After the desolventizing process, the extracted peanut meal can be discharged out of the extractor and the next production cycle can begin in the extractor.

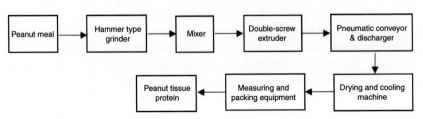

**FIGURE 4.105** Technological chart of peanut tissue protein production.

The specific procedures of the complete set of supercritical extraction are as follows: the $CO_2$ is pumped from the $CO_2$ storage tank to the condenser by means of the pressure pump, and afterward, the condensed $CO_2$ flows into the intermediate storage tank. After the liquefied $CO_2$ is further condensed by the condenser in the intermediate storage tank, it is pumped into the preheater by the high-pressure pump for heating, where the phase of $CO_2$ is changed into a supercritical state. The supercritical $CO_2$ enters the extraction kettle from the bottom of the extraction kettle where the extraction is made. After the extraction, the $CO_2$ that carries the extract is discharged from the top of the extraction kettle and conveyed to the first separation heater, where its temperature rises to the process temperature required by the Grade I separation kettle. In the meantime, by adjusting the adjustment value of the separation kettle, its pressure reaches the separation pressure of the Grade I separation kettle. Afterward, the liquid enters the Grade I separation kettle for the separation of the extract and $CO_2$. The separated products from the Grade I separation kettle are discharged from the bottom of the kettle and received by a container. The $CO_2$ that still carries the extract is discharged from the top of the Grade I separation kettle and is then discharged from the top of the extraction kettle and conveyed to the second separation heater, where its temperature rises to the process temperature required by the Grade II separation kettle. In the meantime, by adjusting the adjustment value of the separation kettle, its pressure reaches the separation pressure of the Grade II separation kettle. Afterward, the liquid enters the Grade II separation kettle for the separation of the extract and $CO_2$. Then the separated products of the Grade II separation kettle are discharged from the bottom of the separation kettle and received by a container. The gaseous $CO_2$, after being condensed by the preheater, flows into the intermediate storage tank systematic circulation.

The peanut meal after efficient extraction is ultra-finely ground by the nonsieve airflow pulverizing grinder to obtain peanut protein powder above 200 mesh. Peanut protein powder products are obtained by means of a microwave sterilizer and measuring and packing equipment (Fig. 4.106).

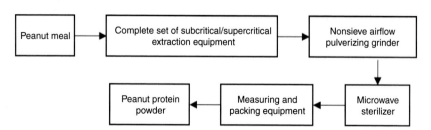

**FIGURE 4.106**   Flow chart of peanut protein powder production equipment.

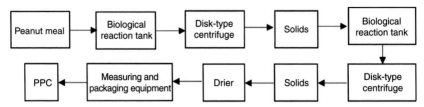

**FIGURE 4.107**  Flow chart of PPC production equipment.

3. PPC
   a. PPC

   The peanut meal is evenly and proportionally mixed with the ethanol solution in the biological reaction tank for the first leaching. After that, the feed liquid is pumped to the disk-type centrifuge for solid–liquid separation. The solids at the slag discharger are then collected to evenly mix with the ethanol solution in the biological reaction tank for the second leaching. After that, the leached feed liquid is pumped to the disk-type centrifuge for solid–liquid separation. The solids at the slag discharger are collected and conveyed to the drier for drying. Finally, the solids are measured and packaged to obtain the PPC products (Fig. 4.107).

   b. Soluble PPC

   The pure water is heated to 70°C by means of a plate-type heat exchanger to determine the water volume in the biological reaction tank. The PPC is evenly and proportionally mixed with the water in the reaction tank by means of the emulsifying mixer for preenzymolysis physical modification. Afterward, enzyme preparations are added for proper enzymatic modification. Then, the enzymolyzed feed liquid is rapidly subject to the enzymatic inactivation and sterilization by means of the ultra-high-temperature instantaneous sterilizer and then pumped to the high-pressure homogenization machine for modification. The modified feed liquid is then pumped to the feed balance cylinder of the double-effect concentrator for preheating, where the liquid is gradually heated to the concentration temperature. Then the feed liquid enters the first evaporator, first separator, second evaporator, and second separator in turn. The concentrated feed liquid is sprayed into the top of the spray drying tower for drying by means of high-pressure spraying. The powder is obtained at the outlet of the Grade I cyclone separator and Grade II bag-type powder collector. Finally, the soluble PPC products are obtained after being measured and packed (Fig. 4.108).

   c. Gel-Type PPC

   The PPC is evenly and proportionally mixed with the water in the reaction tank by means of the emulsifying mixer. After that, enzyme preparations are added for proper enzymatic modification. Afterward, the enzymolyzed feed liquid is rapidly subject to enzyme inactivation,

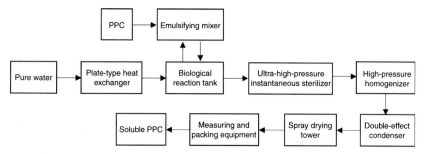

**FIGURE 4.108**  Flow chart of soluble PPC production equipment.

and then pumped to the feed balance cylinder of the double-effect concentrator for preheating, where the liquid is gradually heated to the concentration temperature. Then the feed liquid enters the first evaporator, first separator, second evaporator, and second separator in turn. The concentrated feed liquid is sprayed into the top of the spray drying tower for drying by means of high-pressure spraying. The powder is obtained at the outlet of the Grade I cyclone separator and Grade II bag-type powder collector. Finally, the soluble PPC products are collected measured and packed (see Fig. 4.109).

**4.** PPI

A hammer-type grinder is used to grind the peanut meal into 40–60 mesh, which is then put into the biological reaction tank. It is stirred constantly after water injection and then NaOH solution is used to condition the pH of the leaching solution for the first leaching. The leached feed liquid is pumped to the disk-type centrifuge for solid–liquid separation. After that, leaching solution and solids are collected. The collected solids are used for second leaching in the same way as the first leaching. After two uses for leaching, the leaching solutions are pumped into the acid precipitation tank. While stirring constantly, edible hydrochloric acid solution is slowly added to condition the solution pH, and the stirring is stopped after reaching the isoelectric point. After the solution is left to stand for 20–30 min, the feed

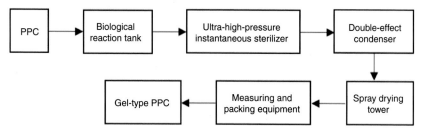

**FIGURE 4.109**  Flow chart of gel-type PPC production equipment.

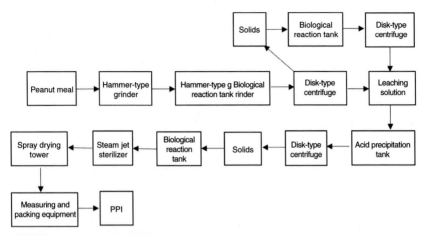

**FIGURE 4.110** Flow chart of PPI production equipments.

liquid is pumped to the disk-type centrifuge for solid–liquid separation. The solids collected at the slag discharger are put into the biological reaction tank for rehydration, pumping, and back-conditioning. Afterward, the solution is sterilized by the steam jet sterilizer and then sprayed into the top of the spray drying tower for drying by means of high-pressure spraying. The powder is collected at the outlet of the Grade I cyclone separator and Grade II bag-type powder collector. Finally, the soluble PPI products are obtained after being measured and packed (Fig. 4.110).

5. Peanut Protein Component (Cold Precipitation Method)

The peanut protein power is evenly and proportionally mixed with the PBS in the reaction tank, and then the solution is stirred constantly. Afterward, the solution is pumped to the disk-type centrifuge for solid–liquid separation, and the supernatant is then transported to the cold precipitation tank for cold precipitation. After the cold precipitation is completed, the supernatant is pumped to the disk-type centrifuge equipped with a water cooling system for low-temperature centrifuge.

The separated supernatant is pumped to the biological reaction tank. While stirring constantly, edible hydrochloric acid solution is slowly added to condition the solution pH, and the stirring is stopped after reaching the isoelectric point. After the solution is left to stand for 20–30 min, the feed liquid is pumped to the disk-type centrifuge for solid–liquid separation. The solids collected at the slag discharger are put into the biological reaction tank for rehydration, pumping, and back-conditioning. Afterward, the solution is sterilized by the steam jet sterilizer and then sprayed into the top of the spray drying tower for drying by means of high-pressure spraying. The powder is collected at the outlet of the Grade I cyclone separator and Grade II bag-type powder collector. Finally, the conarachin products are obtained after being measured and packed.

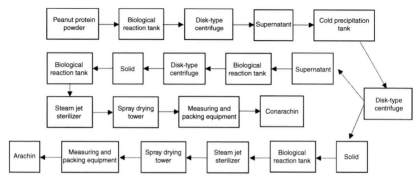

**FIGURE 4.111**   Flow chart of peanut protein component production equipment.

The separated solids, after being put into the biological reaction tank for rehydration and pulping, are sterilized by the steam jet sterilizer. Afterward, the solution is sprayed into the top of the spray drying tower for drying by means of high-pressure spraying. The powder is collected at the outlet of the Grade I cyclone separator and Grade II bag-type powder collector. Finally, the arachin products are obtained after being measured and packed (Fig. 4.111).

## 4.8.2   Key Equipment Introduction

1. Subcritical Extraction Equipment

   The extraction equipment mainly consists of an extraction tank, solvent tank, mixing tank, vaporizing tank, rising film evaporator, condenser, compressor, vacuum pump, and auxiliary buffer tank and filtering tank. All of the vessel equipment are low-pressure vessels (type II). The compressor, according to the type of solvent, can be an ammonia compressor, liquefied petroleum gas compressor, or Freon compressor. Depending on the yield, there are generally one to six extraction tanks. There are vertical tanks and horizontal tanks driven by driving force. The volume of the tank is normally 4–12 $m^3$. The extract liquor is temporarily stored in the mixing tank and the solvent separation will be conducted in the vaporization system, which usually consists of a 2-L climbing film evaporator and a 1-L falling film evaporator. The evaporators are continuously arranged and a 1-L vacuum film evaporator is added for certain individual products. Evaporators are intermittently applied in the evaporation separation of extract liquor in small production units (Qi, 2012).

2. Supercritical $CO_2$ Extraction Equipment

   Supercritical $CO_2$ extraction equipment incorporates many technologies, such as mechanical structures, pressure vessels, pipeline valves, instruments and meters, electric controls, and chemical process, etc. For small and medium sized equipment, the device has a simple structure and small volume. The design and fabrication of laboratory equipment are relatively simple. For large-scale industrialization equipment, the design is complicated and continuous filing materials are required, such as continuous extraction. For the usage of solvent, the recycling of $CO_2$ tail gas is required to cut the cost and increase efficiency.

In supercritical fluid extraction equipment, the extraction tank and separation tank are the key equipment of the system. The pump to supply solvent is the key component for driving the system. The auxiliary system includes solvent storage, pipeline transportation, valves, safety devices, heat exchange systems, and electronic control systems. Currently, the heat cycle system applied in the extraction tank and separation tank meets the international standard, such as a water jacket structure which is installed in the outer shell. Small-scale equipment may use an electrical heating system, such as electric heating tape, to ensure the extraction by isothermal method. A water jacket is used in the cooling circulating system, such as a precooler, an intermediate storage tank to lower the $CO_2$ temperature and keep it in its liquid state. Rubber and plastic thermal insulation material can be applied in the outer layer of a single layer water jacket to increase the effect (Sun et al., 2009).

3. Twin-Screw Extrusion Machine

Extrusion technology, as the new type of processing method, has economical and practical characteristics. It is not only widely used in the food industry, but also used in the feed industry, oil production industry, brewing industry, etc. When producing puffed, half puffed, and textured products, the transportation, compression, mixing, crush, shear, sterilization, puffing, and molding of materials are all completed by the extrusion machine. Different extrusion dies are used to produce products with different shapes. Extrusion technology has the characteristics of simple technology, multifunction, continuous production, high efficiency, low energy consumption, small investment, and rapid return. The food produced has the characteristics of fine and smooth in taste, easily digested and absorbed, small loss in nutritional ingredient, long storage time, antiretrogradation, and easy to eat (Liu, 2006).

4. Disk-Type Centrifuge

The thin layer partition principle is employed to optimize centrifugation processes in disk-type centrifuges, which can be applied in the material separation of liquid–oil, liquid–solid, and liquid–oil–solid in different proportions. As one of the main pieces of equipment for modern oil and fat refining of vegetable oil in continuous refining, it is commonly used in the process of degumming, desoponificating, and washing, etc. (Ma, 2010). According to the slagging mode, a disk-type centrifuge can be divided into the following three types (Zhang, 2010).

a. Manual slagging type, also known as solid reserve type, has characteristics including simple drum structure, continuous operation, and manual unloading. It is used to deal with suspension liquid with solid concentrations of less than 5% and solid volume between 5 and 20 L. It is commonly used in milk degreasing.

**b.** Nozzle type. The drum structure is bicone with slagging nozzles equally distributed around the cones. It could be used for continuous slagging. The drum provides a large space for solids to avoid solid deposition accumulation and blocking and gaining thick mud. The number and diameter of nozzles should be carefully designed (the number is normally 1224 with diameter 0.75–2 mm). This type of centrifuge can be applied in the dehydration and clarification of clay, kaolin, pigment, yeast, oil, tar, and phosphoric acid by wet process. Wear-resistant materials, such as cemented carbide, corundum, and boron carbide, should be used in the nozzle because the fluid flow velocity is high.

**c.** Circle valve type, also known as piston slagging disk-type centrifuge. There are slagging piston devices in the same axis in the drum with circularly grooved slagging holes with different circular arc lengths around the device. This kind of slagging hole can be used to discharge slags which are thicker and stickier. According to the solid deposition thickness, the piston moves up and down by hydraulic pressure and automatically starts and stops the slag outlet to discharge slags. A timer can be used to start and stop the slagging hole.

The disk-type centrifuge mainly consists of a body, cover, drum, horizontal axis, vertical axis, inlet and outlet devices, etc. The key component is the drum which is installed on the vertical axis. There are multilayer disks in the drum, which are orderly installed on the distributor. The distribution holes of disks and distributor are vertically aligned, and form several liquid inlet passageways (neutral hole). The disks are separated by interval bars. In this case, the drum is divided into several thin-layered separated spaces, which greatly shortens the sedimentation distance and improves the separation effect. Under the huge centrifugal force generated by high speed rotation of the drum, the mixture of oil and fat will enter the separation interval through neutral holes after it has entered the drum. However, the relatively large materials (phospholipid, nigre, water, etc.) will be thrown to the inner wall of the drum and the ones with good liquidity will then be discharged into heavy phase exit by a big centripetal pump through the passageway between the drum cover and disk cover. In this case, heavy sediment will gather in the sediment area. Manual cleaning should be conducted after periodical automatic discharge or shutdown. The relatively small materials (light phase) will be discharged into light phase exit by a small centripetal pump after they flow into the middle passageway through the disk inner edge. The design of heavy phase centripetal pump for domestic and imported disk-type centrifuge is normally variable type, whose inner diameter and separation interface can be changed to obtain different separation effects.

# Chapter 5

# Peanut By-Products Utilization Technology

Q. Wang, A. Shi, H. Liu, L. Liu, Y. Zhang, N. Li, K. Gong, M. Yu, L. Zheng

*Institute of Food Science and Technology, Chinese Academy of Agricultural Sciences, Beijing, China*

## Chapter Outline

## 5.1 PEANUT OLIGOPEPTIDES PROCESSING TECHNOLOGY

Biopeptides refer to peptide compounds that are bioactive or beneficial to vital activities of living organisms (Liu and Cao, 2002; Pang and Wang, 2001). Since Brantl (1979) first reported the discovery of small peptides with morphine-like activity in the small intestine of a cavy fed with protein hydrolysates of bovine

**Peanuts: Processing Technology and Product Development.** http://dx.doi.org/10.1016/B978-0-12-809595-9.00005-3

cheese, a variety of bioactive peptides have been isolated from plants, animals, and microorganisms (Gill et al., 1996), among which the peptides with bioactivity that are composed of 2 to 10 or more amino acids are called functional oligopeptides. Functional oligopeptides are biopeptides. Modern nutrition study has found that proteins are absorbed mainly in the form of oligopeptides rather than entirely in the form of free amino acids after being affected by digestive enzyme. The biological significance of functional oligopeptides is mainly that they not only have multiple functional activities but also have a fully independent uptake mechanism.

Currently, peanut protein products on the domestic market are mainly restricted to peanut protein powder and it is mainly used as a basic raw material in food processing. Peanut protein powder has poor functional properties and bioactivity and other disadvantages in processing. Therefore, the preparation of peanut functional oligopeptides with peanut protein powder as the raw material can improve the functional properties of peanut protein, release the functional peptides from proteins, fill the gap in China's deep-processed products of peanut protein, and effectively improve the conversion rate and utilization in peanut deep processing. It is of great importance for optimizing the peanut industrial structure and product mix and promoting the development of the whole peanut protein industry and even the functional food industry.

At present, China has achieved industrialization of soybean protein peptides and milk peptides, with directed enzymolysis, controllable enzymolysis, and microbial fermentation technology mainly used in production. However, China is short of the meticulous research and optimization of the oligopeptide production process and parameters, with the result that most peptide products are crude peptides with low purity and low degree of hydrolysis rather than functional oligopeptides with high purity and high degree of hydrolysis. In addition, the selection system of oligopeptide material with the corresponding functional activity needs to be established. Zhang and Wang (2007a) prepared functional oligopeptides of peanuts using the optimized composite enzymatic method. Wang et al. (2013b) carried out separation and purification of peanut oligopeptides and analyzed the structure–activity relationship between peanut oligopeptides and their structure. On this basis, Wang et al. (2013a) simulated the molecular conformation of hypotensive peanut oligopeptides using Discovery Studio 3.5.

These studies provided a theoretical basis for the production, processing, and application of functional oligopeptides of peanut in China.

## 5.1.1 Peanut Oligopeptide Composition

### 5.1.1.1 Basic Ingredients

Functional oligopeptides refer to the peptides produced with enzymolysis or microbial fermentation technique and with peanut meal or peanut protein as the main material, with a molecular mass of 5000 Da or less. To comply with GB/T 22492-2008 (soy peptide powder), the sensory requirements for peanut peptides powder are as shown in Table 5.1.

**TABLE 5.1** Sensory Quality of Peanut Peptide Powder

| Item | Quality Requirements | Item | Quality Requirements |
|---|---|---|---|
| Fineness | 100% passing through a sieve with aperture size of 0.250 mm | Taste and smell | Unique taste and smell of the product, without any other odor |
| Color | White, light yellow, and yellow | Impurities | No visible foreign matter |

Soybean peptides are divided into three grades in GB/T 22492-2008. The physical and chemical indicators of functional peanut oligopeptides, as the high added value product in peanut peptides, should comply with the standards of Grade I soybean peptides (Table 5.2).

### 5.1.1.2   Amino Acid Composition

Amino acid analysis results (Fig. 5.1) have shown that the free amino acid content of peanut functional oligopeptide is 66.4 mg/g. It can be seen that the product is mainly oligopeptide with lower amino acid content, despite high hydrolysis of complex enzymolysis.

### 5.1.1.3   Molecular Mass Distribution

The molecular mass distribution of peanut functional oligopeptide is mainly determined using HPLC. The basic conditions are: column: TSKgel2000S-WXL 300 mm × 7.8 mm; mobile phase: acetonitrile/water/trifluoroacetic acid, 20/80/0.1 (V/V/V); detection wavelength: 220 nm; flow rate: 0.5 mL/min; detection time: 30 min; column temperature: room temperature; molecular

**TABLE 5.2** Physical and Chemical Indicators of Functional Peanut Oligopeptide

| Item | Peanut Oligopeptide |
|---|---|
| Crude protein (dry basis, N × 6.25) (%) | ≥90.0 |
| Peptide content (dry basis) (%) | ≥80.0 |
| Molecular mass of peptides ≥ 80% (Da) | ≤2000 |
| Ash (dry basis) (%) | ≤6.5 |
| Water (%) | ≤7.0 |
| Crude fat (dry basis) (%) | ≤1.0 |
| Urease (urease enzyme) activity | Negative |

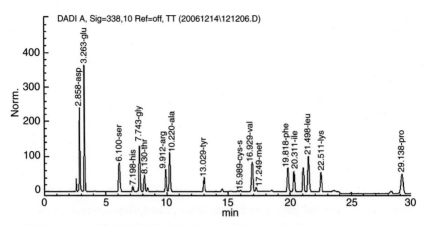

**FIGURE 5.1** Amino acid composition of peanut functional oligopeptide.

mas markers are respectively cytochrome C (MW12500), Bacillus enzyme (MW1450), acetic acid–acetic acid–tyrosine–arginine (MW451), and acetic acid–acetic acid–acetic acid (MW189).

The standard curve of relative molecular mass is obtained as follows: prepare peptide standard solution of the different relative molecular masses with the mass concentration of 1 mg/mL with mobile phase; after mixing them in certain proportion, carry out sample injection after filtering with an organic phase film with aperture of 0.2–0.5 μm, and a standard chromatogram will be obtained. Make a drawing of retention time with the logarithm of relative molecular mass or make a linear regression to obtain the relative molecular mass calibration curve and its equation.

Zhang and Wang (2007b) analyzed the molecular mass distribution of peanut functional oligopeptides with cold-press peanut meal as raw material, and the standard regression equation was lgMw = 7.18 −0.239T, with the relative coefficient $R^2$ = 0.9925. It shows that the logarithm of each molecular mass has good relativity with the elution time of each standard sample, and the molecular mass distribution of peanut peptides can be accurately determined (Table 5.3). Fig. 5.2 is the size-exclusion higher performance liquid chromatogram of peanut functional oligopeptides. It can be seen from the figure that molecular mass of the peanut functional oligopeptide is 126–11,197 Da and mainly in the range of 126–949 Da.

### 5.1.2 Preparation Process and Technique

By using the enzymatic method to hydrolyze protein in mild conditions, the polypeptides produced have a very high nutritional value. Modern nutrition research (Grimble and Silk, 1990; Sheng, 1993; Dun and Chen, 2004; Zhang and Feng, 2004) shows that oligopeptides with molecular mass lower than

**TABLE 5.3** Molecular Mass Distribution of Peanut Functional Oligopeptide

| Retention Time of Each Peak (min) | Molecular Mass (Da) | Peak Area Proportion (%) |
|---|---|---|
| 13.100–14.933 | 11,197–4083 | 2.94 |
| 14.933–16.500 | 4083–1723 | 3.12 |
| 16.500–17.583 | 1723–949 | 4.15 |
| 17.583–19.050 | 949–423 | 20.18 |
| 19.050–20.123 | 423–235 | 26.49 |
| 20.123–21.250 | 227–126 | 34.64 |

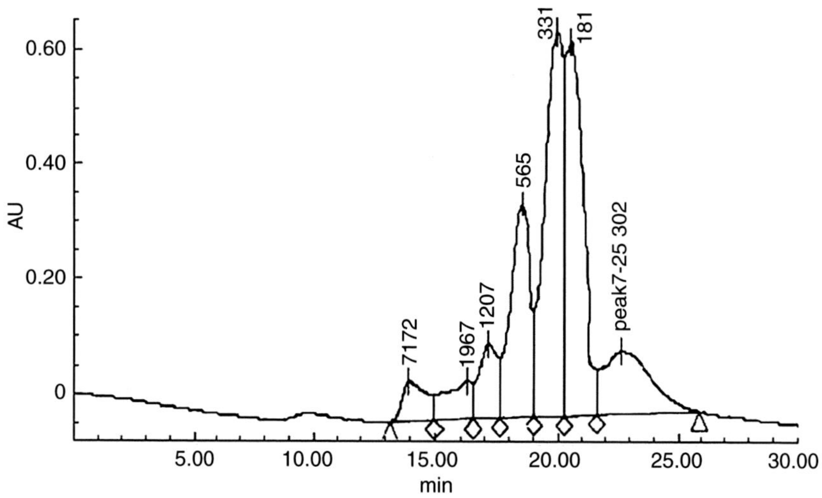

**FIGURE 5.2**   Molecular mass distribution of peanut functional oligopeptide molecular mass standard curve: lgMw = 7.18−0.239 T, $R^2$ = 0.9925.

1000 Da can be easily absorbed by the human body and have a strong functional activity. This indicates that the hydrolysis process selected for oligopeptide preparation should maximize the DH of protein while guaranteeing high yield of oligopeptides.

Currently, in the production process of oligopeptide preparation with protein hydrolysis (Zhang, 2002; Li, 2005; Mo, 1996; Wu and Ding, 2002), the enzymolysis time is generally 360–960 min; the oligopeptide yield is approximately 75%, and the system DH is around 16%. Thus, it can be seen that how to obtain oligopeptides with high DH and high yield in a short time has become a major problem to be solved.

For the problem of low DH and low yield of peptides, in this study, cold pressed peanut protein powder is used as raw material and enzymatic hydrolysis is used for the preparation of peanut oligopeptides, aiming to prepare peanut oligopeptides with high yield and high DH through studying the influencing factors in the enzymolysis process and optimization of the parameters of enzymatic hydrolysis of peanut protein, with the purpose of providing a theoretical basis for the industrialization of functional oligopeptides of peanuts.

### 5.1.2.1 Enzymolysis Conditions

#### 5.1.2.1.1 Single protease

Enzyme species are one of the most important influencing factors in enzymolysis (Deng, 1981). Five species of proteases are selected for peanut protein hydrolysis in this experiment and the result is shown in Fig. 5.3. Under the recommended conditions for each protease, the TCA-NSI of neutral proteases 1398, FM, Protamex, and N120p in peanut protein hydrolysis is 45–53% and DH is 9.4–12.1%; while TCA-NSI and DH of alcalase in peanut protein hydrolysis are respectively $63.80 \pm 0.47\%$ and $15.01 \pm 0.88\%$, significantly higher than those of other proteases ($P < 0.01$). Therefore, alcalase is selected as the tool enzyme for peanut protein hydrolysis in the following experiment.

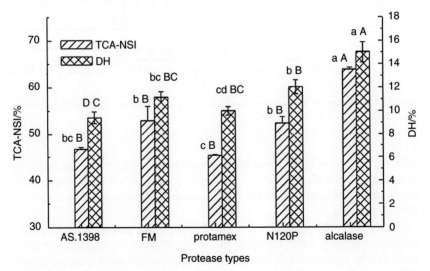

**FIGURE 5.3** Comparison between enzymolysis effects of different proteases. Lowercase letters represent 0.05 significance level; uppercase letters represent the 0.01 significant level; the both lower and upper case letters indicates no significant difference.

### 5.1.2.1.2  Single-factor experiment for peanut protein hydrolysis with alcalase

**1.** Enzyme dosage

In this experiment, the dosage of alcalase is further increased on the basis of 304 U/g to study the effect of enzyme dosage on TCA-NSI and DH (Fig. 5.4).

The results show that when enzyme dosage is 304–2428 U/g, TCA-NSI and DH of enzymatic solution rise sharply with the increase of enzyme dosage; when the enzyme dosage reaches 3642 U/g, TCA-NSI and DH respectively reach 71.63 ± 0.16% and 19.74 ± 0.27%, 10.86%, and 11.45% higher than that at 304 U/g. When enzyme dosage is 3642–6070 U/g, TCA-NSI basically has no change and the increase of DH is also not obvious, being only 0.85%. At the beginning of enzymolysis, base proteins decrease rapidly and small peptides increase rapidly. Meanwhile, peptides, as the substrate of protease, will be further hydrolyzed into peptide molecules with lower molecular mass. They will have competitive binding with proteases in the hydrolytic process. The initial velocity of the system reaction will be increased with the increase of enzyme dosage, which means that most large proteins are degraded into small peptides in a short time. When the amount of large proteins is decreased to a certain level, the probability of reaction between protease and large proteins will be greatly lowered, regardless of whether the protease concentration in the system is high or low. Then the main substrate of protease is peptides. Therefore, when the enzyme dosage reaches 3642 U/g, a further increase of enzyme dosage will have no obvious effect on the peptide yield. Although protease

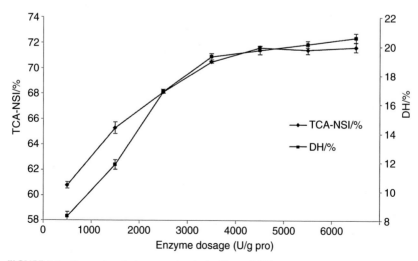

**FIGURE 5.4** Comparison between enzymolysis effects of different proteases.

still has an effect on peptides, DH of the system rises very slowly. That is because the action sites of protease have been greatly reduced. Thus, when the enzyme dosage reaches 3642 U/g, the oligopeptide extraction rate of the system is basically invariable; and the increase of DH with the further increase of enzyme dosage is very small. Therefore, the appropriate enzyme dosage is 3642 U/g.

**2.** Substrate concentration

Fig. 5.5 shows that when the substrate concentration is within the range of 1–10%, there is substrate inhibition in the peanut protein–alcalase system, that is, both TCA-NSI and DH are decreased with the increase of substrate concentration. When the substrate concentration is further decreased to 0.5%, substrate inhibition will disappear, and TCA-NSI and DH are both decreased compared with that at the substrate concentration of 1%. When the substrate concentration is 4%, TCA-NSI and DH are $67.46 \pm 0.31\%$ and $16.95 \pm 0.21\%$, respectively, which is 1.11% and 0.4% lower than those at the substrate concentration of 2%, and 3.23% and 1.79% higher than those at 6%.

**a.** It shows that TCA-NSI and DH are decreased greatly when substrate concentration is increased from 4% to 6%. Theoretically (Tello et al., 1994), a too high substrate concentration of the system will lead to low effective moisture concentration, which reduces the diffusion and movement of substrate and protease and produces inhibition of hydrolysis. A too low substrate concentration could reduce the probability of collision between the protease and acting substrate, resulting in the inhibition of hydrolysis. The experimental results have shown that when substrate concentration is 0.5%, the influence of the reduced probability of protease and substrate collision is greater than that of the high effective water concentration, manifested as the hydrolysis inhibition caused by substrate concentration. In the range of 1–10%, it was manifested as the hydrolysis inhibition caused by high substrate concentration. That is,

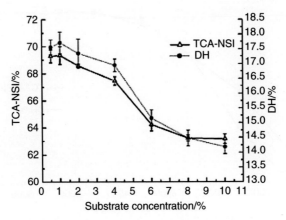

**FIGURE 5.5** Influence of substrate concentration on TCA-NSI and DH.

TCA-NSI and DH are the highest when the substrate concentration is 1%. But in the actual production, the concentration of raw material protein resulted in low absolute yield of product (Deng et al., 2005). With a fixed handling capacity, repeated production was carried out to increase the yield, which led to the increase of energy consumption and low equipment utilization. Therefore, low substrate concentration should not be promoted in actual production. 4% was finally chosen as the optimum substrate concentration.

3. pH

    The effect of pH on enzymatic reactions is mainly reflected in its effect on enzyme activity (Munilla-Moran and Saborido-Rey, 1996). The experiment has analyzed the changes in TCA-NSI and DH when pH changes from 6 to 9, as shown in Fig. 5.6. With the rise of pH, TCA-NSI and DH are rapidly increased. When the pH of the system reaches 8.0, TCA-NSI and DH reach $68.61 \pm 1.23\%$ and $16.36 \pm 0.33\%$, respectively; when the pH of the system reaches 9.0, TCA-NSI and DH are lower than at pH 8.0, with a decrease of 1.59% and 0.84%, respectively. Thus, alcalase is more suitable for an alkaline environment. So 8.0–8.5 is selected as the optimum pH.

4. Reaction temperature

    The experiment has analyzed the changes in TCA-NSI and DH when the temperature changes from 40 to 75°C, as shown in Fig. 5.7. TCA-NSI and DH in the system are gradually increased when the temperature rises from 40 to 55°C and reach their maximums at 55°C, that is, $67.70 \pm 0.58\%$ and $15.96 \pm 0.22\%$, respectively; TCA-NSI and DH show a downward trend

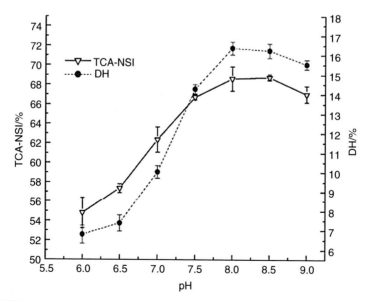

**FIGURE 5.6** Influence of pH on TCA-NSI and DH.

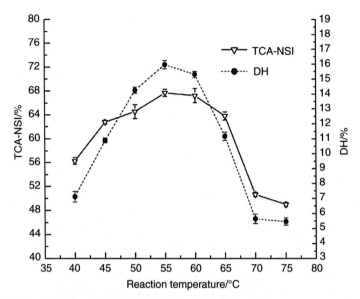

**FIGURE 5.7**    Influence of reaction temperature on TCA-NSI and DH.

when the temperature rises above 55°C and rapidly drop when the temperature rises above 60°C. The temperature for enzymolysis is closely related to
the stability of the protease molecules (Wang, 2001). That is because peptide
bonds of protease molecules have a specific spatial structure; if the reaction
temperature is higher than a certain limit, it will easily lead to dissociation
of secondary bonds, resulting in proteases' wholly or partially loss of catalytic activity. But when the temperature is too low, it will greatly reduce the
intensity of molecular motion in the system, thereby reducing the chance
of protease and substrate collision. Therefore, the suitable temperature for
alcalase reaction is around 55°C.

5. Enzymolysis time
   The experiment has analyzed the influence of different enzymolysis times
   between 0 and 240 min on TCA-NSI and DH. According to Fig. 5.8, TCA-
   NSI and DH show an upward trend at 120 min after hydrolysis is started
   and reach 69.69 ± 0.47% and 15.81 ± 0.26%, respectively, when the hydrolysis lasts 120 min. When the hydrolysis lasts 180 min, DH continues to
   rise by only 0.21% and TCA-NSI is showing a downward trend. That may
   be because alcalase itself is a compound protease composed of an incision
   enzyme and an end-cut enzyme. During hydrolysis, the activity of the end-
   cut enzyme decreased with the decrease in number of incision sites. Some
   peptide molecules in the system are hydrolyzed into amino acids under the
   activity of the end-cut enzyme in the hydrolysis system. In the later period
   of hydrolysis, the hydrolysis rate of the end-cut enzyme is higher than that of
   the incision enzyme, so the peptide extraction rate of the system is slightly

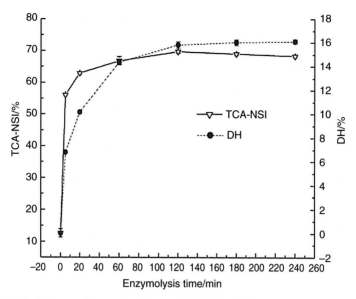

**FIGURE 5.8**  Influence of enzymolysis time on TCA-NSI and DH.

decreased after 180 min. Thus, when the hydrolysis lasts 120 min, the peptide extraction rate of the system reaches its maximum, and the increase of DH with further hydrolysis time is very small. Therefore, the suitable enzymolysis time is around 120 min.

### 5.1.2.1.3 Orthogonal rotation combination experiment for peanut protein hydrolysis with alcalase

The results of TCA-NSI and DH are obtained through the 23 treatment combination experiments of quaternary quadratic orthogonal rotation combination. The results are analyzed and processed with SPSS platform to establish the mathematical model for TCA-NSI and DH with the factors; TCA-NSI is taken as the first index and the oligopeptide yield model is analyzed to get the optimal conditions and determine DH of the system under the optimal conditions with DH model. The model analysis results show that, within the range adopted by the experiment, enzymatic hydrolysis efficiency gradually decreases with the increase of protein concentration. However, as the increase of protein concentration can reduce energy consumption and improve equipment utilization in production, higher substrate concentration can be selected according to the model in practical production. The results of single factor experiments show that 4% is a suitable substrate concentration, so it is selected as the preferred substrate concentration. On the basis of the regression model, it is known that TCA-NSI of the system is 80.35% when substrate concentration is 4% and all other factors are optimal.

#### 5.1.2.1.4  Screening of proteases for complex enzymolysis

Enzyme species are one of the most important influencing factors in enzymolysis. In this experiment, the peanut protein is hydrolyzed with alcalase and then the peanut protein is hydrolyzed with four kinds of proteases. Under the recommended conditions for each protease, the DHs of peanut protein hydrolysis with N120P and FM following alcalase are respectively $20.96 \pm 0.33\%$ and $20.08 \pm 0.37\%$, with no obvious difference ($P > 0.05$) but significantly higher than that with the other two proteases. TCA-NSI of peanut protein hydrolysis with N120P following alcalase reaches $84.6 \pm 0.74\%$, $6.45\%$ higher than hydrolysis with only alcalase and significantly higher than with the other proteases ($P < 0.05$). Thus, it can be seen that both DH and oligopeptide yield are higher in the enzymatic solution system of peanut protein hydrolysis with N120P following alcalase. Therefore, through comprehensive consideration, N120P is selected as the tool enzyme for continuous hydrolysis of peanut protein following alcalase (Fig. 5.9).

#### 5.1.2.1.5  Single-factor experiment for peanut protein hydrolysis with N120P

**1.** Enzyme dosage

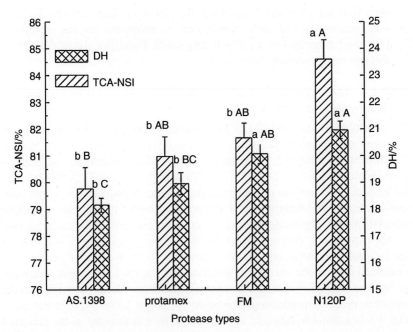

**FIGURE 5.9**  Screening of proteases for complex enzymolysis. Lowercase letters represent significance difference ($P < 0.05$); uppercase letters represent the extremely significance difference ($P < 0.01$); the same letter indicates on significance difference.

This experiment analyzed the changes in TCA-NSI and DH when N120P dosage changed from 300 to 3600 U/g (Fig. 5.10). The results showed that both TCA-NSI and DH in the enzymatic solution rise with the increase of N120P dosage. When N120P dosage is lower than 1800 U/g, TCA-NSI and DH rise more rapidly; when N120P dosage reaches 1800 U/g, TCA-NSI and DH are 54.44 ± 0.42% and 13.89 ± 0.44%, respectively. And when enzyme dosage is within the range of 1800–3600 U/g, the increases of TCA-NSI and DH are only 1.17% and 0.707%, respectively. According to enzyme kinetics, a too high or too low enzyme dosage has a negative effect on hydrolysis, and although too much of the enzyme will achieve hydrolysis, it would be a waste of resources. Therefore, the appropriate enzyme dosage is around 1800 U/g.

2. pH

The influence of pH on enzymatic reaction is mainly reflected in its influence on enzyme activity. This experiment analyzed the changes in TCA-NSI and DH when the pH changes from 5 to 8, as shown in Fig. 5.11. TCA-NSI and DH rise with the increase of pH. When pH is 6.0, TCA-NSI and DH reach their maximum, 51.94 ± 0.42% and 13.83 ± 0.34%, respectively; when pH > 6.0, TCA-NSI and DH begin to decline. Thus, it can be seen that if the system is too acidic or too alkaline it will affect the activity of the enzyme and N120P is more suitable for neutral acidic conditions. Therefore, pH 6.0 is selected as the optimal pH.

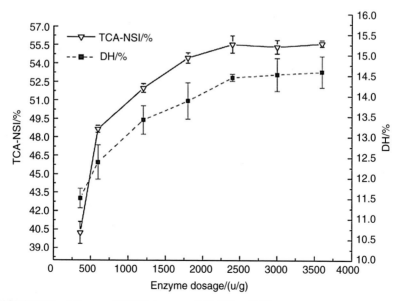

**FIGURE 5.10**    Influence of N120P dosage on TCA-NSI and DH.

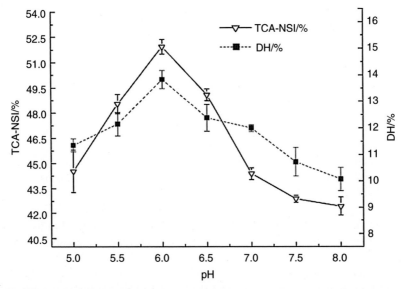

**FIGURE 5.11**    Influence of pH on TCA-NSI and DH.

3. Reaction temperature

   Enzyme activity largely depends on the temperature. This experiment analyzed the changes in TCA-NSI and DH when the temperature increases from
   40 to 70°C, as shown in Fig. 5.12. TCA-NSI and DH are greatly increased
   when the temperature rises from 40 to 55°C and reach their maximum at 55–
   60°C, 53.29 ± 1.48%, and 13.97 ± 0.56%, respectively; TCA-NSI and DH
   start to decline when the temperature is above 60°C. That is because peptide
   bonds of protease molecules have a specific spatial structure; if the reaction
   temperature is higher than a certain limit and the molecules absorb excess
   energy, it will easily lead to dissociation of secondary bonds, resulting in
   proteases' complete or partial loss of catalytic activity (Wang and Gu, 1999).
   Therefore, the suitable temperature for N120P is probably around 55–60°C.

4. Reaction time

   This experiment analyzed the influence of enzymolysis time between 0 and
   240 min on TCA-NSI and DH. According to Fig. 5.13, TCA-NSI and DH
   rise very rapidly in the first 20 min after hydrolysis is started and the rise
   begins to slow down after 20 min; TCA-NSI and DH reach 48.99 ± 1.34%
   and 12.8 ± 0.37%, respectively, when the hydrolysis lasts 60 min. When the
   enzymolysis time is between 60 and 240 min, the increases of TCA-NSI and
   DH are very slow, only 2.92% and 1.21%, respectively. Therefore, a suitable
   reaction time is around 60 min.

5. Determination of DH in peanut protein complex enzymolysis solution

   After 170 min complex enzymolysis of peanut protein under the optimal
   conditions, the DH in the system is 23.76 ± 0.93%, 6.24% higher than that

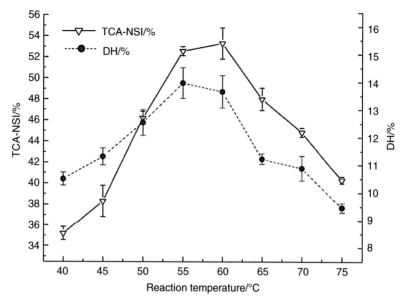

**FIGURE 5.12**   Influence of reaction temperature on TCA-NSI and DH.

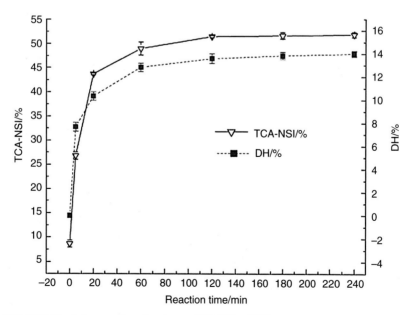

**FIGURE 5.13**   Influence of reaction time on TCA-NSI and DH.

of enzymolysis with only alcalase; the average peptide chain length (PCL) is $4.21 \pm 0.16$, 1.5 lower than that of enzymolysis with only alcalase; the average relative molecular mass is 481.1. Thus, the yield of functional oligopeptides of peanuts can be greatly improved through complex enzymolysis and it will significantly reduce the PCL ($P < 0.01$) of peanut functional oligopeptides.

## 5.1.2.2 Refinement and Classification

### 5.1.2.2.1 Desalination

In peanut protein enzymolysis, a certain proportion of salt in peanut peptides should be maintained by adjusting the pH or keeping it stable. If the salt content is too high there is a negative impact on product applications, so desalination is carried out for improving the product quality. Traditional desalination methods include ion-exchange resin desalination and macroporous resin methods. The former employs anion and cation exchange resins, contributing to a complicated desalination process and low peptide recovery rate. This is because the solution system shows acidity which enabling the positive charge of peptide. When this hydrolysate is filtered by resin cation, a lot of peptide molecules will be absorbed resulting in great loss. The macroporous resin method involves an adsorption and elution process, with a complicated operation and low yield of peptides; organic solvent is required for elution, which makes it difficult to achieve industrialization.

Zhang and Wang (2007b) studied the desalination method of peanut peptides and established acation-anion resin mixed bed desalination method on arachin, the basis of the traditional desalination method, and contrasted the process of mixed bed method, traditional ion exchange method, and macroporous resin method. Experimental results show (Fig. 5.14) that peanut peptide desalination with the mixed bed method has a higher desalting rate and oligopeptide yield. That is because the whole solution system shows acidity when hydrolysate passes resin cation, in traditional is known as anion and cation resin desalination; and therefore causes a lot of peptide molecules in the solution to have positive charge of resin anion which absorbed by the resin cation. In addition, some peanut peptides will also be absorbed on the apertures of resin. Therefore, desalination with resin anion and resin cation will lead to a great loss of oligopeptides. While in the mixed bed desalination, the hydrolysate has a stable pH in desalination, which will greatly reduce the peptide loss in desalination. In addition, the interaction between anions and cations will reduce the resin's adsorption of oligopeptides. It is clear that mixed bed desalination has obvious advantages for oligopeptide yield compared with other two methods.

Zhang and Wang (2007b) also optimized some conditions for the mixed bed desalination process. The results showed that the optimal operating conditions for the anion–cation exchange resin mixed bed are as follows: hydrolysate rate

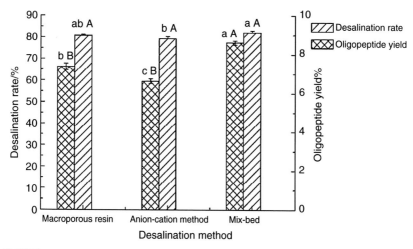

**FIGURE 5.14**  Composition of desalination method. Lowercase letters represent significance difference ($P < 0.05$); uppercase letters represent the extremely significance difference ($P < 0.01$); the same letter indicates on significance difference.

through the column is 5–10 times column volume/h, hydrolysate pH 4.5, and anion and cation resin ratio of 3:2. Under these conditions, both the peanut oligopeptide desalination rate and the protein recovery rate are higher than 80%.

### 5.1.2.2.2  Ultrafiltration

Peanut peptide enzymatic hydrolysate also contains some macromolecular protein, protease, and organic colloids. All of these will restrict its application in the food industry; therefore ultrafiltration and classification is needed for peanut oligopeptides. Ultrafiltration is a system for achieving heterogeneous mixture separation of ingredients with different particle sizes in a solution, using the energy difference on either side of the membrane as the driving force. Its features include: no interphase variation, low energy consumption, easy operation, and small footprint, etc. A major problem of protein hydrolysate separation by ultrafiltration is the decrease of membrane flux with the operation time due to concentration polarization, membrane pore blockage, and the appearance of a gel layer. Therefore, it is of great importance for the long-term, safe, and stable operation of the ultrafiltration system and improvement of product yield to master and implement the appropriate operating parameters.

1. Influence of operating pressure on membrane flux

   The influence of operating pressure on membrane flux mainly lies in the formation of a gel layer on the membrane surface. Under the conditions of specific solute concentration and ultrafiltration temperature, the selection of the appropriate operating pressure will postpone the formation and thickening of the gel layer. In the ultrafiltration process, higher operating pressure will lead to a greater pressure difference across the membrane and greater membrane flux. But the membrane surface will have concentrate polarization as the ultrafiltration proceeds; this will be more serious as the operating pressure increases. Due to the concentrate polarization, a gel layer gradually forms on the membrane surface; a higher operating pressure will lead to thicker gel layer, greater mass transfer resistance, and smaller initial flux increase. Therefore, excessive operating pressure will lead to the increase of energy consumption and the increase of thickness and density of gel layer formed in the ultrafiltration process, thus resulting in reduced efficiency of ultrafiltration and serious membrane pollution. But low ultrafiltration pressure will lead to low initial flux of membrane, resulting in low efficiency of the ultrafiltration process.

Zhang and Wang (2007b) have used an ultrafiltration membrane with molecular mass cutoff (MMCO) of 5000Da for Class-1 ultrafiltration of peanut oligopeptides and a membrane with MMCO of 1000 Da for Class-2 ultrafiltration of the filtered solution, and studied and explored the influence of operating pressure on the ultrafiltration effect (Figs. 5.15 and 5.16). The results show that the most suitable operating pressure for Class-1 ultrafiltration is 0.18 MPa and that for Class-2 ultrafiltration is 0.21 MPa.

2. Influence of feed concentration on membrane flux

Proteins and peptides are amphoteric compounds with strong surface activity and can be easily adsorbed on the polymer surface. Therefore, the influence of feed concentration on ultrafiltration mainly lies in concentrate polarization and the gel layer formation. The appropriate feed concentration will

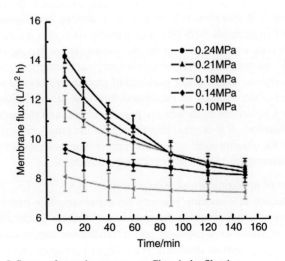

**FIGURE 5.15**    Influence of operating pressure on Class-1 ultrafiltration.

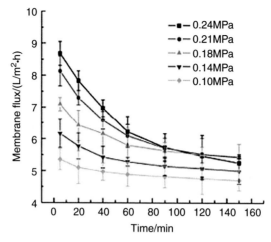

**FIGURE 5.16**   Influence of operating pressure on Class-2 ultrafiltration.

enable more small solutes to pass with low mass transfer resistance, thus improving the overall efficiency of the ultrafiltration.

Zhang and Wang (2007b) studied and explored the influence of feed concentration on the ultrafiltration effect in the ultrafiltration process (Figs. 5.17 and 5.18). The results showed that the optimal feed concentration for Class-1 ultrafiltration is 5–7.5% and for Class-2 ultrafiltration, around 7.5%.

3. Influence of operating temperature on membrane flux

The influence of operating temperature mainly lies in the influence on the feed itself, and thus affects the concentrate polarization on the membrane

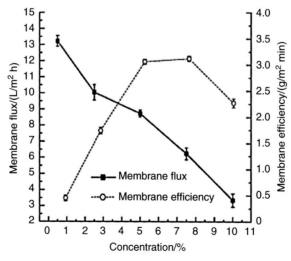

**FIGURE 5.17**   Influence of feed concentration on Class-1 ultrafiltration.

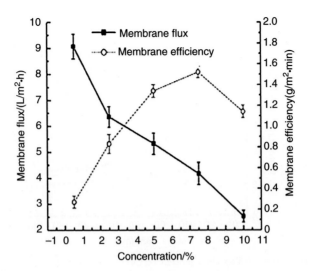

**FIGURE 5.18**   Influence of feed concentration on Class-2 ultrafiltration.

surface. The rise of operating temperature will cause the decrease of feed viscosity, the increase of the diffusion coefficient, and the decrease of the effect on concentrate polarization. Therefore, membrane flux increases with the rise of temperature.

Zhang and Wang (2007b) studied and explored the influence of operating temperature on the ultrafiltration effect in the ultrafiltration process (Figs. 5.19 and 5.20). The results showed that the optimal operating temperature for Class-1 ultrafiltration is 35°C, and for Class-2 ultrafiltration, 40°C.

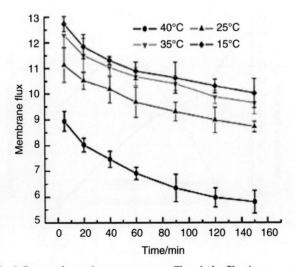

**FIGURE 5.19**   Influence of operating temperature on Class-1 ultrafiltration.

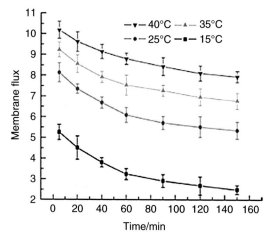

**FIGURE 5.20** Influence of operating temperature on Class-2 ultrafiltration.

### 5.1.2.3 Isolation and Purification

Oligopeptide isolation and purification refers to the process of oligopeptide enrichment, purification, and identification with gel chromatography, ion exchange chromatography, HPLC, and HPLC-MS by adopting a particular functional activity as the evaluating indicator. Gel filtration chromatography is a chromatographic method for gel isolation on the basis of different molecular masses of substances. The material used for gel filtration chromatography is a bead-like particle substance with a 3D network structure and uniform mesh diameter. When solution containing compounds with different molecular masses go through the gel column, the buffer, small compounds, can freely diffuse and permeate through the mesh while macromolecular substances are blocked outside the mesh. Substances with high molecular mass are not able to diffuse so easily through the mesh and thus are eluted earlier than substances with low molecular mass.

#### 5.1.2.3.1 Gel filtration and isolation

Zhang and Wang (2007b) used Sephadex G-15 for gel chromatographic separation of ACE I peptides. The results showed that peanut oligopeptides with a molecular mass smaller than 1000 Da can be separated into three peaks (Fig. 5.21) by Sephadex G-15, in which Peak II (PP-II) has the strongest hypotensive activity, and IC50 can be up to 0.091 mg/mL.

#### 5.1.2.3.2 PP-II separation with semipreparative RP-HPLC

The principle of RP-HPLC is to isolate the sample on the basis of different molecular polarities of ingredients. Due to its high resolution, high speed of analysis, and high recovery, it has become one of the most effective means for

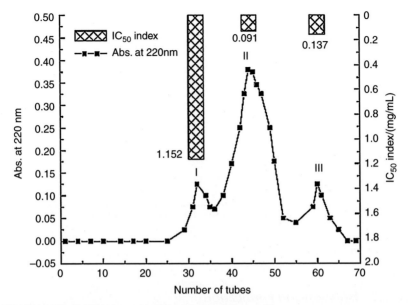

**FIGURE 5.21** Gel filtration and isolation of peanut oligopeptide.

the isolation, preparation, and analysis of polypeptides. After peptide ingredient PP-II isolation with semipreparative RP-HPLC, 27 major ingredients are automatically collected, as shown in Fig. 5.22. The injection is repeated and the same ingredients are combined for freezing and drying, and then dissolved in distilled water to prepare a solution with a peptide concentration of 0.010 mg/mL to

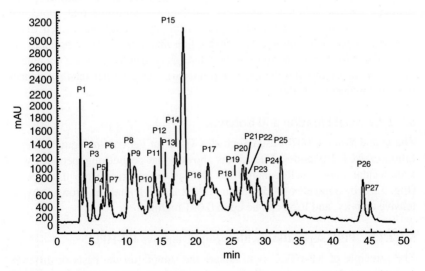

**FIGURE 5.22** Diagram of peanut peptide ingredient PP-II isolation and purification with semi-preparative RP-HPLC.

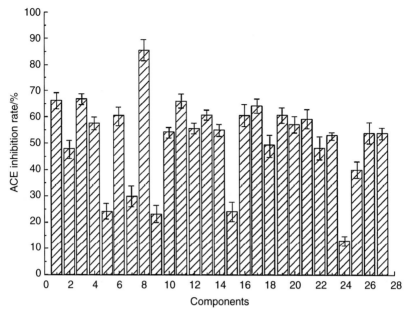

**FIGURE 5.23**    ACE inhibition of each ingredient in peanut peptide PP-II isolation and purification with semipreparative RP-HPLC.

determine the ACE inhibitory activity of each ingredient. As can be seen from the results in Fig. 5.23, ingredient No. 8 (P8) has the strongest activity and its ACE inhibition rate is $85.77 \pm 3.95\%$; next are P3 and P11, whose ACE inhibition rates are $66.79 \pm 2.00\%$ and $66.15 \pm 2.11\%$, respectively. Therefore, P8 is selected for structural identification. Column: Varian C18, 21.2 mm $\times$ 150 mm; column temperature: 25°C; flow rate: 10 mL/min; detected wavelength: 280 nm; injection volume: 1 mL; elution conditions: 0–50 min, 70–95% A, 5–30% B (A: water +0.05% TFA; B: acetonitrile +0.05% TFA).

### 5.1.2.3.3 P8 purity identification after purification

To detect the above isolation effect with RP-HPLC, the purity of P8 is identified in this experiment. As shown in Fig. 5.24, the RP-HPLC diagram of P8 shows a single peak, indicating that P8 may include only one peptide, and further structural identification and analysis can be carried out. Column: Column C18 (4.6 $\times$ 250 mm); flow rate: 0.5 mL/min; detected wavelength: 280 nm; injection volume: 20 uL; column temperature: 30°C; elution conditions: 0–20 min, 70–95% A, 5–30% B (A: water +0.05% TFA, B: acetonitrile +0.05% TFA).

### 5.1.2.3.4 Analysis of amino acid sequence of ACE inhibitory peptides with MALDI-TOF-TOF

Traditional sequencing methods for polypeptide and protein analysis include the C-terminal chemical degradation method, Edman degradation method, and

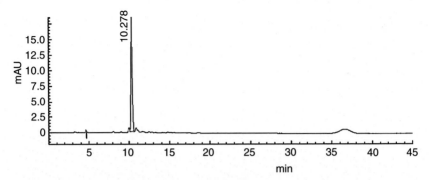

**FIGURE 5.24** RP-HPLC analysis diagram of P8 after purification.

DNA translation method. All these methods have some defects. For example, the Edman degradation method (also known as the PTH method), as the standard method for peptide and protein sequencing, requires high purity and large quantity of samples, has low sequencing speed, often misidentifies modified amino acid residues, and is unable to sequence N-terminal protected peptides; for the C-terminal chemical degradation method, it is difficult to find an ideal chemical probe (Liao et al., 2006). A mass spectrum (MS) is highly sensitive, accurate, easy to operate, and fast. With the birth of fast atom bombardment (FAB), electrospray ionization (ESI), matrix-assisted laser desorption interpretation (MALDI), and other soft ionization methods, MS has been developed rapidly in the biological field. MALDI-TOF-TOF can be directly used for peptide amino acid sequencing. Peptide parent ions generated by a Class A mass spectrum selectively enter Class B; after collision with inert gas, peptide fractures along peptide chain, and the difference between the masses of the peptides obtained is used for inferring peptide sequence. The tandem MS data can be used for analyzing peptide sequences with software or manually with the help of amino acid analysis. Peptide fracture follows certain rules. Generally peptide fracture modes and fragment types are as shown in Fig. 5.25 (Guo, 2009). Peptide bond cleavage mainly produces N-terminal series fragment ions (series b) and C-terminal fragment ions (series y), and these fragments may further form dehydrated and deammoniated ions.

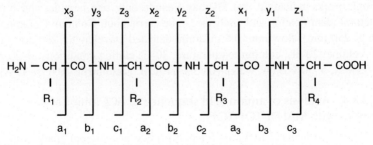

**FIGURE 5.25** Peptide chain backbone fracture and fragment ions naming.

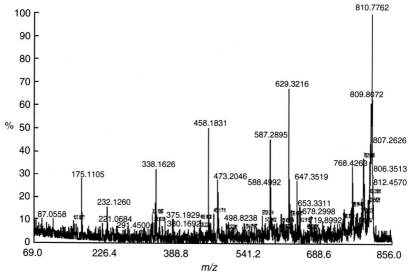

**FIGURE 5.26** MALDI MS/MS spectrogram of oligopeptide P8.

The peanut oligopeptide P8 was analyzed with MALDI-TOF-TOF, and the MS/MS diagram was obtained through class B mass spectroscopy of fragment ions of parent ions (Fig. 5.26). In the MS graph, m/z is the molecular ion peak $[M + 2H]^{2+}$ of 810.8, and P8 has the molecular mass of 808.8 Da (with theoretical molecular mass of 807.7). Using software for analysis from scratch, we obtain that the amino acid sequence of P8 is Lys-Leu-Tyr-Met-Arg-Pro. By determining the ACE inhibitory activity of P8, we see that IC50 is 0.0052 mg/mL (6.42 μmol/L) (Table 5.4).

### 5.1.2.3.5 Verification of ACE inhibitory activity of synthetic oligopeptide ILILLP

To further verify the functional activity of the hexapeptide Lys-Leu-Tyr-Met-Arg-Pro (KLYMRP), solid-phase synthesis is adopted to synthesize oligopeptide

**TABLE 5.4** Amino Acid Sequence and Activity of ACE Inhibitory Peptides of Peanuts

| Peanut ACE Inhibitory Peptide P8 | Analysis Result |
|---|---|
| Theoretical relative molecular mass | 807.7 |
| Measured MS mass | 808.8 Da |
| Amino acid sequence | Lys-Leu-Tyr-Met-Arg-Pro |
| IC50 value | 0.0052 mg/mL (6.42 μmol/L) |

Lys-Leu-Tyr-Met-Arg-Pro with the same amino acid sequence. Through determination of its activity, we find that its ACE inhibitory activity IC50 is 0.0038 mg/mL (4.69 µmol/L). The experiment has verified that oligopeptides with such an amino acid sequence have ACE inhibitory activity in vitro and the activity of the synthesized oligopeptide is higher than that of oligopeptide KLYMRP from peanut protein.

### 5.1.3 Application and Development Prospects

#### 5.1.3.1 Current Situation of Domestic and Foreign Markets

The research on the oligopeptide preparation using the protein enzymolysis method in China started late. However, in recent years, research into the production of functional oligopeptides has been receiving much attention in China, and the major objects of study have mainly included soy protein and milk protein. In terms of soy protein, Shandong Shansong, Wuhan Tallyho, Heilongjiang Hagaoke, and Leneng have prepared soy oligopeptide series products from soybean protein using the enzymolysis method. In terms of milk protein, currently the domestic industrialized bioactive peptide from milk protein is mainly casein phosphopeptide (CPP). There are three to five manufacturers and the CPP content of the products is 25–80%; the molecular mass of CPP is 2000–3000 Da. Guangzhou Institute of Light Industry is the first to achieve industrialized CPP production and its CPP content is the highest (75–82%), with an annual output of 12 tons. In recent years, many enterprises in China have begun the production of peanut peptides, including Shandong Changlin Biotechnology Co., Ltd., Hubei Reborn Biotechnology Co., Ltd., and Shandong Zhongshi Taihao Biochemical Factory. The price of peanut peptides is 100–500 Yuan/kg. But overall, these companies are in their infancy and unable to achieve large-scale production, to say nothing of brand effect. Therefore, China's peanut peptide industry still needs further development.

Compared with developed countries, China still has a long way to go in the study of all aspects of functional oligopeptides, mainly reflected in the following aspects.

#### 5.1.3.1.1 Inadequate research on the production process

China's study on oligopeptides is mainly focused on oligopeptide separation and extraction, structure identification, and their functional activity. China is lacking in detailed study and optimization of the production process and its parameters. For example, there is great blindness in the selection of protease and its action conditions in oligopeptide preparation, leading to low DH and yield of oligopeptide products; a mature system is not yet formed for the oligopeptide purification process, thereby resulting in low purity of oligopeptides and wide distribution of molecular mass, not conducive to the application.

### 5.1.3.1.2  Unclear functional activity of peptides

China's oligopeptide research is seriously divorced from industrialization. Many peptides whose functional activity is deeply studied are not industrialized, while industrialized peptides are lacking in specialized research on functional activity. Therefore, there are hardly any peptide products with specific functional activity.

### 5.1.3.1.3  The raw material protein selection system for peptides with different functional activities needs to be established

Currently, there is still a degree of blindness in enzymatic production of oligopeptides with special functional activity and the materials need to be screened from a lot of raw proteins and protease, with a great workload. As the primary structure of more and more proteins has been elucidated, there is an urgent need to establish a material selection system through research on the relationship between the functional activity and amino acid composition or structural characteristics of the corresponding oligopeptides, to reduce the blindness in raw material selection.

At present, no refined peanut peptides are seen in European and American countries and Japan and there may be two reasons. One is that people in European and American countries are accustomed to processing products directly from peanuts while Japanese are accustomed to raw food. Therefore, there are deemed to be no large amounts of raw materials for peanut peptide production. The other is that many foreign enterprises engaged in polypeptide research and sales, such as Antisera of Yanaihara Institute Inc. in Japan, mainly deal with pure peptides.

### 5.1.3.2  Market Development Prospects

Functional oligopeptides have become a research and development focus for their good processing properties and a variety of biological activities. In China, peanut proteins are mainly by-products of peanut oil production. Currently, proteins and oil can be efficiently and thoroughly isolated in peanut processing with cold press and leaching techniques. Therefore, the preparation of peanut functional oligopeptides with peanut protein powder as the raw material can improve the functional properties of peanut protein in processing and release the functional peptides from proteins. In-depth improvement of China's deep processing of peanut protein will effectively improve the conversion rate and utilization in peanut deep processing, optimize the peanut industrial structure and product mix, and promote the development of the whole peanut protein industry and even the functional food industry. Meanwhile, the increase of product profit margin will make peanut prices biased toward sellers, which will effectively promote agricultural efficiency and rural incomes, and lead farmers to get rich. Therefore, peanut oligopeptides have good application prospects.

Currently, China's peanut peptide producers are all small in scale and in a recession. Therefore, there is an urgent need to integrate existing technologies and raw material resources to establish powerful large enterprises for building brand strategy and in-depth development of the peanut peptides industry. In addition, China still has no standards specifically for peanut peptides, which has also restricted the further development of China's peanut peptides industry.

## 5.2   PEANUT POLYPHENOLS PROCESSING TECHNOLOGY

Polyphenols, also known as tannin or tannic acid, are secondary metabolites in plants and hydroxyl compounds whose molecule contains one or more benzene rings (An et al., 2010). Polyphenols have antibacterial, anti-inflammatory, antioxidant, antiradiation, blood pressure lowering, and other functional activities and are known as "the seventh class of nutrients." They have become a hotspot of research and development and have been widely used in practical production (Zhao and Zheng, 2006). Studies have shown that the polyphenol content of peanut red coat is 90–120 mg/g (Jianmei et al., 2005) and that of peanut shell is 81–371 mg/g (Zhao et al., 2012). Peanut phenols mainly include proanthocyanidins, luteolins, resveratrols, eriodictyol, ferulic acid, phlobaphene, arachidoside, leucocyanidin, and leucodephninidin. The processing techniques of several major peanut phenols are introduced below.

### 5.2.1   Proanthocyanidins

#### 5.2.1.1   Detection Techniques

##### 5.2.1.1.1   Total content determination

1. Vanillin method

    The principle of the vanillin method is that proanthocyanidin flavonoid rings undergo a phenolic condensation reaction with vanillin in a strong acid media to produce colored products (Broadhurst and Jones, 1978), and the total content of proanthocyanidin monomers and polymers can be determined with a spectrophotometer. Hydrochloric acid or sulfuric acid is generally used as a catalyst media, but the hydrochloric acid–vanillin method reaction is not stable easily resulting flocculation (Hong et al., 2009a); in the sulfuric acid–vanillin method, the adding of sulfuric acid will cause a sharp heat production in the reaction system, thereby resulting in oxidative decomposition of proanthocyanidins and a low measured result (Jiang, 2010) with poor reproducibility (Hong et al., 2009a). The specific operation of determining the proanthocyanidin content of peanut skin crude extract with the hydrochloric acid–vanillin method is as follows: weigh some peanut skin crude extract and place it in a test tube; add 3.0 mL of 4% vanillin–methanol solution and 1.5 mL of concentrated hydrochloric acid (36–38%); immediately mix them well and keep color development for approximately 15 min, and determine the absorbance at 550 nm (Tong and He, 2011).

2. Butanol-hydrochloric acid method

   The principle of the butanol–hydrochloric acid method is as follows: proanthocyanidins are oxidized and degraded in the presence of inorganic acids and heating; extended units produce red anthocyanin (Sun, 2010), with the maximum absorbance at 546–550 nm, and the relationship between the proanthocyanidin content and absorbance complies with Lambert–Beer Law. The butanol–hydrochloric acid method has high selectivity and the determination result is subject to proanthocyanidin structure, degree of polymerization, and the storage period after preparation, and it is also easily affected by reddish brown polymers produced by side reactions (Li and Li, 2007).

3. UV spectrophotometry

   Proanthocyanidin molecules contain benzene rings, which have a strong UV absorption in the vicinity of 280 nm, and the relationship between its concentration and absorbance complies with the Lambert–Beer Law. According to this feature, UV spectrophotometry can be used to determine the proanthocyanidin content (Wang et al., 2011). UV spectrophotometry is susceptible to other substances with similar structures and the determination result is rough. The minimum detectability of procyanidins of UV spectrophotometry is 8.23 ug; proanthocyanidin concentration has a linear correlation with absorbance within the range of 10.4–52.0 ug/mL ($R^2 = 0.9992$) (Fu et al., 2002).

4. Catalytic colorimetry with ferric ions

   The principle of catalytic colorimetry with ferric ions is as follows: proanthocyanidins are hydrolyzed by hot acid to produce anthocyaninin; anthocyanin reacts with ammonium ferric sulfate to produce a deep-color compound; the compound has the maximum absorption peak at 550 nm; its absorbance is determined by colorimetry to get the proanthocyanidin content. The features of catalytic colorimetry with ferric ions are easy operation, high accuracy, and good reproducibility (Sun et al., 2004), but it has long reaction time and requires n-butyl alcohol with high toxicity as the reaction medium (Fu et al., 2002). The optimal conditions for the determination of proanthocyanidin content using catalytic colorimetry with ferric ions are: use a mixture of n-butyl alcohol and 5 mL of concentrated hydrochloric acid as the reaction medium and 0.415 mol/L of Fe3+ as a catalyst, keep reaction for 60 min at 60°C (Hong et al., 2009a).

5. Other methods

   Atomic absorption spectrometry and the electrochemical method can also be used to determine proanthocyanidin content. Atomic absorption spectrometry is based on complexation reactions between procyanidins and basic lead acetate; the reaction produces water-insoluble brown precipitate; after filtration and separation, the excess lead ions in the filtrate are determined with atomic absorption spectrometry to indirectly determine the proanthocyanidin content. The electrochemical method is a new method of determining the proanthocyanidin content using its resistance to oxidation. The features of this method include a simple and rapid operation, high sensitivity, and wide linear range (Zhao et al., 2006).

## 5.2.1.1.2 Determination of the components

1. TLC

   TLC is a rapid and simple chromatography developed in recent years. It can be used for rapid separation and qualitative analysis of small amounts of substances and is a form of solid–liquid adsorption chromatography. Its procedure is as follows: evenly coat the clean glass plate with a layer of absorbent or support agent; after drying and activation, drop the sample solution with a capillary tube onto the starting line about 1 cm from one end of the thin-layer plate; after air drying, place the thin-layer plate in a developing tank with solvent, with an immersion depth of 0.5 cm. When the solvent front is approximately 1 cm from the end, take out the chromatographic sheet; after drying, spray chromogenic agent or develop color in UV light. Use a 0.2-mm-thick silica gel plate; use toluene: acetone: formic acid = 3:3:1 (V/V/V) as a solvent and 10% (m/V) vanillin/hydrochloric acid as a chromogenic reagent to determine the procyanidins (Ricardo et al., 1991).

2. HPLC

   HPLC is currently the most commonly used method of proanthocyanidin analysis and a superior separation and analysis method for simultaneous separation and determination of multiple components of complex samples, featuring high separation efficiency, high sensitivity, and good reproducibility. However, as it is difficult to get a standard sample of proanthocyanidin (PC), coupled with incomplete separation between various isomers, a lot of PC isomers are combined together and the analysis duration is long, seriously affecting the accuracy of the results of quantitative analysis (Li, 2010).

   Currently, in some literature, HPLC is used for quantitative analysis of PC in grape seeds (Chen et al., 2007c), apple (Hao et al., 2005), blueberry (Gu et al., 2002), tea polyphenols (Mo et al., 2008), and other samples. In GB/T 8313-2008, HPLC is also used to detect catechin in tea; acetonitrile of different concentration gradients and water are mobile phase and the determination is carried out at room temperature at a flow rate of 1 mL/min (GB/T 8313-2008).

   With the rapid development of MS technology, HPLC is also used to identify the proanthocyanidin structure. It can be determined using HPLC-MS that proanthocyanidin is mainly composed of catechin, epicatechin, and a mixture of catechin and epicatechin, which has laid a good foundation for further development and utilization of proanthocyanidin. In addition, NP-HPLC-APCI-MS can be used to carry out separation and identification of procyanidins of polymerization degrees lower than 15 (Ying and Ming, 2000).

3. Capillary electrophoresis

   Capillary electrophoresis is a liquid separation technique using a high-voltage electric field as the driving force and a capillary tube as the separation channel to achieve separation based on different velocities of movement and distribution coefficients of the components in the sample. Its features are short analysis time, high degree of separation, and low cost (Zhao et al., 2006).

Kreimever et al. (1998) separated flavan-3-ol from procyanidins of similar structure using a pH = 7.0 phosphate buffer and surfactant SDS and capillary electrophoresis. Cifuentes et al. (2001) separated procyanidin B1, procyanidin B2, and procyanidin B3 in samples from food sources from monomers constituting proanthocyanidins with capillary electrophoresis.

4. Other methods
   Fluorescence detection, paper chromatography, and other methods can also be used for isolation and identification of proanthocyanidins. When detecting procyanidins with a fluorescence detector and under the conditions of excitation wavelength of 276 nm and emission wavelength of 316 nm, procyanidin monomers to straight nonamers in the samples can be identified (Adamson et al., 1999). Paper chromatography can be used to identify catechin, epicatechin, and procyanidins in plants using two-dimensional chromatography (Thompson et al., 1972).

## 5.2.1.2   Preparation Process and Technique

### 5.2.1.2.1   Extraction methods

The commonly used methods of proanthocyanidin extraction are the organic solvent extraction method, supercritical $CO_2$ extraction method, and ultrasonic and microwave assisted extraction method. The proanthocyanidin extraction rate and composition of extract are affected by the type and grinding degree of raw materials, extraction solvent, temperature, and other factors.

1. Water extraction method
   The specific operation of proanthocyanidin extraction with water is as follows: raw material crushing ($\leq$15 mm) $\rightarrow$ Deoxy hot water extraction (atmospheric pressure at 60–100°C or high pressure at 100–125°C) $\rightarrow$ filtration (ultrafiltration or reverse osmosis) $\rightarrow$ filtrate concentration $\rightarrow$ vacuum spraying or freeze-drying.

   Using the water extraction method, the product is mainly water-soluble proanthocyanidins with molecular mass not higher than 5000 Da; the yield is 0.5–10.0%, generally 6.5–9.6% (Duncan and Gilmour, 1999). The water extraction method features simple equipment, safety, and cleanness, but it has time-consuming leaching, and low product purity and yield, which has limited its application (Tian, 2008). And during the extraction with cold water, inorganic salt needs to be added to precipitate tannin, which will result in a loss of oligomeric components; hot water leaching will oxidize the product, resulting in a lower yield of proanthocyanidins (Wei et al., 2003).

2. Organic solvent extraction method
   Currently, the organic solvent method is mainly used for proanthocyanidin extraction, including reflux, percolation, and thermostatic water bath. Solvents with strong polarity are generally used, such as methanol, ethanol,

and ethyl acetate. The key to the solvent extraction method is to select an effective solvent. It requires both good solubility of active ingredients and avoiding dissolution of large amounts of impurities, and solvent recovery, environmental pollution, technical processes, and other factors should also be considered (Tian, 2008). In peanut proanthocyanidin extraction with organic solvent, the major factors to be considered include particle size, extraction time, temperature, solid–liquid ratio, pH, and equipment conditions. The technical process of organic solvent extraction technique is as follows: raw materials → crushing → adding organic solvent → extraction → filtration → concentration under reduced pressure → vacuum drying → extract. The optimal conditions for proanthocyanidin extraction from peanut red coat with ethanol are: solid–liquid ratio 1:21, ethanol concentration 56%, extraction time 59 min, temperature 62°C. The yield of proanthocyanidin under such conditions was 7.94% (Liu et al., 2010d).

3. Supercritical extraction method

Supercritical extraction is a process of separating and extracting mixture from the sample using a supercritical fluid as a solvent by taking advantage of its high permeability and high solubility. Currently, $CO_2$ is generally used as the solvent. Supercritical extraction features high extraction efficiency, good selectivity, no solvent residual, recycling of solvent, and low temperature operation. It is especially suitable for substances with poor thermal stability and is widely used for the extraction of active ingredients of natural plants. However, this method requires equipment of high cost (Tian, 2008) and is difficult to use. In the proanthocyanidin extraction with supercritical extraction method, the pressure, temperature, time, flow rate, and other factors need to be considered. Currently, there are few reports on the application of supercritical extraction method in peanut skin proanthocyanidin extraction.

4. Ultrasound-assisted extraction method

The ultrasound-assisted extraction technique mainly involves mechanical action, cavitation, thermal effects, and media interaction. Compared with conventional extraction methods, ultrasound-assisted extraction method is fast, economical, safe, and efficient (Yuan and Cao, 2011). The technical process of ultrasound-assisted extraction method is as follows: raw materials → crushing → adding organic solvent → ultrasonic treatment → filtration → concentration under reduced pressure → vacuum drying → extract.

The optimal conditions for proanthocyanidin extraction from peanut red coat with ethanol are: ethanol concentration 60%, temperature 55–70°C, extraction time 20–30 min, solid–liquid ratio 1: 6–25, pH = 3, 1 time of extraction. The yield of proanthocyanidin under such conditions was 3.48–4.96% (Tong and He, 2011; Liu et al., 2005).

5. Microwave-assisted extraction method

The principle of microwave-assisted extraction is to achieve rapid turning and oriented arrangement of polar molecules to be extracted in a microwave electromagnetic field, which causes tearing and friction heat and rapid transfer of energy, accelerating the dissolution and release of substances. Microwave-assisted extraction method features low consumption of solvent, high selectivity and yield, and has been successfully used in crude drug leaching and active ingredients extraction (Yuan et al., 2002). The technical process of proanthocyanidin extraction from peanut with microwave-assisted extraction method is as follows: raw materials → crushing → adding organic solvent → microwave treatment → filtration → concentration under reduced pressure → vacuum drying → extract.

The factors to be considered in proanthocyanidin extraction from peanut red coat with the microwave-assisted method include microwave power, extraction solvent type, solid–liquid ratio, extraction temperature and time. The optimal conditions include output power 600 W, the volume fraction of ethanol 40%, solid–liquid ratio 1:24.5, extraction time 91 s, and extraction temperature 40°C. Proanthocyanidin extraction rate under such conditions was 12.59% (Zhu et al., 2009).

### 5.2.1.2.2  Purification technique

Proanthocyanidin extract contains many impurities, so further separation and purification is needed to improve the purity and the content of proanthocyanidin (Zhang et al., 2006). Currently, the major methods for proanthocyanidin separation and purification include solvent extraction method, membrane separation method, and chromatography.

1. Solvent extraction method

   In the solvent extraction method, ethyl acetate, toluene, methylene chloride, and ether are usually used for purification. This method has low yield and purification of proanthocyanidin and the content is typically only 50–70%; it requires a lot of organic solvent, easily pollutes the environment, and causes toxic organic solvent residues in the product; the complex steps involve repeated heating and distillation; the method also involves difficult solvent recovery and high production costs (Tian, 2008).

2. Membrane separation method

   Membrane separation method is the method of separating, classifying, purifying, and concentrating the solute and solvent in a mixture with the promotion of outside energy or chemical potential difference by making use of the selective osmosis effect of a membrane. This method uses low energy consumption, no chemical reagents, does not contaminate the product and has good separation effect, but it has a narrow range of applications, immature technology, and low productivity (Tian, 2008; Liu et al., 2010c). In the purification of proanthocyanidins from peanut red coat with membrane separation method, the optimal purification membrane is HPS-1 ultrafiltration membrane module; the operation pressure is 0.54 MPa and the operation temperature is 25°C. The yield of proanthocyanidins under such conditions was 13.3% and the purity was 85.8% (Liu et al., 2010b).

3. Chromatography

Chromatography is a technique of separating mixed components on the basis of different distribution coefficients of substances in the mobile phase and the stationary phase. Chromatography includes TLC, gelchromatography, HPLC, and gel exclusion chromatography. TLC is generally used for basic analysis; it has low accuracy of purification and low efficiency, and products cannot be obtained directly, so it is not suitable for large-scale industrial production. Gel chromatography features a simple process, low energy consumption, no pollution, and high extraction efficiency. HPLC has limited application on the separation of proanthocyanidins of higher degrees of polymerization because proanthocyanidins of high degrees of polymerization have low solubility in organic mobile phase and strong adsorption in stationary phase. Gel exclusion chromatography can be used for the classification of proanthocyanidins of high degrees of polymerization, and the classification gradient can be controlled through change of elution gradient, which is convenient and easy; but the chromatographic column is expensive and acetone is needed for eluting proanthocyanidins of high degrees of polymerization, so a UV detector cannot be used for the detection (Tian, 2008). High-speed countercurrent chromatography features high handling capacity, low solvent consumption, easy operation, high separation efficiency, and product purity, and involves no adsorption and pollution of the sample by the carrier (Li and Li, 2008).

### 5.2.1.3 Application and Development Prospects

#### 5.2.1.3.1 Current situation of application

In the past 30 years, the application basis, development, and utilization of proanthocyanidins have attracted wide attention, and a greater depth of knowledge and understanding of such natural products has been achieved. Proanthocyanidins are applied in not only traditional industries, such as agriculture, silvichemicals, and chemical products, but also in medicine, health products, cosmetics, and polymer functional materials. There have been more than 60 kinds of related commodities on the world market (Shi and Du, 2006).

1. Application in food industry

Proanthocyanidins are present in many natural foods and have a variety of chemical and physical activities. Therefore, they are widely used in the food industry. Proanthocyanidins contribute to the color and luster of many fruits and vegetables and can be used as a natural pigment for food; they show "convergency" in drinks and contribute to the "addiction" of wine, tea, and coffee (Huang et al., 1996); they can also be used as a fining and stabilizer for fruit juice, wine, and beer (Liu and Xi, 1994); they can be added to food as an antioxidant (Lv, 2002) to extend the shelf life.

In recent years, European countries, the United States, Australia, and other countries have developed proanthocyanidins into health foods, such as antiaging, antiradiation, antitumor, and brain-invigorating foods. For example, Pycnogenol (proanthocyanidin ≥85%) of Horphay Research, Switzerland, Pana-life Grape Seed Antioxidant of America, OPC of Italy, Tigra Pill of France, and China's "Shimingbao," "Lino Beautifying Tablet," and "Leizhenzi Health Care Capsule", etc. have been put on the market. According to reports, to prevent radiation damage during space flight, astronauts of the former Soviet Union used a PC-rich vegetable drink (Zhao and Yao, 2000).

2. Application in medicine industry

   Studies have shown that proanthocyanidin can enhance and regulate the immune activity (Lin et al., 2002), prevent cancer, lower cholesterol, and inhibit platelet aggregation; it can also be used to treat diarrhea (Oshima and Ueno, 1993), gastric ulcer (Park et al., 2000), peripheral venous insufficiency (Aliou, 1995), and hypertension (Pruss, 1997). Proanthocyanidins have also been widely used for the treatment of eye diseases, periodontal disease, hay fever, anaphylaxis, and alcoholism (Shi and Du, 2006). For example, Bonnaure et al. in France invented proanthocyanidin oligomer and monomer preparation for periodontal disease treatment and obtained patent protection; Berkhman et al. in Germany developed proanthocyanidin preparation for the treatment of alcoholism; Endotelon proanthocyanidin preparation of Romania has been put on the market for the treatment of capillary disease (Wan et al., 2001).

3. Application in chemical industry

   The plant proanthocyanidins initially used by people were mainly polymeric proanthocyanidins. People extracted tannin extract, whose main ingredient is polymeric proanthocyanidin, from bark and other plant materials for leather tanning, and later for the production of adhesives and drilling mud thinners, etc. (Shi and Du, 2006).

   In addition, proanthocyanidins are widely used in the cosmetics industry. Studies have shown that proanthocyanidin can prevent skin allergies and inflammation, reduce the damage of ultraviolet radiation to the skin (Garini et al., 2000), and reduce blackened skin and brown spots (Kakgawa et al., 1985); it can also restore skin elasticity, skin elasticity, prevent wrinkles, and aging (Wei, 1997), and promote growth of hair (Takahashi et al., 1999). Therefore, proanthocyanidin has been used as a functional cosmetic factor for the production of efficient cosmetics for speckle removing and whitening.

### 5.2.1.3.2 Market development prospects

Proanthocyanidins, as "green" renewable resources and a natural functional product, have good development and utilization prospects. Currently, commercialized proanthocyanidins are mainly the extracts of maritime pine bark and grape seed. The proanthocyanidin products from these sources have a high price and it is difficult to meet market demand. Therefore, it is very important to find raw material resources from a wide range of sources and at a low price. Peanut red coat, as a screened material, has attracted attention from researchers. The preparation of proanthocyanidin with peanut red coat can not only improve its comprehensive utilization value and economic value but also have the social value of promoting the development of China's food and pharmaceutical industry.

In addition, proanthocyanidins with different degrees of polymerization and molecular masses have different chemical and biological activities. The preparation of proanthocyanidin fractions with different molecular masses by degradation, modification, and certain separation method can effectively improve the biological activity of the proanthocyanidin and expand its range of applications. With the increase in relevant research, refined utilization of proanthocyanidin is an inevitable trend.

## 5.2.2 Luteolin

### 5.2.2.1 Detection Techniques

Luteolin determination methods mainly include colorimetry, coulometric titration, fluorescence spectrometry, UV spectrophotometry, TLCS, weighted pulse polarography, and HPLC. Most of these methods have troublesome operation and large uncertainties and are seldom used, while HPLC is widely used for luteolin determination because of its quick operation, high sensitivity, high precision, and accuracy. Several commonly used luteolin determination methods are briefly introduced below.

#### 5.2.2.1.1 UV spectrophotometry (UV method)

Luteolin has a pyrocatechol structure and has a similar reaction with ammonium molybdate to produce yellow molybdic acid ester. In a nearly neutral medium, yellow molybdic acid ester has an absorption peak at the wavelength of 360 nm. Therefore, molybdate UV spectrophotometric determination of peanut shell luteolin has high sensitivity and is simple, rapid, and reliable (Xu et al., 2008b).

1. Experimental method

   Take two 25-mL colorimetric tubes and add 1 mL of 100 g/L ammonium molybdate solution into them; add a certain volume of luteolin standard solution or test solution into one tube, shake it up and then add HOAc-NaOAc buffer solution to the specific scale; directly add HOAc-NaOAc buffer into another tube to the specific scale. Place the first colorimetric tube in the sample cell and the second in the reference cell; determine their absorbance at the wavelength of 360nm.

   With this method, HOAc-NaOAc is selected as the buffer and the pH of the solution is controlled to be 6.5; 100 g/L ammonium molybdate solution is selected as the reaction solution; luteolin at the mass concentration of 4–40 mg/L shows good linear relationship, and the minimum detection limit is 0.5 mg/L.

2. Sample determination

Take 5 g of peanut shells and add 50 mL of water; leach it in a 60°C water bath for 40 min; carry out vacuum filtration; add 50 mL of water to the filter residue and leach it in a 60°C water bath for 40 min; carry out vacuum filtration; combine the filtrate; take 2 mL of filtrate and dilute it with water to 100 mL in a volumetric flask as luteolin test solution; take 2 mL of the solution and carry out determination with the experimental method.

### 5.2.2.1.2    Flow injection chemiluminescence

Under acidic conditions, luteolin reacts with Ce (IV) and chemiluminescence occurs; rhodamine 6G can enhance the chemiluminescence, thereby establishing a new chemiluminescence analysis method for simple and rapid luteolin determination (Wang et al., 2006).

As shown in Figure 5.27, cerous sulfate solution and rhodamine 6G is pumped into the flow route through channels b and c, respectively; after mixing, it is mixed with a sample solution pumped from channel a to react and generate chemiluminescent signals; detection is carried out by the photomultiplier (PMT) of the flow injection chemiluminescence analyzer, and all data acquisition and processing is carried out by Remax on Windows XP.

It is found through the determination that, under optimized experimental conditions, chemiluminescence intensity within the range of $5.0 \times 10^{-8}$–$2.0 \times 10^{-6}$ g/mL shows a good linear relationship with the concentration of luteolin; the calculation shows that the minimum detectability is $1.0 \times 10^{-8}$ g/mL; the relative standard deviation is 0.87%, with accurate results.

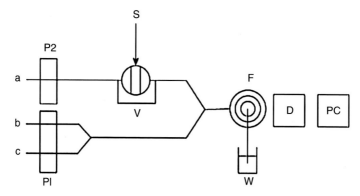

**FIGURE 5.27**    IFFL-D-type flow injection chemiluminescence analyzer. a. $H_2O$; b. Ce(IV) + $H^+$; c. Rh6G; S. sample; P1 and P2: preistaltic pump; V. eight-way value; F. flow cell; W. waste liquid; D. Luminescence detector; PC. computer.

### 5.2.2.1.3 Indirect atomic absorption spectrometry

Basic lead acetate is used as a luteolin complexation precipitating agent to produce yellow insoluble precipitates; determine the $Pb^{2+}$ concentration in the supernatant with indirect atomic absorption spectrometry, thus obtaining a method of indirect determination of luteolin in peanut shells (Xu et al., 2007).

1. Operating parameters of atomic absorption spectrometer
   Wavelength 217 nm; passband width 0.4 nm; lamp current: 3.0 mA; burner height 5 mm.
2. Experimental method
   Transfer a specific amount of luteolin standard solution with a pipette to a 10-mL centrifugal tube; add 1 mL of 1.0 g/L basic lead acetate solution and dilute it with water to the specific scale; keep the reaction for 20 min in a 80°C water bath; carry out centrifugal separation after cooling and draw 2 mL of supernatant to a 10-mL colorimetric tube and dilute it with water to the specific scale; determine the absorbance.
3. Sample determination
   Take 10 g of peanut shells and add 100 mL of water; leach it in a 50°C water bath for 60 min; carry out vacuum filtration; add 50 mL of water to the filter residue and leach it in a 50°C water bath for 30 min; carry out vacuum filtration; combine the filtrate; take 2 mL of filtrate and dilute it with water to 100 mL in a volumetric flask to produce the luteolin test solution; take 2 mL of the solution and carry out determination with the experimental method.

   The linear range of this method is 3.0–25.0 mg/L; the minimum detectability is 1.2 mg/L; pH is controlled in the neutral range; reaction time is 15 min; select 1 mL of 1.0 g/L basic lead acetate solution as the precipitating agent. This method features simple and fast operation, high sensitivity, and low reagent consumption; it can be used for actual sample determination and has accurate results.

### 5.2.2.1.4 HPLC

HPLC is mostly used for the determination of luteolin and is notable for speediness and simplicity. In the Chinese Pharmacopoeia of 2005, this method was established for Duyiwei Capsule and honeysuckle, etc. This method is notable for convenience, reliability, high sensitivity, easy operation, short duration, good degree of separation, and accurate results and can be used as one of the quality control methods of pharmaceutical preparation (Tang et al., 2005).

1. System adaptation experiment of chromatogram
   Column: the most commonly used columns for HPLC are C18 bonded silica gel column and Kromasil ODS column; the method established is verified by C1 bonded silica gel columns of different brands and specifications, and has similar separation effects.

Detection wavelength: as scanned by UV spectrum, luteolin has absorption peak at the wavelength of 254 nm, so 254 nm is selected.

Column temperature: the graph system assessment is basically consistent within the temperature of 25–35°C, so 30°C is selected.

Mobile phase: using methanol-water as the mobile phase will cause serious tailing of the luteolin peak. Considering the acidity of luteolin, we select a methanol–water–acetic acid system and the peak shape is significantly improved. Through a comparative study, when methanol–water–acetic acid (50:50:1) was used as the mobile phase, the peak shape was good at the flow rate of 1 mL/min and the retention time was approximately 9 min; the tailing factor (1.038) reached the requirements for analysis; the calculation with the luteolin peak showed that the theoretical plate number was greater than 5000.

Under such conditions, the luteolin peak tailing factor in the sample chromatogram was 0.987; the theoretical plate number was greater than 5000; the degree of separation from other peaks was higher than 2.0; chromatographic conditions complied with the requirements for content determination.

2. Methodological study

Precision: draw the same volume of luteolin reference solution and determine it six times; the retention time RSD = 0.12%; peak area RSD = 0.96%, complying with the requirements.

Stability: draw the same test sample solution and carry out determinations at the time points of 0, 2, 4, 6, 12, 24, and 48 h, respectively; according to the calculation on the basis of peak area integral value of test sample, RSD = 0.87% (n = 7); the test sample solution was stable within 48 h.

Reproducibility: take six portions of peanut shell samples from the same origin to prepare test sample solution and determine the content; according to the calculation based on the sample content, RSD = 0.63% (n = 6), which complies with the requirements.

Recoveries: accurately weigh luteolin reference to prepare approximately 1 mg/mL solution; accurately draw tow portions each of the solution of 3, 5, and 10 mL, and place them in triangular flasks; then dry them and add accurately weighed peanut shell samples to prepare the test sample solution; determine the content; the recoveries of luteolin were 100.5–102.1%; the average recovery was 101.2%; RSD = 0.72% (n = 6).

3. Sample determination

Crush dry peanut shells and sieve with a 60-mesh sieve; keep for later use. Take peanut shell sample of approximately 2 g; accurately weigh it and place it in a triangular flask; accurately add 50 mL of methylate; weigh it and leave it standing overnight; carry out ultrasonic extraction for 30 min; cool it down and supplement the lost weight; filter it and take the subsequent filtrate; filter it with 0.45 μm microporous membrane to get the solution for test.

## 5.2.2.2 Preparation Process and Technique

### 5.2.2.2.1 Extraction methods

The methods for luteolin extraction mainly include water extraction, ethanol–water solvent extraction, dilute alkaline alcohol extraction, microwave method, ultrasonic method, and extraction by supercritical fluid. Since luteolin is generally present in the form of aglycone, it can be extracted with an organic solvent with high polarity, such as methanol or ethanol. Considering safety, environmental, and other factors, laboratories usually extract luteolin with high-concentration ethanol or water. 60% acetone, 70% methanol, 70% ethanol, and water were used as extracting agents to determine and compare the luteolin content of the extract with HPLC. It was found that luteolin content of extract with organic solvents, especially with ethanol, is higher and that of extract with water is the lowest (Li, 2005). Table 5.5 shows the process of luteolin extraction with different methods. It can be seen that the solvent extraction

**TABLE 5.5** Comparison Between Different Methods of Luteolin Extraction Process

| Extraction Method | Raw Material | Extraction Process | Extraction Rate (%) | Extraction Amount (mg/mL) | Content (%) | References |
|---|---|---|---|---|---|---|
| Solvent extraction | Peanut shells | 70% ethanol; solid–liquid ratio 1:10; three times' extraction and 2 h each time | 90 | — | 0.354 | Huang et al., 2008 |
| Ultrasonic method | Peanut shells | 70% ethanol; solid–liquid ratio 1:10 (g/mL); three times' extraction and 15 min each time | 3.7 | — | 0.31 | Xiao et al., 2008 |
| Ultrasonic method | Peanut shells | 30 times 70% ethanol; 30 min's extraction at 60°C | — | 1.601 | — | Li, 2010 |
| Microwave method | Peanut shells | Solid–liquid ratio 1:20; 80% ethanol; microwave power 900 W and extraction duration 70 s | — | — | 0.969 | Yang et al., 2009b |

Note: "—" means no data.

method, ultrasonic method, and microwave method have been studied more, and there are large differences between the yields of luteolin extracted with the three methods.

#### 5.2.2.2.2  Purification methods

There are many methods for luteolin separation and purification, including column chromatography, TLC, lead salt precipitation, boric acid complexation, gradient extraction method, and solvent extraction, as well as HPLC, droplet countercurrent chromatography, gas chromatography method, and microemulsion TLC, etc. that have been more frequently used in recent years. In the separation of luteolin from peanut shell methanol extract using the polyamide chromatography method, the optimal conditions for adsorption and elution of luteolin on the polyamide column is 70–75% ethanol–aqueous solution elution. The analysis shows that luteolin in peanut shells is present in its completely free form and its content is approximately 0.3% (Wang et al., 1997); if peanut shell alkaline extract is extracted with ethyl acetate and concentrated under reduced pressure, the luteolin purity (in mass fraction) in crude luteolin can be increased from 2.47% to 19.50% (Ding et al., 2005); 95% methanol peanut shell extract solution is purified with column chromatography, and pure luteolin is obtained through recrystallization; peanut shell luteolin is purified through solid phase extraction with molecular imprinting technique, and its purity can be nearly 20 percentage point higher than that of luteolin extracted with the silica gel column separation method (Xiao et al., 2010).

### 5.2.2.3  Application and Development Prospects

#### 5.2.2.3.1  Current situation of application

Today, the pharmacological activity of luteolin has been recognized and Chinese patent medicines have been developed, such as Duyiwei Capsule, Maishu Capsule, and Xiaochunhua Oral Solution, etc.

1. Duyiwei Capsule

   It is made from common lamiophlomis and the medicinal ingredients are flavonoids, saponins, sterols, amino acids, and trace elements, among which the luteolin content shouldl not be lower than 0.80 mg per capsule. Functions and indications: invigorating the blood circulation and relieving pain; removing extravasated blood; mainly used for incision pain and bleeding after surgical operations, trauma fractures, sprained muscles, rheumatism, uterine bleeding, dysmenorrhea, sore gums, and bleeding.
2. Maishu Capsule

   It is made from peanuts and the medicinal ingredients are flavonoids, mainly luteolin. Functions and indications: blood-lipid lowering and for hyperlipidemia treatment.
3. Xiaochunhua Oral Solution

   It is a compound prescription made from Xiaochunhua and other herbal medicines according to folk traditions and the active ingredient is luteolin.

Functions and indications: removing heat from the liver, removing rheumatism, and detoxifying; clinically for the treatment of respiratory diseases.

### 5.2.2.3.2 Market development prospects

It is found in practice that luteolin extracted from plants clinically has antitussive, expectorant, and anti-inflammatory effects, and antibacterial, antiviral, anticancer, and cholesterol-lowering effects in vivo. It is a material used for both food and medicine with a great potential for development. In addition, the antitumor mechanism of luteolin can not only directly act on tumor cells to interfere with cell metabolism and inhibit cell growth, induce or sensitize apoptosis, but can also exert antitumor effects by enhancing the immune function of the body. Therefore, luteolin has a multitarget antitumor effect and a clear structure; it is a very promising potential anticancer drug and can be used as a new drug for in-depth study.

Peanut shell is currently a renewable plant resource with high luteolin content and rich sources. On one hand, it is an ideal raw material for luteolin production but not yet effectively developed as an agricultural by-product; on the other hand, it is an important issue to accelerate the comprehensive development and deep processing of agricultural products, increase their added value, and establish a recycling green ecological agriculture. We can extract luteolin or produce luteolin-containing products from crop waste peanut shells, and improve the experimental yield and reduce cost by improving the extraction, separation, and purification methods. That will effectively improve the added value of peanut shells and contribute to better economic and social benefits.

## 5.2.3 Resveratrol

### 5.2.3.1 Detection Techniques

#### 5.2.3.1.1 HPLC

HPLC is currently the most commonly used method for resveratrol detection and is suitable for the detection of resveratrol in parts of peanut plants and products. This method is featured by high sensitivity, good reproducibility, and reliability. The current national standard GB/T24903-2010 adopts HPLC in the peanut resveratrol determination. The specific conditions are column C18 (150 mm × 3.9 mm, 4 μm); mobile phase acetonitrile: water: acetic acid = 25:75:0.09; flow rate 0.7 mL/min; UV light detection wavelength 306 nm. Figs. 5.28 and 5.29 show the peaks when the author measures the standard product and the sample with column C18 of 150 mm × 4.6 mm (5 μm) by referring to the national standard.

#### 5.2.3.1.2 Gas chromatography

Gas chromatography is a method using gas as the mobile phase. Gas chromatography-mass spectrometry (GC-MS) is generally used for qualitative and

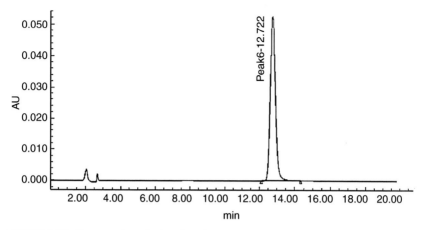

**FIGURE 5.28** Resveratrol standard product chromatogram.

quantitative analysis and determination of resveratrol. Gas chromatography-features high sensitivity and accuracy, simple and rapid operations, and low sample consumption, but the sample pretreatment is complicated. Generally, a silanizing derivation is required and the instrument is expensive, which has limited its application (Li et al., 2011). Specific steps of the GC-MS method for the determination of resveratrol in peanuts are as follows: extract $N_2$ purged to dry → add derivatization reagent (BSTFA) → reaction after shaking → draw sample solution and inject it into the sample. Operating conditions for peanut resveratrol determination with GC-MS method: chromatographic condition: inlet temperature 280°C; temperature programming for column; beginning temperature 10°C; keep it for 10 min and raise the temperature to 300°C at the rate of 10°C/min; keep the temperature constant for 10 min; use helium as the carrier

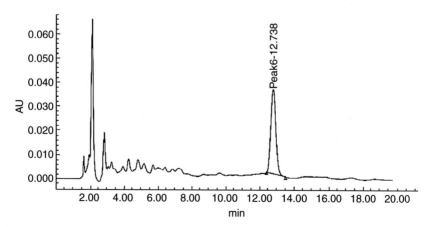

**FIGURE 5.29** Resveratrol sample hromatogram.

gas; flow rate 1.0 mL/min; split ratio 30:1; MS conditions: inlet temperature 280°C; E1 ion source 70 eV; ion source temperature 230°C; quadrupole temperature 106°C; electron multiplier voltage 1500 V; mass scanning range 30–50 0u (Liang et al., 2006).

### 5.2.3.1.3 TLC

The principle of TLC is to use the fluorescence intensity of TLC spots (components) or fluorescent quenching degree of dark spots on fluorescent thin layer plate for quantification (Huang, 2007), including TLCS and micellar thin-layer chromatography (Huang et al., 2010b). Resveratrol and some derivatives show blue and purple light in UV light with a wavelength of 366 nm. By taking advantage of this property, TLC can be used to separate and determine resveratrol and its derivatives (Pezet and Perret, 2003). TLC is simple, rapid, sensitive, and low cost. The conditions for the determination of resveratrol in peanut roots and stems using thin-layer fluorescence scanning method are as follows: using chloroform: ethyl acetate: formic acid = 8:2:(0.2–0.35) as developing agent; slit size 3 mm $\times$ (0.145–0.45) mm; carry out determination by scanning at 335 nm; resveratrol showed a good linear relationship within the range of 21.6–108.0 ng; the average recovery rate was 103.46% and the relative standard deviation was 0.66% (Han et al., 2006, 2010).

### 5.2.3.1.4 Fluorescence spectroscopy

The principle of fluorescence spectroscopy is that resveratrol reacts with $Al^{3+}$ to produce complex compounds and undergoes a color reaction, and the intensity of the emitted fluorescence is proportional to the concentration of resveratrol. According to this principle, fluorescent spectrometry can be applied in resveratrol determination. Features of this method are easy operation, low cost, good accuracy, high sensitivity, and good reproducibility (Wang et al., 2009c). The optimal conditions for peanut resveratrol determination with fluorescence spectrophotometer are as follows: excitation wavelength and emission wavelength are respectively 324.06 and 400.00 nm; the fluorescence intensity of resveratrol shows a good linear relationship with the concentration within the range of 0–1.68 $\times$ $10^{-5}$ mol/L, and the minimum detection limit is 8.14 $\times$ $10^{-10}$ mL/L (Zhang, 2006a, 2006b).

### 5.2.3.1.5 Other detection methods

In recent years, some new detection methods have been applied in the detection of resveratrol and its derivatives, such as the electrochemical luminescence method (Tang et al., 2002), the quadratic differential simple oscilloscope voltammetry (Zheng et al., 2002), and the in vivo fluorescence method (Hollecker et al., 2009). With an improvement in analysis techniques and equipment updates, the detection methods of resveratrol and its derivatives will become simpler, faster, and more accurate.

## 5.2.3.2 Preparation Process and Technique

### 5.2.3.2.1 Preparation methods

1. Organic solvent extraction method

   The principle of organic solvent extraction is that "substances with similar polarity are compatible," that is, organic solvent with similar polarity is used for the extraction of organic substances from vegetable cells. Organic solvents commonly used for peanut resveratrol extraction are methanol, ethanol, and ethyl acetate. Resveratrol has high polarity and is soluble in high-polarity solvent, so methanol or ethanol will have better extraction effect. Commonly used organic solvent extraction methods are Soxhlet extraction, reflux method, and percolation method. Organic solvent extraction method is widely used and its features include low investment in equipment, simple techniques, and low cost, but it has many steps, the extract has a complex composition, and separation and purification are required to obtain a finished product (Hu, 2006). Table 5.6 shows the comparison between the effect of different solvent extraction methods, and it can be seen that ethanol solution is mostly used as the solvent in practical application.

2. Enzymatic method

   Enzymatic extraction involves destroying plant cell walls by enzymolysis to fully release the cell contents, which are conducive to effective extraction with organic solvent. A commonly used emzyme is cellulase, as shown in Table 5.7. In addition, the enzymatic method can convert plant polydatin into resveratrol, thus improving the extraction rate of resveratrol (Cai, 2010). Enzymatic extraction features mild conditions, short duration, and high yield, but has high cost and requires rigorous conditions for reaction (Zhang et al., 2009d). The specific operation of enzymatic extraction is as

**TABLE 5.6** Effect of Solvent Extraction Methods

| Raw Material | Extraction Solvent | Extraction Volume (Extraction Rate) | References |
|---|---|---|---|
| Peanut roots | 65.00% ethanol | 0.012% (extraction rate) | Zhang et al., 2009b |
| Peanut roots | 61.48% ethanol | 0.320% (extraction rate) | Shan et al., 2008 |
| Peanut roots | 95.00% ethanol | 179.500 µg/g | Chen, 2004 |
| Peanut vine | 60.00% ethanol | 855.050 µg/g | Wang et al., 2009a |
| Giant knotweed | 60.00% ethanol | 0.860% (extraction rate) | Zhao et al., 2010 |
| Giant knotweed | 80.00% ethanol | 0.430% (extraction rate) | Lan, 2007 |
| Grape vines | Ethyl acetate | 10.676 µg/g | Sun and Zhao, 2011 |
| Wine grape pomace | 80.00% ethanol | 94.200 µg/g | Bi, 2006 |

**TABLE 5.7 Enzymatic Extraction Effect**

| Raw Material | Enzyme Variety | Extraction Rate (%) | References |
| --- | --- | --- | --- |
| Peanut red coat | Cellulase | 1.25 | Wang et al., 2009b |
| Peanut roots | Cellulase | — | Zhang et al., 2009d |
| Giant knotweed | Enzyme | — | Yan, 2010 |
| Giant knotweed | Cellulase | 0.92 | Lan, 2007 |

Note: "—" means no data.

follows: raw material peanuts → cleaning → drying → crushing → adding water/buffer → adjust pH → enzymolysis → enzyme inactivation → add solvent for extraction → filtration → resveratrol extract.

3. Microbial fermentation extraction method

   The microbial fermentation method uses peanut tissue as the culture medium and uses microbial fermentation to produce enzymes and extract resveratrol by enzymolysis, and further induces the generation of resveratrol as a phytoalexin. This method can not only greatly reduce production cost, improve extraction volume, and product quality, but also can provide a new way to resveratrol industrialization. Resveratrol was extracted by fermenting peanut roots with *Aspergillus niger* under the conditions of ethanol concentration 64%; extraction temperature 55°C, extraction duration 60 min, and solid–liquid ratio of 1:8. The yield of resveratrol was 0.191% (Zhang et al., 2009c).

4. Alkali extraction method

   Resveratrol is weakly acidic and can produce salt under alkaline conditions; and its water solubility increases significantly. The principle of the alkali extraction method is to take advantage of this property of resveratrol and make it react with certain inorganic bases or basic salts to form phenolate and then dissolve in the system; it is then precipitated by adusting the pH of the solution, to achieve the purpose of resveratrol extraction. Alkaline extraction does not require large amounts of organic solvent and has good selectivity, thus producing a higher content of active ingredient (Hu, 2006).

5. Supercritical extraction method

   The supercritical extraction uses supercritical fluid as the solvent and extracts and separates a mixture by taking advantage of its high permeability and high solubility. Currently, $CO_2$ is mostly used as the solvent. This method has no solvent residue, and the extraction solvent can be recycled; high temperature operation is avoided, so it is suitable for heat-sensitive substances (Hu, 2006). However, supercritical fluid extraction requires high one-time investment and it has high maintenance costs and is not suitable for mass production applications, with a low extraction efficiency. Extraction duration, temperature, pressure, and other factors need to be considered in peanut resveratrol extraction using the supercritical extraction method.

6. Microwave-assisted extraction method

   The principle of microwave-assisted extraction method is to use a magnetron to generate UHF rapid vibrations of hundreds of millions of times per second to cause collision and extrusion between molecules creating thermal effects; cell expansion and rupture will speed up resveratrol leaching from peanut cells (Lan, 2007). The microwave-assisted extraction method features a short cycle, low energy consumption, and low organic solvent consumption; it can also improve the yield and extract purity; it is conducive to the extraction of thermally unstable substances and can prevent sample decomposition due to prolonged high temperature (Miao, 2008). The pecific operation of this method is as follows: raw material peanuts cleaning → drying → crushing → adding extraction solvent → processing with microwave → centrifugation and obtaining the supernatant. The optimal conditions for resveratrol from peanut roots with microwave-assisted extraction method are as follows: output power 400 W, microwave irradiation duration 4.5 min, extraction temperature 80°C, ethanol volume fraction 60%, and solid–liquid ratio 1:14 (g/mL) (Zhang et al., 2009a).

   Table 5.8 shows results from the relevant literature on resveratrol extraction using microwave-assisted extraction method in recent years.

7. Ultrasound-assisted extraction method

   The principle of ultrasound-assisted extraction is to use ultrasound to generate strong disturbance, crushing, mixing, and other mechanical effects, cavitation effect, and thermal effect, to increase the frequency and speed of molecular movement of substances, to increase the penetrating power of solvent and accelerate the breakdown of the cell walls, thus promoting resveratrol extraction (Li et al., 2006). Ultrasound assisted extraction can shorten the extraction duration, reduce organic solvent consumption, and improve the yield of resveratrol. The ultrasound-assisted extraction process is as follows: raw material peanuts → washing → drying → crushing → adding extraction solvent → ultrasonic extraction → filtration → resveratrol extract. Peanut roots, stems, and leaves were used as raw materials to extract resveratrol with ultrasound-assisted extraction method under the conditions

**TABLE 5.8** Microwave-Assisted Extraction Effect

| Raw Material | Microwave Rate (W) | Extraction Rate (%) | References |
|---|---|---|---|
| Peanut red coat | 300 | 0.038 | Zhou et al., 2010 |
| Peanut roots | 400 | 0.376 | Zhang et al., 2009a |
| Giant knotweed | 510 | 1.02 | Lan, 2007 |

of ultrasonic frequency 15.33 kHZ, duty cycle 1:2, extraction duration 6 min, solid–liquid ratio 1:25, and ethanol concentration 70%; the resveratrol extraction volume was 885.96 µg/g (Zou et al., 2011). Table 5.9 shows the results from the relevant literature on resveratrol extraction using the ultrasound-assisted method in recent years, and it can be seen that this method is notable for a short production cycle and high extraction rate.

8. High-pressure microwave-assisted extraction method

High-pressure microwave-assisted extraction is a new extraction method combining high temperature and microwaves. The principle is to apply a certain pressure on raw material at a normal temperature or low temperature (<100°C) to destroy the tissues of raw material by taking advantage of the fact that macromolecular substances (such as protein and starch, etc.) are sensitive to pressure and easily destroyed while micromolecular substances are not affected; in addition, the heating effect of microwaves will further accelerate the leaching of resveratrol in peanut tissue cells (Liao et al., 2010; Xu et al., 2008a). The features of this method are low energy consumption, short reaction period, and high extraction rate. The optimal conditions for resveratrol extraction from peanut roots with high-pressure microwave-assisted extraction method are as follows: 72% ethanol concentration; solid–liquid ratio 1:30 (m/V); pressure 208 MPa; dwell time 4 min; microwave power 320 W; microwave duration 90 s. The extraction rate of resveratrol under these conditions is 73.76% (Chen et al., 2013).

### 5.2.3.2.2 Purification technique

1. Column chromatography

Column chromatography is the most commonly used method for resveratrol separation and purification. This method achieves separation by taking

**TABLE 5.9** Ultrasound-Assisted Extraction Effect

| Raw Material | Extraction Duration (min) | Extraction Volume (Extraction Rate) | References |
|---|---|---|---|
| Peanut shells | 7 | 12.990 µg/g | Liu et al., 2009 |
| Peanut stems | 20 | 0.11% | Xu et al., 2009 |
| Peanut roots, stems and leaves | 6 | 885.960 µg/g | Zou et al., 2011 |
| Peanut red coat | 20 | 0.04% | Liu et al., 2005 |
| Peanut roots | 5.5 | 3.96% | Li et al., 2009 |
| Peanut kernels | 40 | 2.630 µg/g | Liu et al., 2006b |
| Giant knotweed | 10 | 1.12% | Lan, 2007 |

advantage of different adsorption, distribution, and affinities of mixture components in different phases. Stationary phases for column chromatography include silica gel and polyamide gel. Silica gel column chromatography has a low cost and high separation capacity, but it has complex operations, low yield, many impurities in product, and recrystallization and further purification is needed. Macroporous resin separation has good adsorption selectivity, reusability, mild desorption conditions, and low cost (Cai, 2010). Resins that have good resveratrol separation effect include NKA-9 (Cai, 2010), H1020 (Yang et al., 2009a), AB-8 (Ma, 2006), and HP500 (Cao, 2001). The author's research team studied the physical properties, adsorption, and desorption properties of different models of resin and determined polyamide to be the best adsorbent (Table 5.10).

Su et al. (2004) purified crude extract of resveratrol with medium-pressure column chromatography. After recrystallization, the resveratrol purity reached above 99%; the resveratrol extract was purified with a molecular imprinting technique; after going through the chromatographic column with added imprinted polymer, and elution with a methanol solution with 5% acetic acid, product with high purity could be obtained (Xiang et al., 2005); the purity of resveratrol extracted with macroporous resin–silica gel column chromatography reached 87% (Ma, 2006).

2. High-speed countercurrent chromatography

High-speed countercurrent chromatography is a liquid–liquid partition chromatography without any solid support or carrier. Its principle is to achieve separation by taking advantage of different partition coefficients of mixture components in two phases (Xin et al., 2009). High-speed countercurrent chromatography shows good separation effect, high product purity, no carrier adsorption and contamination of samples, large preparation capacity,

**TABLE 5.10** Physical Properties, Adsorption, and Desorption Properties of Resin

| Resin Model | Particle Size (mm) | Specific Surface Area (m²/g) | Appearance | Polarity | Adsorption Rate (%) | Desorption Rate (%) |
|---|---|---|---|---|---|---|
| AB-8 | 0.30–1.25 | ≥480 | Milky | Weak polarity | 87.64 ± 1.50 | 55.09 ± 2.11 |
| S-8 | 0.30–1.25 | ≥100 | Yellowish | Weak polarity | 96.49 ± 0.64 | 66.30 ± 0.87 |
| NKA-9 | 0.30–1.25 | ≥250 | Milky | Polarity | 95.05 ± 1.66 | 54.09 ± 2.53 |
| Polyamide | 30–60 (mesh) | 120–360 | White | Polarity | 96.84 ± 0.13 | 69.67 ± 1.19 |

and low solvent consumption (Cai, 2010). The purity of resveratrol extracted with countercurrent chromatography can be up to 99% (Chen et al., 2000), making it a new separation method with good application prospects. In resveratrol extraction from crude peanut extract with high-speed countercurrent chromatography, mobile phase system, flow rate, speed, and other factors need to be considered.

3. Membrane separation method

   Membrane separation method is solution separation by taking pressure difference on both sides of the membrane as the driving force and mechanical gradation as the basis. It can be used to isolate substances in a solution on the basis of molecular mass, thereby achieving the purpose of classification, separation purification, and concentration. This method has low production cost, involves no hazardous solvents, and is clean and safe (Chen, 2011). It can reduce resveratrol degradation under normal temperature, with a simple process, easy operation, and it is easy to achieve in industrial production. Membrane flux, operating pressure, temperature, duration, and other factors need to be considered in the resveratrol extraction from peanut roots using the membrane separation method.

### 5.2.3.3 Application and Development Prospects

#### 5.2.3.3.1 Current situation of application

Resveratrol has many functional activities, such as anticancer, anti-inflammation, blood fat lowering, and cardiovascular disease prevention, and is listed as one of the "100 most effective antiaging substances" by "Anti-aging Holy Scriptures" of the United States (Bi, 2006). It is natural, nontoxic, and has been widely used in food, health, medicine, cosmetics, and other industries. Resveratrol can be used as an additive to food and health care products to improve health, or as a raw material for the treatment of skin diseases, hepatitis, cardiovascular diseases, or as a natural antioxidant in the cosmetic industry to effectively keep moisture, remove wrinkles, and prevent radiation. For example, Platinum Ultimate Revitalizing Serum of Lander and INNER LIGHT new makeup series of the world's 15th skin care brand AVEDA contain resveratrol (Li et al., 2008).

In the United States, resveratrol is used as a dietary supplement and the recommended daily intake for adults is 4 mg; in Japan, resveratrol extracted from plants is used as a food additive (Shu and Chen, 2003) and Natrol Resveratrol of Canada has developed resveratrol health care products. In China, resveratrol-containing plant extracts are made into lipid lowering, beautifying, slimming, and anticancer capsules, such as "Tiens Huolikang Capsules," "Zijin Capsules," and "Nabei Probiotic Capsules." Resveratrol is also added to various wines to prepare new types of low-alcohol and high-resveratrol healthy table wines that have good cardiovascular disease preventive effects (Hu and Zhang, 2002). Please refer to Table 5.11.

**TABLE 5.11** Chinese and Foreign Resveratrol Producers

| Sample | Dosage (ug/mL) | HepG2 Inhibition Rate (%) | IC50 Value | AGS Inhibition Rate (%) | IC50 Value |
|---|---|---|---|---|---|
| Resveratrol nanoliposome | 100 | 73.4882 | 29.769 | 57.3416 | 2.8897 |
| | 50 | 75.5436 | | 72.6429 | |
| | 25 | 77.3607 | | 68.6244 | |
| | 12.5 | 34.3461 | | 64.9665 | |
| | 6.25 | 25.6181 | | 55.383 | |
| Resveratrol +blank nanoliposome | 100 | 74.8585 | 52.3977 | 54.7656 | 57.2414 |
| | 50 | 69.9434 | | 51.3138 | |
| Resveratrol phosphate buffer solution | 25 | 31.2184 | 74.844 | 46.9861 | 146.4766 |
| | 12.5 | 5.0045 | | 32.0969 | |
| | 6.25 | 9.5025 | | 17.7743 | |
| | 100 | 59.5174 | | 45.6981 | |
| | 50 | 42.5082 | | 44.9768 | |
| | 25 | 25.0521 | | 35.188 | |
| | 12.5 | 19.4519 | | 31.6847 | |
| | 6.25 | 12.7495 | | 29.83 | |

| Product | Purity (%) | Enterprise (Origin) |
|---|---|---|
| Resveratrol | 50, 98 | Xi'an Dongju Biotechnology Co., Ltd. |
| Resveratrol | 50, 98, 99 | Shanghai DND Pharm-Technology Co., Inc. |
| Resveratrol | >98 | Shanghai Lanyuan Biotechnology Co., Ltd. |
| Resveratrol | 50, 98 | Baozetang Pharmaceutical Co., Ltd. |
| Resveratrol | >98 | Xi'an Tianrui Biological Technology Co., Ltd. |
| Resveratrol | 50, 98, 99 | Xi'an Tenghua Biotechnology Co., Ltd. |
| Resveratrol | 50–98 | Xi'an Sanwei Biotech Co., Ltd. |
| Resveratrol | 50, 98, 99 | Xi'an Xiaocao Botanical Development Co., Ltd. |
| Resveratrol | 10, 20, 50, 98 | West Kaimeng Biotechnology Co., Ltd. |
| Resveratrol | 98, 99 | Shaanxi Ciyuan Biotechnology Co., Ltd. |
| Resveratrol | 98 | Xi'an Guanyu Biotechnology Co., Ltd. |
| Resveratrol | 50, 98, 99 | Beijing Krohesin Technology Co., Ltd. |
| Resveratrol | >98 | Chengdu Must Bio-Technology Co., Ltd. |

*(Continued)*

**TABLE 5.11 Chinese and Foreign Resveratrol Producers (*cont.*)**

| Product | Purity (%) | Enterprise (Origin) |
|---|---|---|
| Resveratrol | 20, 35, 99 | Chengdu Jule · Zhuoyue Herbal Consortium |
| Resveratrol | 20–99 | Sichuan Qilin Biotechnology Co., Ltd. |
| Resveratrol | 20, 50, 80, 98 | Chengdu Weihui Biotechnology Co., Ltd. |
| Resveratrol | ≥ 50, 80, 98 | Hubei Yuancheng Pharmaceutical Co., Ltd. |
| Resveratrol | 50, 90, 98 | Wuhan Shuyuan Technology Co., Ltd. |
| Resveratrol | 50, 98 | Wuhan Xianbao Bio-Chemical Co., Ltd. |
| Resveratrol | 98 | Wuhan Jinnuo Chemical Co., Ltd. |
| Resveratrol | 50, 98, 99 | Hunan Hongjiang Huaguang Biological Co., Ltd. |
| Resveratrol | 20, 30, 40, 50, 98 | Zhangjiajie, Xianghui Biological Co., Ltd. |
| Resveratrol | — | Changzhou Songhong Import and Export Co., Ltd. |
| Resveratrol | 50, 98 | Yongzhou Yidong Biotechnology Co., Ltd. |
| Resveratrol | ≥98 | Taizhou Auto Chemical Co., Ltd. |
| Resveratrol | 99 | Taizhou Time Biochemical Technology Co., Ltd. |
| Tiens Huolikang Capsule | ≥0.75 | Daqing, Heilongjiang Province |
| Resveratrol Complex Nutritional Capsule | — | Sail Health Natural Products Inc. |
| Paradise Herb Resveratrol Life Extension | — | United States |
| Natrol Resveratrol | — | Canada |

Note: "—" means no data.

At present, there are more than 10 domestic producers of natural pure resveratrol and plant extracts, among which Hunan Hongjiang Huaguang Biological Co., Ltd. is the largest dealer in China. It has an annual resveratrol productivity of more than 10 tons, including more than 40 resveratrol-based products, which are mostly for export (Rao, 2008).

### 5.2.3.3.2 Market development prospects

At present, there are more than 1000 high-end resveratrol products (including drugs and health products) approved in the European and American market. The price of resveratrol with a purity of above 95% is approximately 8000–10,000 US dollars/kg (Lan, 2007). In the next 8 years, the sales volume of resveratrol products is expected to reach 500–800 million US dollars, forming a huge industry (Huang et al., 2006). According to relevant statistics, currently the global resveratrol production capacity is apprximately 50–60 t, but the market demand is more than 100 t and will continue to increase in the coming years. Therefore, there is a large gap in the market demand of resveratrol-related products and this industry has very good market prospects (Weidong Ma, 2006).We studied the in vitro antitumor effect of resveratrol nanoliposome from peanut roots and found that resveratrol of peanut roots at low concentration can significantly inhibit the proliferation of HepG2 and AGS cells, with good antitumor effect (Table 5.12). This has widened the way to a further market development of resveratrol.

The production of resveratrol from peanut roots, stems, and leaves can not only reduce environmental pollution, promote waste utilization, and improve product added value but also can regulate the health level of the population.

**TABLE 5.12** In-Vitro Antitumor Effect of Resveratrol Nanoliposome

| Sample | Dosage (ug/mL) | HepG2 Inhibition Rate (%) | IC50 Value | AGS Inhibition Rate (%) | IC50 Value |
|---|---|---|---|---|---|
| Resveratrol nanoliposome | 100 | 73.4882 | 29.769 | 57.3416 | 2.8897 |
| | 50 | 75.5436 | | 72.6429 | |
| | 25 | 77.3607 | | 68.6244 | |
| | 12.5 | 34.3461 | | 64.9665 | |
| | 6.25 | 25.6181 | | 55.383 | |
| Resveratrol + blank nanoliposome | 100 | 74.8585 | 52.3977 | 54.7656 | 57.2414 |
| | 50 | 69.9434 | | 51.3138 | |
| Resveratrol phosphate buffer solution | 25 | 31.2184 | 74.844 | 46.9861 | 146.4766 |
| | 12.5 | 5.0045 | | 32.0969 | |
| | 6.25 | 9.5025 | | 17.7743 | |
| | 100 | 59.5174 | | 45.6981 | |
| | 50 | 42.5082 | | 44.9768 | |
| | 25 | 25.0521 | | 35.188 | |
| | 12.5 | 19.4519 | | 31.6847 | |
| | 6.25 | 12.7495 | | 29.83 | |

Therefore, it will have good economic and social benefits to make great efforts to develop various products rich in peanut resveratrol. In future studies of resveratrol, on the one hand, resveratrol content of peanut can be improved by genetic engineering, microbial fermentation, and other means; on the other hand, resveratrol yield can be improved posttreatment by the improvement of extraction techniques and equipment.

## 5.3 PEANUT POLYSACCHARIDE PROCESSING TECHNOLOGY

Carbohydrates are the main source of energy for maintaining the vital activities of organisms. They not only are nutrients but also have a special physiological activity. Carbohydrates in peanuts mainly include active polysaccharides, starch, oligosaccharides, and dietary fiber. These components are distributed in peanut by-products, such as peanut shells and peanut meal. Han et al. (2011) have found through determination that the oligosaccharide content of peanut meal is 12.67%; the dietary fiber content of peanut shell is the highest, accounting for 30% of the peanut mass, followed by XOS content, 13%. The data suggest that peanuts and their by-products contain a lot of carbohydrates. At present, because of its low production costs and rich sources, peanut meal is mainly used as fuel and feed (Liu and Ma, 2005), and has not been fully developed and utilized. Therefore, making clear the composition and contents of carbohydrates in peanut meal is of great significance for the further development and utilization of peanut meal for good economic benefits.

China's annual production of peanut meal from oil extraction is about 3.3 million tons. Peanut polysaccharides are the second largest component in peanut meal and are not yet developed and utilized. According to the relevant literature, peanut polysaccharides have the functional activities of liver protection and oxidation resistance. Systematic research on peanut polysaccharide preparation, functional assessment, and structural identification will help make full use of the by-products of peanut processing, improve product added value, and promote the technological progress of the peanut industry and the functional food industry.

We carried out an in-depth study of the composition of carbohydrate from peanut meal and optimized, using response surface methodology, the optimal process of peanut meal polysaccharide extraction with hot water and further conducted separation and purification to get purified components, in the hope of providing theoretical guidance for the industrial production of peanut meal polysaccharide and lay a solid foundation for subsequent studies of functional activity.

### 5.3.1 Composition of Carbohydrates in Peanut Meal

The total sugar content of peanut meal is determined to be 32.45% using the sulfuric acid–phenol method. In addition, national standard methods were used to determine the monosaccharide, oligosaccharide, and crude fiber contents of peanut meal. According to standard AOAC996.11, AOAC996.11 the Megazyme

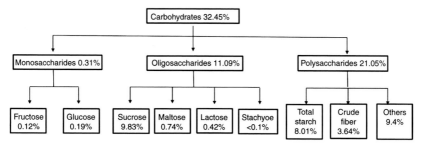

**FIGURE 5.30**    Composition of carbohydrates in peanut meal.

kit method was used to determine the total starch content. The results are shown in Fig. 5.30. The monosaccharide content of peanut meal is low, only 0.31%. The monosaccharides mainly include fructose and glucose, with the contents of 0.12 and 0.19%, respectively. Oligosaccharide is the major carbohydrate in peanut meal and mainly includes sucrose, maltose, lactose, and stachyose. Among these, sucrose is the major sweetening substance, with a content of 9.83%. This is consistent with the literature (Bryant et al., 2004). Maltose and stachyose content is low, only 0.74% and 0.42%, respectively; the lactose content is less than 0.1%. Polysaccharides show great variety and have complex compositions. Results show that the total starch content of peanut meal is 8.01% and crude fiber content is 3.64%. These two substances are polysaccharides with no biological activity. Taking the total sugar content minus the content of monosaccharide, oligosaccharide, starch, and cellulose, we can determine that the content of the biologically active heteropolysaccharides is 9.4%. This provides a direction for further study of the biological activity of peanut meal polysaccharides.

## 5.3.2    Optimization of the Preparation Process of Peanut Meal Carbohydrates

Cold-press peanut meal is the by-product of cold-press peanut oil. Low-temperature oil extraction technique causes little damage to the active ingredients, such as polysaccharides and protein, in peanut meal, so cold-press peanut meal has good functional characteristics and high utilization value. Polysaccharides with natural activity usually refer to nonstarch polysaccharides, which have a wide range of biological functions and have been successfully used in medicine, and clinical and health foods (Tu, 2012). According to existing research reports, peanut meal polysaccharides have hypoglycemic, liver protection, and antioxidant effects (Wei Yang and Chengming Wang, 2010; Yao et al., 2011; Han, 2010). Therefore, it is of practical significance and helps improve the utilization rate of cold-press peanut meal to conduct in-depth extraction and purification of peanut meal polysaccharides.

As peanut meal has a high protein content, protein should be removed from peanut meal before polysaccharide extraction. At present, the protein removal methods used during the refining process of plant polysaccharides mainly include sevag method, TCA precipitation method, isoelectric point method, and enzymatic method. To avoid contamination of hazardous reagents and for the purpose of large-scale industrial production, we selected the protease enzymolysis method for protein removal.

There are many methods for plant polysaccharide extraction, such as hot water extraction, dilute alkali extraction, dilute acid extraction, enzymatic extraction, ultrafiltration method, microwave-assisted extraction, ultrasound-assisted extraction, ultrafine grinding technique, and supercritical fluid extraction. The hot water extraction method is low cost, with a simple operation, has mild functioning conditions for polysaccharides, low equipment requirements, and no solvent contamination. More importantly, it is suitable for large-scale industrial production. Considering the above advantages, we used the hot water extraction method for peanut meal polysaccharide extraction and used RSM to analyze and optimized the process parameters.

### 5.3.2.1 Optimization of Peanut Meal Polysaccharide Extraction With Hot Water

For the full development and utilization of the potential commercial value and nutritional value of peanut by-products, cold pressed peanut meal is used as the raw material to extract peanut polysaccharides using hot water extraction method.

The process is as follows: Defatted peanut meal powder 5 g → enzymolysis with neutral protease → enzymolysis with compound protease → hot water extraction → centrifugation for 20 min at 4500 rpm to take supernatant → rotary evaporation → precipitate in thrice-volumed absolute ethanol → drying → redissolve the polysaccharide sample and determine its content.

On the basis of studying the influence of temperature, duration, and solid–liquid ratio on polysaccharide extraction rate, Box-Behnken is used for establishing a mathematical model; select three significant levels for each factor and use the peanut meal polysaccharide extraction rate as a response value to optimize the preparation process of peanut meal polysaccharide extraction.

The formula for the polysaccharide extraction rate is as follows:

$$\text{Polysaccharide extraction rate (\%)} = \text{extract mass} \times \text{polysaccharide content} / \text{peanut meal mass} \times 100\% \tag{5.1}$$

### 5.3.2.2 Single-Factor Experiment

#### 5.3.2.2.1 Influence of temperature on the extraction rate of peanut polysaccharides

The influence of temperature on the extraction rate of peanut meal polysaccharides is shown in Fig. 5.31. The extraction rate of peanut meal polysaccharides increases with the increase of temperature. With the restrictions of equipment,

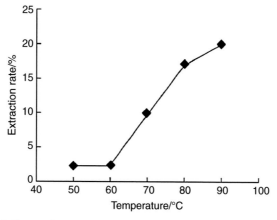

**FIGURE 5.31** Influence of temperature on extraction rate of peanut meal polysaccharides.

energy consumption, and other factors during production, however, 90°C is selected as the optimal temperature, and the extraction rate of peanut meal polysaccharides is 20.01%.

### 5.3.2.2.2 Influence of time on the extraction rate of peanut meal polysaccharides

The influence of time on the extraction rate of peanut meal polysaccharides is shown in Fig. 5.32. It can be seen that, with the extension of leaching duration, the extraction rate of peanut meal polysaccharides first drops and then rises. The polysaccharide extraction rate is 3.55% when the leaching duration is 3 h,

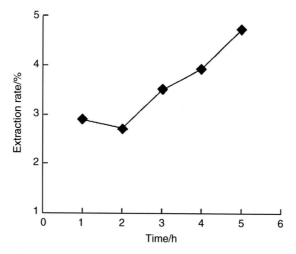

**FIGURE 5.32** Influence of time on the extraction rate of peanut meal polysaccharides.

3.93% when the duration is 4 h, and 4.74% when 5 h. The increase of polysaccharide extraction rate is not significant, possibly because the leaching time is too long and most polysaccharides have been extracted. Considering the time consumption in actual production, 3 h is selected as the optimal duration.

### 5.3.2.2.3 Influence of solid–liquid ratio on the extraction rate of peanut meal polysaccharides

The influence of solid–liquid ratio on the extraction rate of peanut meal polysaccharides is shown in Fig. 5.33. According to the mass transfer law, the continuous increase of solid–liquid ratio during extraction can increase the release of polysaccharides, but it also reduces the concentration of polysaccharides to a certain extent. This is not conducive to the polysaccharide precipitation when a certain volume of ethanol is added. It can be seen that, with the increase of solid–liquid ratio, the extraction rate of peanut meal polysaccharides first gradually rises to the peak when the solid–liquid ratio is 1:15 (g/mL) and then drops slowly. Therefore, 1:15 (g/mL) is selected as the optimal solid–liquid ratio and the extraction rate of peanut meal polysaccharides is 3.83%.

### 5.3.2.3 Response Surface Optimization Experiment

To optimize the process conditions obtained from the single factor experiment, quadratic regression rotation design is adopted to optimize the main influencing factors of the peanut polysaccharide preparation process. Extraction temperature (X1), extraction time (X2), and solid–liquid ratio (X3) are selected as three influencing factors, and three significant levels are selected for each factor. It can be seen that all the three factors, namely extraction temperature, duration, and solid–liquid ratio, have a significant effect. There is significant interaction

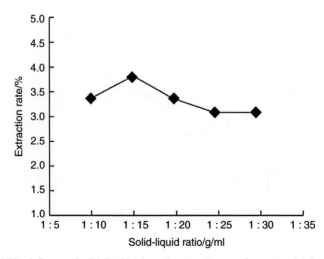

**FIGURE 5.33** Influence of solid–liquid ratio on the extraction rate of peanut meal polysaccharides.

effect between extraction temperature and duration, extraction temperature, and solid–liquid ratio, but not between extraction duration and solid–liquid ratio.

SAS 9.1 is used to establish the regression equation of hot water extraction of peanut polysaccharides as follows:

$$Y = 14.11667 + 9.23625X1 + 0.695X2 - 0.47625X3$$
$$+6.570417X1\ 2 + 0.735X1X2 - 5.5625X1X3 \tag{5.2}$$
$$-8.252083\ X2\ 2\ 0.36X2X3 - 0.934583\ X3\ 2$$

To accurately calculate the optimum process parameters, calculate the partial derivatives of the three independent variables in the regression equation for peanut polysaccharide extraction with the hot water extraction method and make them equal to 0, then the equation can be solved to get the extreme values, which are the optimal parameters of the relevant factors: extraction temperature 83.7°C, extraction duration 2.8 h, solid–liquid ratio 1:18 (g/mL); the extraction rate under these conditions is 34.49%.

To test the reliability of the response surface method for the optimization of the peanut polysaccharide extraction process, an experiment is carried out with the optimized process conditions for verification. The process parameters are set as follows: extraction temperature 85°C, extraction time 3 h, solid–liquid ratio 1:20 (g/mL). The final peanut polysaccharide extraction rate is 34.49%, consistent with the forecast value of model.

### 5.3.3 Separation and Purification of Peanut Meal Polysaccharides

Polysaccharides prepared using general methods are a mixture of carbohydrates with different properties. Classification of polysaccharides is the process by which a polysaccharide mixture is separated into different carbohydrates to obtain single polysaccharides after noncarbohydrate impurities have been removed. Separation and purification is a key link related to the structural identification and functional evaluation of active substances and are also the basic requirements of the structure–activity relationship studies. At present, there are many methods of polysaccharide classification, such as fractional precipitation method, metal salt precipitation method, salting out method, quaternary ammonium salt precipitation, ion exchange resin chromatography, cellulose anion-exchange column chromatography, gel column chromatography, preparative high performance liquid chromatography, preparative zone electrophoresis, and ultrafiltration method, etc. Different methods have their advantages and disadvantages and the selection of a method depends on the specific cases. This research team used α-amylase to remove starch from peanut meal polysaccharides and used dextran gel column chromatography to conduct separation and purification of nonstarch polysaccharides to obtain the corresponding purified components.

## 5.3.3.1 Enzymolysis of Peanut Meal Polysaccharides With Amylase

Peanut meal contains complex carbohydrates and the peanut meal polysaccharides obtained through extraction have a high starch content, which greatly reduces the purity of polysaccharide and restricts its physiological activity. Therefore, the removal of starch from peanut meal polysaccharides is a key step to peanut polysaccharide purification. It is better to use thermostable α-amylase and diastatic enzyme for peanut meal polysaccharide enzymolysis to remove starch. The process is as follows: peanut meal polysaccharides 5 g → redissolve with water → treatment with α-amylase and glucoamylase → enzyme inactivation → centrifugation for 10 min at 4500 rpm to take supernatant → thrice-volumed absolute ethanol → drying → redissolve the polysaccharide sample and determine its content.

Enzymolysis conditions with α-amylase and glucoamylase are shown in Table 5.13. Starch will be enzymolyzed by α-amylase into a lot of dextrin and oligosaccharides (Ramesh and Lonsane, 1990). The dextrin cannot be removed during alcohol precipitation and dialysis, so glucoamylase needs to be added to further hydrolyze dextrin into glucose and a small amount of maltose as well as some other micromolecular carbohydrates (Li et al., 2011), and thereby remove it in the subsequent purification process. Therefore, glucoamylase needs to be used for further treatment.

The total sugar, starch, and protein contents of peanut meal polysaccharides before and after treatment with amylase are shown in Table 5.14. As can be seen, the starch content of peanut meal polysaccharides before treatment with amylase is as high as 61.92%; after enzymolysis with amylase, most starch is

**TABLE 5.13 Enzymolysis Conditions of Amylase**

|  | Enzymatic Activity (U/mg) | Temperature (°C) | pH | Time (min) |
|---|---|---|---|---|
| α-amylase | 20 | 90 | 6.5 | 20 |
| Glucoamylase | 100 | 60 | 4.5 | 120 |

**TABLE 5.14 Index Contents of Peanut Meal Polysaccharides Before and After Treatment With Amylase**

| Amylase | Total Sugar Content (%) | Starch Content (%) | Protein Content (%) |
|---|---|---|---|
| Before treatment | 80 | 61.92 | 17.65 |
| After treatment | 50.13 | 7.86 | 40.15 |

removed and the content is only 7.86%. Most starch is removed, so the total sugar content is reduced to 50.13%. Nonstarch polysaccharides of peanut meal also include a certain amount of protein. As protease has been used to enzymolyze macromolecular protein into micromolecular substances early in the process, dextran gel column chromatography is used for further separation and purification in subsequent experiments to get purified components.

### 5.3.3.2  Dextran Gel Column Chromatography

Take a small amount of nonstarch polysaccharides of peanut meal and dissolve them in distilled water at a concentration of 10 mg/mL; take a Sephadex G-100 gel column with the specification of 60 cm × 2.6 cm; sample volume 5 mL; eluent flow rate 0.4 mL/min; elute with distilled water, and collect 4 mL in each tube. Use the phenol-sulfuric acid method to track and detect the polysaccharide content in each tube of eluent at 490 nm until no elution peak is detected. Collect and combine eluent of major elution peaks, concentrate it, and then dialyze it with distilled water for 48 h; dry it by freezing and finally collect it to get a purified component, as shown in Fig. 5.34. Determination shows that the molecular mass of the purified component is $2.383 \times 105$ Da.

## 5.3.4  Development Prospect of Peanut Polysaccharides

Research on peanut polysaccharides has just started, but some other plant polysaccharides have achieved industrial production and are widely used in food, medicine, cosmetics, chemicals, and other fields. Biological activity and gel properties are two important properties of polysaccharide macromolecules. For their special physiological activities, such as immunoregulation, antitumor activity, anti-AIDS, and antioxidation effect, polysaccharides are applied

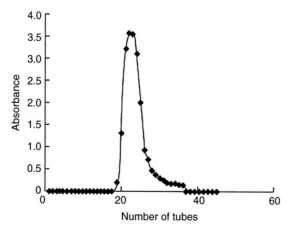

**FIGURE 5.34**  Sephadex G-100 column elution curve of nonstarch polysaccharides of peanut meal.

in health foods as raw materials or ingredients, such as lentinan, Ganoderma lucidum, and dendrobium polysaccharides, etc. Furthermore, polysaccharides also have good gel property and are used as thickeners, emulsifiers, and stabilizers in goods, such as carageen and konjac gum.

Studies have shown that peanut polysaccharides have good oxidation resistance and a protective effect on alcohol-induced liver injury in mice, which means that it is the trend to investigate more functional activities of peanut polysaccharides. As more and more scholars are concerned about the development and utilization of peanut polysaccharide, it will be soon widely used in food, pharmaceutical, chemical, and other industries.

China has a very large annual production of peanut meal, but it is mainly used as vegetable protein feed, or for the extraction of peanut protein, and the residuals are discarded, resulting in a waste of resources and environmental pollution. Peanut meal is rich in total sugar, which is second only to protein. Research has shown that peanut polysaccharides exhibit good oxidation resistance, liver protection, and other biological activities. Therefore, the extraction of polysaccharides from peanut processing by-products and the development of peanut polysaccharides products is of great importance for the full utilization of resources, improving peanut added value, increasing economic efficiency, and environmental protection

## 5.4   DIETARY FIBER IN PEANUTS

Peanut and its by-products are rich in dietary fiber. Peanuts dietary fiber is mainly from peanut shells, peanut meal, and stem and leaves, of which peanut shells have the highest dietary fiber content, approximately 65–80%. Peanut dietary fiber is a complex mixture and mainly contains cellulose, hemicellulose, pectin, and lignin. Peanut dietary fiber has good water holding capacity, cation binding and exchange, adsorption, and other physical and chemical properties. It also has many physiological functions, such as blood sugar regulation, cholesterol lowering, blood pressure lowering, obesity prevention, as well as cancer prevention and oxidation resistance, etc.

### 5.4.1   Detection Techniques

With the in-depth study of dietary fiber in peanuts, the demand for peanut dietary fiber detection methods is increasing. The dietary fiber content is also an important indicator for evaluating the purity and functional characteristics of dietary fiber extracted. The methods of detecting peanut dietary fiber include the enzyme-gravimetric method, enzyme-chemical method, detergent method, and near-infrared spectroscopy.

#### 5.4.1.1   Enzyme-Gravimetric Method

The enzyme-gravimetric method is the most important method to determine the dietary fiber content. It is used to determine total dietary fiber (TDF), soluble dietary fiber (SDF), and insoluble dietary fiber (IDF). The enzyme-gravimetric method was developed in the 1980s and has become an AOAC-approved analytical method. AOAC 991.43 is an internationally recognized measurement and evaluation criteria. China's GB/T 5009.88 _2008 has provided the method of determining dietary fiber in food and modified AOAC 991.43 "Enzyme-gravimetric Method for the Determination of Total, Soluble and Insoluble Dietary Fiber in Food." The method of determination of dietary fiber in food in GB/T 5009.88 _2008 is described in detail as follows.

### 5.4.1.1.1 Definition

Dietary fiber refers to carbohydrates with the degree of polymerization not lower than 3 in the edible part of the plant; it cannot be digested or absorbed by the small intestine but is beneficial to human health. It includes cellulose, hemicellulose, pectin, and inulin, etc.

### 5.4.1.1.2 Principle

Take some dry sample and remove protein and starch by enzymolysis with $\alpha$-amylase, protease, and glucosidase; then precipitate the sample solution with ethanol and filter it; wash the residue with ethanol and acetone and dry it; weigh the substance obtained to get the weight of TDF residue; take another portion of sample and carry out enzymolysis with the above three enzymes; directly filter the solution and wash the residue with hot water and dry it; weigh the substance obtained to get the weight of IDF residue; precipitate the filtrate with 95% ethanol of four times the volume of filtrate and filter it; dry and weigh the residue to get SDF residue. After drying and weighing the above residues, determine their protein and ash content respectively. After deducting protein, ash, and blank from the TDF, IDF, and SDF residues, the TDF, IDF, and SDF contents of the sample will be calculated.

The TDF determined with this method refers to the carbohydrates that cannot be hydrolyzed by $\alpha$-amylase, protease, or glucosidase, including cellulose, hemicellulose, lignin, pectin, partially retrograded starch, fructan, and Maillard reaction products, etc. As some small-molecular SDF (degree of polymerization 3-12), such as fructooligosaccharides, galacto-oligosaccharides, polydextrose, resistant starch, resistant maltodextrin, etc., are partly or completely soluble in ethanol solution, they cannot be accurately measured with this method.

### 5.4.1.1.3 Analysis steps

1. Preparation of samples
   If the fat content is unknown in the treatment sample, dietary fiber should be defatted before determination.
   a. After mixing the sample well, dry it at 70°C in vacuum overnight; then cool it in the desiccator; after dry sample is crushed, sieve it through a 0.3–0.5 mm sieve.

    **b.** If the sample cannot be heated, carry out freeze-drying before crushing and sieving.

    **c.** If the fat content of the sample is higher than 10% and normal crushing is difficult, the sample can be defatted with benzine; use 25 mL benzine for a gram of sample each time for three times and then carry out drying and crushing. Record the sample loss caused by benzine and make corrections in the final calculation of dietary fiber content.

    **d.** If the sample has high sugar content, it should be delactosed before determination. Treat every gram of the sample two to three times with 85% ethanol of 10 mL, and dry it at 40°C overnight. After crushed and sieved, keep the sample in the desiccator for later determination.

**2.** Sample enzymolysis

Make two reagent blanks for every analysis of sample.

    **a.** Accurately weigh duplicate samples (m1 and m2) of $1.0000 \pm 0.0020$ g and put them in 400 mL or 600 mL tall beakers respectively; add 40 mL of pH 8.2 MES-TRIS buffer and stir it using magnetic stirring until the sample is fully distributed in the buffer (prevent formation of lumps, otherwise the enzyme cannot fully contact the sample).

    **b.** Enzymolysis of thermally stable $\alpha$-amylase: add 50 µL thermally stable $\alpha$-amylase solution and stir it slowly; then cover the beaker with aluminum foil; out the beaker in a 95–100°C thermostat shaking bath for continuous shaking; start timing when the temperature rises to 95°C. The total reaction duration is generally 35 min.

    **c.** Cooling: take the beaker out of the water bath and cool it to 60°C; remove the aluminum foil and scrape off the annular substance on the beaker wall and jelly at the bottom with a spatula; flush the beaker wall and spatula with 10 mL distilled water.

    **d.** Protease enzymolysis: add 50 mg/mL protease solution of 100 µL in each beaker and cover them with aluminum foil; continue the bath shaking; start timing when the water temperature rises to 60°C; and keep the reaction for 30 min at $60 \pm 1$°C.

    **e.** pH determination: after 30 min, remove the aluminum foil and add 5 mL of 3 mol/L acetic acid solution while stirring it. When the solution temperature reaches 60°C, adjust the pH to approximately 4.5 (with 0.4 g/L bromocresol green as an external indicator).

      Note: Be sure to adjust the pH at 60°C; pH will rise if the temperature is lower than 60°C. Measure pH of the blank every time; if the measured value exceeds the scope required, the pH of the enzyme solution should also be checked.

    **f.** Amyloglucosidase hydrolysis: add 100 µL amyloglucosidase solution while stirring; cover the beaker with aluminum foil and continue shaking; start timing when the water temperature rises to 60°C; and keep the reaction for 30 min at $60 \pm 1$°C.

3. Determination
   a. TDF determination
      - Precipitation: add 225 mL (volume after preheating) of 95% ethanol preheated to 60–95°C to each portion of sample; the volume ratio of ethanol to sample solution is 4:1; take out the beakers and cover them with aluminum foil; keep precipitation for 1 h at room temperature.
      - Filtration: moisten the weighed diatomite in a crucible with 15 mL of 78% ethanol and smooth it out; filter to remove ethanol solution to make the diatomite in the crucible form a plane on the sintered glass filter. Pour an ethanol-precipitated enzymatic solution of the sample onto the crucible for filtration; transfer all residues to the crucible with a spatula and 78% ethanol.
      - Washing: wash the residue twice respectively with 78% ethanol, 95% ethanol, and 15 mL acetone; after filtering and removing the washing liquid, dry the crucible and residue at 105°C overnight. Cool the crucible in the desiccator for 1 h; weigh it (including the crucible, dietary fiber residue, and diatomite) to an accuracy of 0.1 mg. Deduct the dry weight of the crucible and diatomite to calculate the mass of residue.
      - Determination of protein and ash: determine nitrogen content of the weighed sample residue respectively according to GB/T 5009.5 and calculate the protein mass with N × 6.25 as the conversion coefficient; determine the ash content according to GB/T 5009.4. That is, ash the residue at 525°C for 5 h, accurately weigh the total mass of crucible (to an accuracy of 0.1 mg); deduct the mass of the crucible and diatomite to calculate the ash content.
   b. IDF determination
      - Weigh the sample according to ① in step (2) and carry out enzymolysis of the sample according to step (2); transfer the enzymatic solution to the crucible for filtration. Before filtration, moisten the diatomite with 3 mL water and smooth it out; remove the water and make the diatomite in the crucible form a plane on the sintered glass filter.
      - Filtration and washing: transfer all sample enzymatic solution to the crucible for fitration; wash the residue twice with 10 mL of distilled water at 70°C; combine the filtrate and transfer it to another 600 mL tall beaker for later determination of dietary fiber. Wash the residue twice respectively with 78% ethanol, 95% ethanol and acetone of 15 mL each; filter to remove the washing liquid and wash, dry, and weigh the residue according to c in ① of step (3); record the residue mass.
      - Determine the protein and ash content according to d in ① of step (3).
   c. SDF determination
      - Filtrate volume calculation: collect the filtrate after IDF filtration into a 600-mL tall beaker; weigh the total mass of beaker and filtrate and deduct the mass of beaker to estimate the filtrate volume.

- Precipitation: add into the filtrate 95% ethanol preheated to 60°C four times the volume of the filtrate and precipitate for 1 h at room temperature. The following determination should be carried out according to b and d in "① TDF determination" in step (3).

4. Calculation result

Blank weight should be calculated according to formula (5.3):

$$m_B = \frac{m_{BR1} + m_{BR2}}{2} - m_{PB} - m_{AB} \tag{5.3}$$

where $m_B$ is blank weight, in mg; $m_{BR1}$ and $m_{BR2}$ are residue mass of blank determination of two portions, in mg; $m_{PB}$ is mass of protein in residue, in mg; and $m_{AB}$ is mass of as in residue, in mg. The dietary fiber content is calculated according to formula (5.4):

$$X = \frac{[(m_{R1} + m_{R2})/2] - m_P - m_A - m_B}{m_1 + m_2} \times 100 \tag{5.4}$$

where $X$ is dietary fiber content, in g/100 g; $m_{R1}$ and $m_{R2}$ are mass of the two portions of sample residue, in mg; $m_P$ is mass of protein in sample residue, in mg; $m_A$ is mass of ash in the sample residue, in mg; $m_B$ is blank weight, in mg; $m_1$ and $m_2$—mass of the sample, in mg.

The results should be accurate to two decimal places.

The masses of TDF, IDF, and SDF are calculated with formula (5.4).

The enzyme-gravimetric method is simple to operate and can be used to simultaneously determine the TDF, IDF, and SDF and is widely used in the determination of dietary fiber. However, it cannot be used to determine some SDF with the degree of polymerization of 3 to 12, such as inulin, polydextrose, resistant dextrin, resistant oligosaccharides, and other dietary fiber with low molecular mass. This is because they can be partly or completely dissolved in 78% ethanol solution, which will lower the determination results (Yin, 2005). To improve the accuracy and precision of enzyme-gravimetric determination, studies can be carried out to develop reagents, enzymes, and methods for completely removing protein, fat, and starch, and to simulate the conditions of the human digestive system as accurately as possible.

### 5.4.1.2 Enzyme-Chemical Method

The enzyme-chemical method was first proposed in 1969 by Southgate. It starts from the angle of chemical analysis and determines the content of every component of dietary fiber to get the amount of dietary fiber. Different from the enzymatic-gravimetric method, the enzyme-chemical method can be used to determine every neutral sugar and total acid glycoprotein (uronic acid); it can also be used to determine lignin separately, but is not suitable for conventional dietary fiber analysis due to the restrictions of instruments and equipment.

Enzyme-chemical methods mainly include the enzyme-chromogenic method, enzyme-gas chromatography, and enzyme-HPLC. The principle of the enzyme-chemical method is to first carry out enzymolysis of the sample; after acid hydrolysis, determine the components and content of hydrolysis products using color comparison, GLC, or HPLC. Features of the enzyme-chemical method are its flexible operation and provision of extensive information. It can be used for not only the determination of dietary fiber content but also studies of composition and structure.

Although the enzyme-chemical method has detailed and accurate detection results, it is based on the definition of dietary fiber as a nonstarch polysaccharide and cannot be used to determine dietary fiber such as polyphenolic compounds and inositol; it is also time-consuming and difficult to operate, not suitable for detection and analysis of conventional dietary fiber, and has been gradually replaced by the enzyme-gravimetric method. Compared with AOAC, however, it can provide more detailed data of monosaccharide composition which is of great importance in terms of nutrition; it is also of significance in guiding more in-depth studies on the influence of different dietary fiber structures on physiological function.

### 5.4.1.3   Detergent Method

The detergent method removes protein and starch and other digestible ingredients in the sample to the maximum with acids, alkalis, and detergents without destroying the nonstarch carbohydrates and lignin (dietary fiber); the difference between the residue and ash after detergent treatment is "detergent fiber." According to different detergents and extraction conditions, the detergent method can be also divided into analytical methods of neutral detergent fiber (NDF) and acidic detergent fiber (ADF) (Chai, 2003). NDF is the residue from neutral detergent (the main component is lauryl sodium sulfate) leaching and is mainly composed of cellulose, hemicellulose, lignin, cutin, and silica. ADF is the fiber determined after carbohydrate is digested with acidic detergent (with the main component being hexadecyl trimethyl ammonium bromide). The neutral detergent method can effectively remove protein from cells and the results of SDF determination have high repeatability, wide applicability, and low cost. Its main disadvantage is that it can only be used for IDF determination and cannot be used for SDF determination; meanwhile, it cannot remove starch completely and is difficult to filter (Wang et al., 2007). China's GB/T 5009.88-2008 "Determination of Dietary Fiber in Food" provides the method of determining IDF in food with neutral detergent method. The detailed description is as follows:

#### 5.4.1.3.1   Principle

Under the digestive effect of neutral detergent, the sugar, starch, protein, and pectin in the sample are dissolved and removed; the indigestive residue is IDF, mainly including cellulose, hemicellulose, lignin, cutin, and silica, as well as insoluble ash.

### 5.4.1.3.2 Analysis steps

1. Sample handling
   a. Grain: wash the sample three times with water; place it in a 60°C oven to dry the surface; grind it to powder and sieve the powder with a 30–30 mesh sieve (1 mm); store the sieved powder in a plastic bottle, add a small packet of camphor fine, and tightly plug the bottle for later use.
   b. Vegetables and other plant foods: take the edible part and wash it three times with water; absorb water droplets with gauze; chop it and take some well mixed sample and dry it at 60°C; weigh it and calculate the water content; grind and sieve it with a 20–30 mesh sieve; keep it for later use. Or absorb water of fresh sample with gauze; chop and well mix it for later use.
2. Determination
   a. Accurately weigh 0.5–1.00 g of the sample and place it in a spout-free beaker. The sample should be defatted if the fat content of the sample is over 10%. For example, extract 1.00 g of sample three times with ben-zine (30–60°C), 10 mL for each time.b. Add 100 mL of neutral detergent solution and then add 0.5 g of anhydrous sodium sulfite.
   c. Heat it with an electric stove to make it boil within 5–10 min; move it to the hot plate and keep weak boiling for 1 h.
   d. Spread 1–3 g glass wool in a heat-resistant glass filter; put it in an oven for drying for 4 h at 110°C; take it out and place it in a desiccator to cool it to room temperature; weigh it to get $m_1$ (keep four decimal places).
   e. Pour the boiling sample in the filter and leach it with a pump. Wash the beaker and filter for several times with 500 mL of hot water (90–100°C); keep leaching until it is dry. Clean the liquid and foam at the bottom of filter and plug it with a rubber stopper.
   f. Add enzyme solution in the filter and the fiber needs to be covered by liquid; remove the bubbles with a thin needle; add a few drops of toluene and cover the filter with a watch glass; place and keep it in a 37°C ther-mostat overnight.
   g. Take out the filter and remove the bottom sieve; leach out the enzymatic solution and wash out the residual enzymatic solution several times with 300 mL of hot water; check whether there is starch residue with iodine; if yes, continue to add enzyme for hydrolysis; if starch has been removed completely, smoke it dry and wash it twice with acetone.
   h. Place the filter in an oven to dry for 4 h at 110°C; take it out and place it in a desiccator to cool to the room temperature; weigh it to get $m_2$ (keep four decimal places).
3. Calculation result

$$X = \frac{m_2 - m_1}{m} \times 100\%$$ (5.5)

where $X$ is IDF content of sample, %; $m_2$ is mass of filter, glass wool, and sample fiber, in g; $m_1$ is mass of filter and glass wool, in g; $m$ is mass of the sample, in g.

The results should be accurate to two decimal places.

Since the detergent method has the disadvantages of the inability of determining SDF, low precision, and low accuracy, it can be improved. The alcohol insoluble residue method (AIR) and cold neutral detergent fiber extraction are two well-known improved methods. The AIR method (Selvendran and Robertson, 1994) is as follows: remove water-soluble saccharides with methanol and defat the sample with ether or acetone; remove starch through enzymolysis with glucoamylase, and then extract the residue with hot water; hydrolyze the extract with sulfuric acid to obtain water-soluble noncellulosic polysaccharides. This method is easy to operate and suitable for determining the dietary fiber content of fruit and vegetable samples that have low content of starch and protein. Cold neutral detergent fiber extraction (CNDF) is a more complete method of dietary fiber extraction and has many advantages, such as mild extraction conditions, high extraction rate, and low loss, etc. (Selvendran and Robertson, 1994). In recent years, Chinese scholars have studied and improved CNDF method, enhanced its practicality and convenience, and made the results of dietary fiber determination more scientific and accurate (Zheng and Geng, 1997; Pan et al., 2002).

### 5.4.1.4 Near-Infrared Spectroscopy

Near-infrared light (NIR) refers to the electromagnetic waves between the visible light and mid-infrared light (MIR), with the wavelength range of 780–2526 nm. The principle of NIR spectrum detection is based on the consistence between cofrequency and double-frequency absorptions of NIR spectrum and organic molecules' hydrogen-containing groups (–CH, –OH, –NH, and –SH) vibration and frequency-doubled absorption at all levels, to scan the NIR spectrum of sample to obtain the characteristic information of hydrogen-containing groups of organic molecules in the sample. Currently, the enzyme-gravimetric method and the enzyme-chemical method are mainly adopted for the detection of dietary fiber composition and content. Traditional methods of dietary fiber detection have high accuracy, but have complicated steps, are time-consuming, labor-consuming and require a lot of reagent, and cannot meet the needs of screening and identification of large quantities of samples. Sample analysis with NIR spectroscopy has many advantages, such as convenient, fast and easy operation, high accuracy, good stability, no pollution, low cost; and it will not destroy the sample, consumes no reagent, and the determination can also be carried out without unpacking and in real time. Therefore, the detection of dietary fiber with NIR spectroscopy has become a new focus of dietary fiber detection. Currently, the crude fiber content of soybean meal can be determined with NIR spectroscopy, which proves that NIR spectroscopy can replace chemical methods to determine the crude fiber in soybean meal (Zhang et al., 2003).

Currently, as NIR spectroscopy has not been extensively developed, there are some limitations in dietary fiber determination and continuous improvement is needed. First, near-infrared bands have weak absorption and low sensitivity. That has limited the application of NIR spectroscopy in the detection of samples with low dietary fiber content, so the dietary fiber content of samples should be higher than 1%. In the future, photoelectric technology needs to be further improved to increase the spectral signal strength to expand the detection range of the near-infrared spectrum. Second, the near-infrared spectrum has different levels of double-frequency absorption and different combinations of cofrequency absorption, with complicated bands and rich information, and useful information will be covered by a lot of invalid information. Therefore, the development of the techniques of the near-infrared spectrum characteristic information extraction and processing should be strengthened. Finally, if new changes occur to the sample set in time and space, the mathematical model established before is not suitable for this new condition, and the model needs constant maintenance and optimization (Zhuang, 2006).

## 5.4.2 Preparation Process and Technique

According to different solubilities in water, peanut dietary fiber can be divided into water-soluble dietary fiber (SDF) and insoluble dietary fiber (IDF). Dietary fibers with different solubilities have different physical and chemical properties; therefore different methods are needed for extraction.

### 5.4.2.1 SDF Preparation

Among peanut dietary fiber, IDF has high functional activity, so it is very important to study the extraction method of SDF in peanuts. Currently, the extraction methods of SDF are mainly acid hydrolysis, composite enzymatic method, membrane filtration method, ultrasonic method, microwave method, and microorganism solid fermentation method.

#### 5.4.2.1.1 Acid hydrolysis

Acid hydrolysis is a simple method of peanut SDF extraction and currently is commonly used for the extraction of SDF from peanut shells. After acid treatment of peanut shells, SDF is hydrolyzed; after filtration, carry out concentration, alcohol precipitation, and drying of the filtrate to get SDF; insoluble residue can be used as the raw materials for the preparation of IDF.

The optimal condition for the preparation of peanut shell SDF with acid hydrolysis method is as follows: using citric acid as the extracting agent; temperature 90°C, citric acid mass fraction $w$ (citric acid) = 3%; liquid–solid ratio $m$ (raw material):$V$ (extract) = 1:12; duration 120 min. Under such conditions, the extraction rate of SDF was 7.36% and the mass fraction of nonstarch polysaccharide was 45.40% (Yu et al., 2008). The yield of SDF increased with the increase of acid extraction times. The yield of three times peanut shell SDF extraction with the acid extraction method was significantly higher than with only one extraction. The process is as follows: Peanut shells → washing → drying → crushing → sieving → extract treatment → filtration → filtrate → concentration → absolute ethyl alcohol → precipitation → centrifugation → precipitation → drying → product (Yu et al., 2009).

### 5.4.2.1.2 Composite enzymatic method

Composite enzymatic method is the method of preparing peanut SDF by using two or more of proteases, amylase and cellulase, etc. Due to the complexity of peanut ingredients, the extraction rate is low and the effect is not ideal with only one enzyme, whereas composite enzymes can simultaneously enzymolyze different ingredients of peanut and improves the extraction rate and purity.

Enzymes selected for the composite enzymatic extraction of peanut shell SDF are papain, α-amylase, and cellulase. First pretreat peanut shell powder with papain and α-amylase; then enzymolyze peanut shells with cellulase to prepare SDF. The optimal condition for peanut shell SDF with composite enzymatic method is as follows: after pretreatment with papain and α-amylase, the peanut shell powder is hydrolyzed for 4.0 h in 0.5 mg/mL cellulase solution with pH of 5.2 at the constant temperature of 45°C; then carry out ethanol precipitation, centrifugation, and drying of the enzymatic solution to get SDF product. The enzymolysis rate of peanut shells in SDF preparation with the composite enzymatic method can reach 19.80%; the nonstarch polysaccharides content of SDF prepared is 38.32%. The process is as follows: pretreated peanut shell power → cellulase solution → filtration (residue) → pretreated peanut shell power → buffer solution → filtration (residue) → half-mass pretreated peanut shell powder → half-concentration cellulase solution → filtration → residue weighing → concentration of filtrate from three filtration → ethanol precipitation → filtration → precipitate dissolved in boiling water → centrifugation → supernatant concentration → ethanol precipitation → filtration → precipitate drying → SDF product (Yu et al., 2010b).

### 5.4.2.1.3 Membrane filtration method

Membrane filtration method is the method of separating and extracting dietary fibers with different molecular masses with membranes with different sizes of pores. It is an effective method of SDF preparation. Using this method, dietary fibers with different molecular masses are prepared by changing cutoff molecular masses of membranes. It can effectively avoid the organic solvent residue and damage to dietary fiber structure that occurs in the chemical separation method. Membrane filtration method has easy operation and high product purity, but it cannot be used for IDF preparation and has high requirements for equipment; the cleaning and maintenance of membrane before and after use is also complicated (Chen and Xu, 2008). There has been no report on using the membrane filtration method for peanut SDF preparation. In the future, membranes with different apertures could be used to separate SDF with different molecular masses.

### 5.4.2.1.4 Ultrasonic method

Using only the ultrasonic method, the yield of peanut SDF preparation is low. This method needs to be used together with acid hydrolysis to improve the yield. According to the literature, scholars have prepared peanut shell SDF with the ultrasonic method together with acid hydrolysis. Before dietary fiber extraction, peanut shells should be pretreated through acid hydrolysis. Take a certain amount of peanut shells and put them in 3% citric acid solution; carry out oscillation extraction for 15 min in a water bath at the constant temperature of 90°C and then carry out ultrasonic treatment for 15 min under the following conditions: ultrasonic frequency 40 kHz; ultrasonic wave power 150 W; normal temperature; then carry out alternate treatment with water bath and ultrasonic wave four times for 2 h in total; filter the extracted mixture in the reactor and retain the filtrate. The filtered residue needs to be retreated with ultrasonic wave. Specifically, add it to citric acid solution and carry out alternate water bath and ultrasonic treatment four times without changing the conditions for ultrasonic treatment; carry out filtration and combine the filtrate from the two filtrations to get the extract. Through rotary evaporation in vacuum, the extract will become a concentrated solution; add to the concentrate absolute ethyl alcohol four times the volume of the solution and precipitate it overnight; filter the mixture after absolute ethyl alcohol precipitation and dry the filter residue at 60°C; then crush it to get SDF products. The ultrasonic method can improve the yield of SDF from peanut shells (Yu et al., 2009).

### 5.4.2.1.5 Microwave method

The raw materials for peanut dietary fiber preparation with the microwave method include peanut shells and stems. As they have a compact structure, the effect of extraction using only the microwave method is not ideal, and this method is often used together with citric acid hydrolysis method to extract SDF from peanut shells and stems.

The optimal condition for SDF extraction from peanut shells using the microwave method is as follows: the mass fraction of citric acid 5%, solid–liquid ratio of 1:20; microwave power 320 W; processing duration 30 s. The yield of peanut shell SDF extraction was 17.25% under such conditions. The SDF obtained is high quality, with good water holding capacity, swellability, water bonding capacity, and certain cation exchange capacity. The process is as follows: peanut shell powder → water bath oscillation extraction (citric acid solution) → microwave extraction → filtration → rotary evaporation in vacuo → concentrate → absolute ethyl alcohol with four times volume → precipitation overnight → filtration → filter residue → drying → SDF product (Wen and Yang, 2011).

The optimal conditions for SDF extraction from peanut shells using the microwave method are as follows: soaking time 55 min, microwave time 4 min, microwave power 800 W, microwave temperature 90°C, solid–liquid ratio 1:14 (g/mL); the yield of peanut stem SDF under such conditions was 6.0% and NSP content was 94.68%; the comprehensive rating of SDF reached 65.12%. The SDF extracted was light yellow with light sugar fragrance and no other odor. The development of SDF extraction from peanut stem can improve the added value of peanut stem, avoid resource waste and environmental pollution, and increase the economic efficiency of the peanut industry (Li et al., 2010a, 2010b). Both the microwave power and time needed for SDF extraction from peanut shells were lower than from peanut stem, and the final yield of SDF extracted from peanut shells was 17.25%, significantly higher than that from peanut stem (6.0%). Thus, it can be seen that peanut shells are better than peanut stem for SDF extraction using the microwave method.

### 5.4.2.1.6 Microorganism solid fermentation method

It has become a trend to extract functional substances with enzymes produced by microorganism fermentation. The microorganism solid fermentation method has many advantages, such as low equipment investment, low operation costs, easy promotion, and it uses agricultural and sideline products as raw materials.

The optimal condition for SDF extraction from peanut shells using the *Aspergillus niger* solid fermentation method is as follows: culture temperature 27°C, strain age 2.9 d, inoculum size 16 mL, and culture time 9.1 d. Under such conditions, the enzymolysis rate of peanut shells using *Aspergillus niger* fermentation broth reached 11.03%. The SDF prepared is light yellow, with a light sugar fragrance. For the specific operation, sterilize 8 g of pretreated peanut shell powder; add *Aspergillus niger* under aseptic condition and keep it growing for 7–11 d at 21–33°C. After adding 75–100 mL of water, carry out shaking culture for 4 h at 45°C; filter it and keep the filtrate, that is, fermentation broth. After sterilization of the fermentation broth, concentrate it by rotary evaporation in vacuo; add to the concentrate absolute ethyl alcohol of four times volume of the concentrate; keep it standing overnight and then filter it; concentrate the filtrate and add absolute ethyl alcohol of four times volume of the concentrate; collect the precipitates and dissolve them in boiling water; carry out centrifugation and concentrate the supernatant; add absolute ethyl alcohol of four times volume of the concentrate; filter and precipitate it; dry the precipitate to get SDF extracted from peanut shell powder using *Aspergillus niger* solid fermentation (Li et al., 2010a, 2010b).

### 5.4.2.2   IDF Extraction

The methods of peanut IDF extraction are mainly the alkali extraction method, acid-base complex method, and composite enzymatic method.

### 5.4.2.2.1  Alkali extraction method

The alkali extraction method can be used for the extraction of IDF from peanut shells and peanut meal. Peanut shells are rich in dietary fiber, so the yield of IDF

**TABLE 5.15** Process Condition and Yield of Alkali Extraction Method

| Raw Material | NaOH Concentration | Temperature (°C) | Time | Yield (%) |
|---|---|---|---|---|
| Peanut shells | 1 mol/L | 40 | 4 h | 66.9 |
| Peanut shells | 4% | 50 | 4 h | 95.6 |
| Peanut meal | 15% | 40 | 50 min | 34.2 |

extracted from peanut shells is significantly higher than that from peanut meal. Since the alkali extraction method is a chemical separation method, the IDF prepared is yellowish brown. If it is added to bread or noodles, the food will have a dark color. Therefore, bleaching is needed. Please refer to Table 5.15 for the process conditions and yield of IDF extraction from peanut shells and peanut meal using the alkali extraction method.

Leach peanut shells for 4 h with 1 mol/L NaOH solution at 40°C; then bleach it for 6 h at 40°C with 6% H2O2 solution with pH of 12. The peanut IDF extracted under such conditions was light yellow, with light scent; the yield was 66.9% and the expansion force was as high as 5.73 mL/g; water holding capacity was 615.86%. The process is as follows: peanut shells → pretreatment → drying (blast drying at 105°C) → crushing → sieving → NaOH hydrolysis → filter washing → HCl hydrolysis → H2O2 bleaching → filtration and full washing → drying (blast drying at 105°C) → crushing → sieving → finished product (Chen et al., 2006).

Leach peanut shells for 4 h with 4% NaOH solution at 50°C; then bleach the liquid with 6% $H_2O_2$ solution at a temperature of 6°C for 2.5 h at a pH of 10.0 to obtain light yellow peanut shell IDF; the yield was 95.6%; the protein content of this dietary fiber can be lowered to 1.68%. The process is as follows: peanut shells → pretreatment → drying (80°C) → crushing → sieving (60 mesh) → defatting → NaOH hydrolysis → rinse to neutral → bleaching → filtration → drying (80°C) → crushing → sieving (100 mesh) → dietary fiber finished product (Wu and Chen, 2008).

Hong et al. (2009b) extracted IDF from peanut meal using the alkali extraction method under the optimal condtions as follows: NaOH concentration of 15%, temperature 40°C; leaching duration 50 min; the IDF yield was 34.20% under such conditions. The process was as follows: peanut meal → NaOH treatment → filtration → filter residue → washing to neutral → IDF product.

Although the yield of IDF extraction from peanut meal using the alkali extraction method is lower than from peanut shells, it can be improved by optimizing the alkali extraction process and pretreatment of peanut meal before alkali extraction. The main component of peanut meal is crude protein, and the crude fiber content is much lower than that of peanut shells. If peanut meal after protein extraction is used for IDF extraction, it will increase the utilization of by-products and improve economic benefits.

### 5.4.2.2.2 Acid-base complex method

This method combines alkali extraction and acid hydrolysis to extract IDF from peanut shells and peanut cake. The acid-base complex method adds an acid hydrolysis process after alkali extraction, so the yield and purity of dietary fiber extracted is slightly higher than that of the alkali extraction method, and the alkali extraction time is also shortened. The specific conditions for peanut IDF extraction with an acid-base complex are shown in Table 5.16.

Take defatted peanut cake as raw material, soak it in pH 12 NaOH solution for 50 min; filter it and wash it to neutral with water; then soak it with pH 2 hydrochloric acid solution for 90 min; wash it to neutral; filter and dry it; grind it to obtain peanut cake IDF. The yield of peanut cake IDF extraction using the acid-base complex method was 67.4%; the water holding capacity of IDF extracted was 3.25% and its expansion force was 3.59 mL/g (Pan et al., 2005).

To extract IDF from peanut shells using the acid-base complex method, NaOH is generally used for alkali treatment and HCL or another acidic detergent is used for acid treatment.

The optimal conditions for IDF extraction from peanut shells using the acid-base complex method with NaOH and HCl as base and acid reagents are as follows: treat 3 g peanut shell powder for 30 min in 40°C water bath with NaOH solution with a mass fraction of 4%; filter the peanut shell powder and wash it to neutral; add 60 mL of HCL solution; treat the solution for 90 min in a 60°C water bath; filter and wash it to neutral; dry it to obtain peanut shell IDF product. Under such conditions, the IDF product extracted was beige, with a good smell; the yield was 86.44% and purity 91.13%. The process is as follows: peanut shells → rinsing → drying → crushing → sieving → SDF extraction → drying and crushing → alkali extraction → filtration and washing → acid extraction → filtration and washing → drying → IDF product (Yu et al., 2010c).

To extract IDF from peanut shells using the acid-base complex method, an acidic detergent can be used as an acid treatment reagent. The optimal conditions for the extraction are as follows: leach peanut shell particles of 40-mesh size for 4 h at 40°C with 1 mol/L NaOH solution; filter and wash it to neutral; then add to filter residue acidic reagent (28 mL sulfuric acid, 1000 mL distilled water, and 10 g cetyl trimethyl ammonium bromide) according to the solid–liquid ratio of 1:25 to leach for 3 h at 70°C. The yield of IDF extracted with this process was 75.8%; it is light yellow and a porous fibrous powder, with an expansion force of 3.20 mL/g and water holding capacity of 399.15%, as well as good functional characteristics. The process is as follows: peanut shells → rinsing → drying → crushing → sieving → NaOH extraction → filtration → washing → acidic detergent leaching → filtration → washing → $H_2O_2$ treatment → filtration → drying (for 3 h at 105°C) → products (Feng et al., 2011).

**TABLE 5.16 Process Parameters and Yield of IDF Extraction With Acid-Base Complex Method**

| Raw Material | Alkali Treatment (NaOH) | Alkali Treatment Time | Alkali Treatment Temperature (°C) | Acid Treatment | Acid Treatment Time | Acid Treatment Temperature (°C) | Yield (%) |
|---|---|---|---|---|---|---|---|
| Peanut cake | pH = 12 | 50 min | — | HCl (pH = 2) | 90 min | 60 | 67.4 |
| Peanut shells | 4% | 30 min | 40 | HCl | 90 min | 60 | 86.4 |
| Peanut shells | 1 mol/L | 4 h | 40 | Acidic detergent | 3 h | 70 | 75.8 |

Note: "—" means room temperature.

Peanut shells are rich in fiber, so the yield of IDF extraction from peanut shells using the acid-base complex method is higher than that from peanut cake, and both the yield and purity of IDF extracted with hydrochloric acid as an acid treatment reagent are higher than that with an acidic detergent, and the treatment time is also greatly shortened. Therefore, peanut shell is the best raw material for peanut IDF extraction using acid-base complex method, and the use of NaOH and HCl as treatment reagents can not only improve the yield and purity of IDF but also shorten the extraction time and improve resource utilization. As the impurities in peanut shells are mainly protein and other alkali-soluble substances, the effect of alkali treatment is greater than that of acid treatment in acid-base complex treatment. The increase of IDF extraction yield using the acid-base complex method can be achieved by the optimization of the alkali treatment process.

### 5.4.2.2.3    Composite enzymatic method

The composite enzymatic method is a method of preparing dietary fiber by removing protein, fat, reducing sugar, starch, etc in raw materials using two or more enzymes. Compared with commonly used chemical methods of peanut dietary fiber preparation, such as acid hydrolysis and alkali hydrolysis, the composite enzymatic method can be used to produce dietary fiber with higher purity. That is because the chemical method cannot thoroughly remove protein, starch, and other impurities in dietary fiber. Enzymes used in the composite enzymatic method mainly include cellulase, hemicellulase, amylase, protease, and arabanase, etc. This method has many advantages, for example, mild extraction conditions, low energy consumption, easy operation, no acidic or alkali reagent required, and environmental friendliness (Li, 2005).

The raw materials for peanut IDF preparation using the composite enzymatic method are peanut shells and peanut meal. The optimal process conditions for IDF preparation with peanut shells as the raw material and using the papain and cellulase hydrolysis method are as follows: cellulase dosage 0.4%, papain dosage 0.4%, pH 6.0; hydrolyzing peanut shells for 2.5 h at 50°C. Under such conditions, the protein hydrolysis rate reached 70.2% and the yield of peanut shell IDF reached 81.5%; the water holding capacity of peanut shell IDF was 3.94%; the expansion force was 5.05 mL/g; water bonding force was 2.87%, with good physiological activity. The composite use of cellulose and papain can overcome the disadvantages of poor protein removal effect by using only cellulase and semicellulose degradation during the chemical method, thus preparing dietary fiber with high purity and good physiological activity. The process is as follows: peanut shells → pretreatment → hydrolysis with cellulase and protease → high-temperature enzyme inactivation → filtration → bleaching → rinsing and filtration → drying → crushing and sieving (100 mesh) → finished product (Wu, 2008).

The optimal conditions for peanut meal IDF preparation with α-amylase and papain as composite enzymes included α-amylase dosage 2%, pH 4.0, temperature 60°C, time 30 min; papain dosage 11%, pH 7.0, temperature 80°C, time 2 h. The yield of peanut meal IDF extracted under such conditions was 37.72%. The process is as follows: peanut meal → drying → crushing → sieving → defatting → gelation → starch enzymolysis → deactivation → protein enzymolysis → deactivation → water washing → filtration → drying → IDF (Chen et al., 2011). Peanut shells and peanut meal have different components, so they need different composite enzymes for hydrolysis. As peanut shells are rich in crude fiber, the yield of peanut IDF prepared using the composite enzymatic method is as high as 81.5%, much higher than that of peanut meal IDF, 37.72%. And the composite enzyme dosage for peanut shell enzymolysis is much smaller than for peanut meal, but the enzymolysis time of the former is longer than that of the latter. Therefore, considering the economy and yield, peanut shell is the optimal raw material for peanut IDF extraction using the composite enzymatic method.

## 5.4.3   Application and Development Prospects

### 5.4.3.1   Current Situation of Application

Similar to other dietary fibers, peanut dietary fiber can be used as a food additive to improve the sensory and nutritional quality of functional foods. Currently, peanut dietary fiber is mainly applied in dairy products, meat products, beverage products, and bakery, etc.

#### 5.4.3.1.1   Dairy products

SDF added to milk powder, fermented yogurt, flavored milk drinks, lactic acid bacteria beverages, ice cream, and other dairy products can improve the taste and storage stability and will not have physical and chemical reactions with any ingredient of dairy products that would be detrimental to human health. Long-term drinking of the products can soothe the intestines, prevent constipation, lower cholesterol, regulate blood lipids and blood glucose, etc. Dietary fiber added to lactic acid bacteria beverages can serve as the nutrition source for active bacteria to maintain their activity, extend the shelf life, and improve the taste and flavor of dairy products (Dello et al., 2004). Peanut SDF extracted from peanut cake can be added to ice cream as a stabilizer, and it can significantly improve the dispersion degree of fat and fat-containing solid particles, taste, internal structure and appearance of ice cream; it can also improve the dispersion stability and resistance to melting of ice cream (Feng and Zhong, 2009). In addition, SDF can also prevent the formation of lactose ice crystals in ice cream.

#### 5.4.3.1.2   Meat product

Currently, dietary fiber has been applied in the production of ham, sausage, jerky, and other meat products. Adding dietary fiber to sausage takes advantage of its water holding capacity and water swelling capacity to increase the plasticity of water, support water in the structure and improve the toughness and the organizational structure of the meat. Adding dietary fiber to jerky takes advantage of dietary fiber's characteristic of water holding capacity, oil holding capacity, and ability to hold water after baking to improve the quality of jerky and make it delicate. Meanwhile, it can also improve the production rate of jerky and lower the costs. In addition, due to its physiological functions of lowering blood lipid and regulating blood sugar, dietary fiber added to meat products can increase their physiological functions and make them suitable for hyperlipidemia and diabetes patients, broadening the consumer groups.

### 5.4.3.1.3 Beverages

SDF is added to beverages. On one hand, its unique physical and chemical properties can make beverage particles uniformly distributed, prevent precipitation and stratification, and improve the stability and dispersion of beverages; on the other hand, due to so many physiological functions of dietary fiber, the beverages can increase satiety, make the drinker reduce the intake of other calory foods, and help to control weight. Long-term drinking can soothe the intestines, prevent constipation, lower cholesterol, regulate blood lipids and blood glucose, etc. Adding peanut SDF extracted from peanut cake to orange juice drink can improve the stability of fruit pulp during storage and prevent stratification and precipitation; it can also improve the taste of the drink. Results of studies have shown that 0.5% of peanut SDF can maintain good stability of orange juice drink and will not have any negative impact on the taste and flavor of the drink (Feng and Zhong, 2009).

### 5.4.3.1.4 Bakery

Dietary fiber applied in bakery products is mainly added to bread, cakes, and biscuits. Due to its water holding capacity, dietary fiber added to bakery products can increase the water content, improve its organizational structure and taste, and prevent dehydration and hardening of bakery products during their shelf life. Moreover, dietary fiber can undergo Maillard reactions with other ingredients during baking to make the products have a golden color and improve their sensory quality. After adding dietary fiber to bread, it can significantly improve cellular tissue, taste, and color of bread, and improve the dietary fiber content and the nutritional quality of bread. Dietary fiber added to cakes can prevent collapse, keep them fresh, and extend their shelf life. In addition, dietary fiber health biscuits can be made by adding a lot of dietary fiber.

### 5.4.3.1.5 Other foods

Apart from the above types of foods, dietary fiber can also be added to infant foods. The number of bifidobacteria in the body of an infant is sharply reduced after weaning, which may lead to diarrhea, anorexia, growth retardation, and low absorption and utilization of nutrients. Adding IDF to infant foods can improve the nutrition utilization of infants and promote the absorption of iron, zinc, and other trace elements. In addition, as dietary fiber has the physiological functions of regulating blood sugar, lowering blood pressure and cholesterol, helping to control weight, and anticancer and oxidation resistance, it can be made directly into health foods, diet foods, or snack foods, etc., for patients with diabetes, hypertension, high cholesterol, obesity, and also for healthy people. Daily intake of a certain amount of dietary fiber is very good for human health.

### 5.4.3.2 Market Development Prospects

China is one of the main peanut producing countries. Peanut by-products, such as peanut hulls, peanut stems, and leaves, etc., contain large amounts of dietary fiber, but they are often directly discarded or burnt as fuels, which pollutes the environment and wastes resources. The diet of Chinese people has been focused on plant-based foods for a long time, so people have a low demand for additional dietary fiber and the research, development, and utilization of dietary fiber is far behind that of developed countries. However, with the improvement of people's living standards, the diet of Chinese people is gradually changing, with a substantial increase in the intake of high-fat and high-calory foods, such as meat, eggs, and dairy products. Consequently, the number of people suffering from obesity, hypertension, and hyperlipidemia, etc., is growing day by day. Therefore, China urgently needs to adjust its daily diet, increase the intake of dietary fiber in daily meals, and strengthen the research on dietary fiber. As a high-quality dietary fiber with nutritional, health, and other functional properties, peanut dietary fiber has many advantages, such as rich material sources and a wide range of applications, etc., and can be added as an additive to dairy products, meat products, bakery products, and beverages, etc., to improve the sensory quality and nutritional value of products, thus having broad development prospects.

To vigorously develop peanut dietary fiber, China also needs to strengthen the research and development of peanut dietary fiber in the following aspects. First, China needs to carry out in-depth study on the relevant underlying theory of peanut dietary fiber to lay a solid theoretical foundation for the promotion of the development of China's peanut dietary fiber industry. Second, China needs to carry out extensive studies of the SDF and IDF extraction methods, to find dietary fiber extraction methods, with high yield and good maintenance of dietary fiber physiological activity, suitable for industrialized production and achieve large-scale production of peanut dietary fiber. Third, China needs to devise an accurate, rapid, and efficient method for peanut dietary fiber determination by drawing on other methods of measuring dietary fiber, which is essential for the development of the entire peanut industry. Finally, China should also expand the range of applications of peanut dietary fiber and strengthen the propaganda of its physiological functions and features, and make full use of the unique features of peanut dietary fiber to establish a wide range of consumer groups.

The development of the peanut dietary fiber industry will not only increase the added value of peanut by-products, such as peanut shells, peanut meal, peanut stems and leaves, but also promote the vigorous development of the whole dietary fiber industry, with huge economic and extensive social benefits.

## 5.5 PROCESSING AND UTILIZATION TECHNIQUES OF BY-PRODUCTS FROM PEANUT OIL PRODUCTION

### 5.5.1 Phospholipids

Peanut phospholipids are by-products of the peanut oil refining process and mainly contain lecithin (35%) and cephalin (64%) (Zhang et al., 2008b). They have good physiological functions and high nutritional value. Phospholipid products contain phosphorus, choline, and unsaturated fatty acids and other functional materials necessary for the human body. A lot of research has been carried out on the functions and features of soybean phospholipids in China, and industrialized production of soybean phospholipids has been achieved. China is very rich in peanut resources and more than 50% of peanuts are used for oil production. Therefore, in-depth research and development of peanut phospholipids are needed to effectively utilize the by-products of the peanut oil refining process.

#### 5.5.1.1 Detection Techniques

Phospholipids are an important phosphorus-containing lipid in vivo and also a high added-value by-product of the oil industry. Oil residue can be obtained by hydration and degumming of peanut oil; phospholipid concentrate can be obtained by concentration and drying of oil residue. According to the reports of Xu et al. (2001), phospholipid concentrate can be further processed into other phospholipid products, such as phospholipid powder, phospholipid with high phosphatidyl choline (PC) content, hydroxyl and acylated phospholipids, etc. In the oil industry, the phospholipid content of degummed oil is a very important indicator reflecting hydration and degumming efficiency, so the determination of phospholipid content in peanuts and peanut oil is an essential method in the peanut industry. The methods used mainly include molybdenum blue colorimetry, gravimetric method, TLC, HPLC, ultraviolet spectrophotometry, infrared spectrometry, and nuclear magnetic resonance (NMR).

##### 5.5.1.1.1 Molybdenum Blue Colorimetry

The basic principle of molybdenum blue colorimetry (GB5537-2008) is: phospholipid is oxidized to phosphorus pentoxide after burning, and then changed to phosphoric acid by hot hydrochloric acid, and produces sodium phosphomolybdate with sodium molybdate; this is reduced to molybdenum blue with hydrazine sulfate, and then the absorbance at the length of 650 nm is measured using a spectrophotometer; the phosphorus content is found according to the standard curve and is then converted to phospholipid content.

The operating procedure is as follows.

1. Sampling
   According to GB/T5524.
2. Preparation of samples

According to GB/T15687.

3. Drawing standard curve

   Take six colorimetric tubes and number them 0, 1, 2, 4, 6, and 8. Pour 0, 1, 2, 4, 6, and 8 mL of standard solution into the respective colorimetric tubes and then add 10, 9, 8, 6, 4, and 2 mL of water, respectively. Then add 8 mL hydrazine sulfate solution and 2 mL sodium molybdate solution into each of the six colorimetric tubes. Stopper the tubes and shake them three to four times; then put the colorimetric tubes in a boiling water bath for 10 min heating; take them out and cool them to room temperature. Dilute them with water to the mark and shake well; let them stand for 10 min. Draw the solution to a dry and clean cuvette; adjust zero point with reagent-blank in the place of 650 nm and measure the absorbance with a spectrophotometer. Draw a standard curve with absorbance as the ordinate and phosphorous content (0.01, 0.02, 0.04, 0.06, 0.08 mg) as the abscissa.

4. Preparation

   According to the phospholipid content of the sample, weigh the samples prepared and put them in a crucible. Weigh 10 g of refined oil sample and 3.0–2.0 g (to an accuracy of 0.001 g) of crude oil and degummed oil. Add 0.5 g of zinc oxide; first slowly heat the sample on an electric stove to make it thick, and gradually heat to fully carbonize it; send the crucible to a 550–600°C muffle furnace to fully ash (white) it by burning for about 2 h. Take the crucible out and cool it to room temperature; dissolve the ash with 10 mL hydrochloric acid solution and heat it to a state of weak boiling for 5 min and then stop heating; wait until the solution is cooled to room temperature, and then filter and pour it into a 100 mL volumetric flask; flush the crucible and filter paper with about 5 mL hot water for three to four times. After the filtrate cools to room temperature, neutralize the turbidity with liquor kalli hydroxidi; slowly drip hydrochloric acid solution to fully dissolve zinc oxide precipitates and then add two drops. Finally, dilute it with water to the scale and shake it. Please prepare a portion of blank control while preparing the test solution.

5. Colorimetry

   Draw 10 mL of the test solution and inject it into a 50 mL colorimetric tube. Add 8 mL hydrazine sulfate solution and 2 mL sodium molybdate solution. Stopper the tubes and shake them three to four times; then put the colorimetric tubes in a boiling water bath for 10 min heating; take them out and cool them to room temperature. Dilute them with water to the mark and shake well; let them stand for 10 min. Draw the solution to a dry and clean cuvette; adjust spectrophotometer to below 650 nm; adjust the zero point with reagent-blank to measure the absorbance.

6. The phospholipid content of the sample is calculated according to the formula (5.6):

$$X = \frac{P}{m} \times \frac{V_1}{V_2} \times 26.31 \tag{5.6}$$

where $X$ is phospholipid content, in mg/g; $P$ is phosphorous content in test solution find from the standard curve, in mg; $m$ is mass of the sample, in g; $V_1$ is diluted sample volume after ashed, in mL; $V_2$ is volume of test solution taken for color comparison, in mL; 26.31 is the weight of phospholipids (in milligrams) equivalent to 1 mg phosphorus.

When the absorbance of the test solution is greater than 0.8, the volume of test solution drawn needs to be properly reduced to ensure that its absorbance is below 0.8.

A parallel test should be carried out twice for each portion of sample. When the test results are in line with the precision requirements, take their arithmetic mean as the result by keeping three decimal places.

### 5.5.1.1.2 Gravimetric method

The basic principle of the gravimetric method (GB5537-2008) is: phospholipids in vegetable oil swell after absorbing water with an increase of density, which will make them change from flocculent suspension to sediment. After hydration of the sample, wash and filter it repeatedly with acetone; since phospholipids are insoluble and oil is soluble in acetone, phospholipids will be isolated from oil. Weigh phospholipids and calculate the content. The sediment filtered with this method is not all phospholipid but also contains other lipids insoluble in acetone.

The operating procedure is as follows.

1. **Sampling**
   According to GB/T5524.
2. **Preparation of samples**
   According to GB/T15687.
3. **Method**
   Take uniform sample of approximately 100 mL and put it in a conical flask; heat it to around 90°C and filter it. Weigh sample of approximately 25 g ($m0$) and put it in a beaker; heat it to 80°C; add 2.0–2.5 mL of water and fully hydrate it by stirring; allow it to stand overnight at room temperature or carry out centrifugation. Pour the supernatant and filter it with a filter paper of known constant weight ($m_1$) (or suction). After all filtrate is filtered, flush the residual sediment in the beaker to the filter paper with cold acetone and continue to wash the filter paper and sediment with acetone until there is no grease mark. After all the acetone in the filtrate and sediment evaporates, send the sediment to a 105°C oven and dry it to constant weight; accurately weigh it ($m_2$).
4. The phospholipid content of the sample is calculated according to formula (5.7):

$$Y = \frac{m_2 - m_1}{m_0} \times 1000 \qquad (5.7)$$

where $Y$ is phospholipid content, in mg/g; $m_2$ is mass of sediment and filter paper, in g; $m_1$ is mass of filter paper, in g; $m_0$ is mass of the sample, in g.

A parallel test should be carried out twice for each portion of sample. When the test results are in line with the precision requirements, take their arithmetic mean as the result by keeping three decimal places.

Repeatability: the absolute difference between the two independent measurement results obtained under repeated conditions should not be greater than 10% of their arithmetic mean.

### 5.5.1.1.3 Thin layer chromatography (TLC)

Thin layer chromatography (TLC) adsorbs each component of the phospholipids with normal phase silica gel, and then elutes and desorbs the adsorbed sample with developing solvent (mobile phase), thus achieving separation and detection according to the differences in polarity of each component (Wang et al., 2003). Before the 1990s, TLC was a common method of detecting phospholipid outside of China, but with the popularity of HPLC, TLC has been less used in in the last 10 years and has been replaced by HPLC and nuclear magnetic resonance (NMR).

TLC is also divided into one-dimensional and two-dimensional development. Generally, one-dimensional development can isolate some major components of phospholipids, but there are still some ingredients that have close rates of flow. Then another developing agent needs to be chosen to isolate these components for further analysis.

Ma and Duan (1999) prepared a developing agent with chloroform–methanol–glacial acetic acid–acetone–water (4:25:7:4:2, V/V/V/V/V) to develop phospholipids. It can not only isolate seven major phospholipids well but also can isolate PS and PI on the thin layer. They are then coated with molybdic acid and hydrazine sulfate by spraying; color development is carried out with sulfuric acid in methanol (10%) after all solvent has evaporated; the content of each component is measured by scanning at 700 nm. Xu and Chang (1998) carried out two-dimensional development of soybean lecithin with developing solvent A: chloroform: methanol: acetic acid: acetone: water (35:25:14: 15:2, V/V/V/V/V), and B: n-hexane–ether: (4:1, V/V); color development is carried out with phosphomolybdic acid ethanol solution (5–10%, W/V) and Dittmer; it is scanned at the detected wavelength of 650 nm and the reference wavelength of 400 nm to determine the PC content.

### 5.5.1.1.4 High performance liquid chromatography (HPLC)

High performance liquid chromatography (HPLC) is the most commonly used method to quickly and accurately detect the content and composition of the phospholipids and it is less affected by human factors. The column of HPLC is generally stationary phase silica gel, but there are many types of mobile phase, generally divided into two categories, including acetonitrile (chloroform)–methanol–water system and n-hexane–isopropanol–water (methanol) systems. The detectors are UV, fluorescent, IR, and differential refractometer detectors. The presence of carbon–carbon double bonds, carbonyl groups, phosphate groups, amino groups, and other unsaturated base groups and functional groups in phospholipid enables it to have strong absorption at the wavelength of 200–214 nm. Therefore, a UV detector is more widely used.

According to the reports of Singleton and Stikeleather (1995), samples should be processed before peanut phospholipids are detected with HPLC. Extract total phospholipid from peanut for 1 min with mixed solution of chloroform and methanol with the volume ratio of 1:2 in a mixing vessel. Then, filter the mixture; continue to dissolve the residue with chloroform mixture and then filter it again; mix the filtrate obtained in the two filtrations and add some saturated NaCl water solution; draw nonfat components from the mixed solution. Wash the filtrate three times and then remove the solvent by flash evaporation; then determine the phospholipid content with HPLC.

### 5.5.1.1.5 UV spectrophotometry

In recent years, lecithin has been determined mostly with UV spectrophotometry. Phospholipids are insoluble but other lipoid substances are soluble in acetone; lecithin is soluble in n-hexane and other nonpolar organic solvents but less soluble in polar organic solvents, such as methanol and ethanol. Acetone can be used for removing fats and n-hexane for dissolving lecithin, thus achieving the isolation of phospholipids. According to relevant reports (Guan et al., 2008), phosphatidyl choline in lecithin has the feature of absorbance at specific wavelengths, in line with the Lambert–Beer's law. That is, when light lecithin solution is presented with monochromatic light of the appropriate wavelength, its absorbance $A$ is proportional to the product of the solution concentration $C$ and the translucent liquid layer thickness $L$. Therefore, UV spectrophotometry is used to determine phospholipids. The biggest advantage of this method is that determination can be carried out directly without isolation, so it is easy.

### 5.5.1.1.6 Infrared spectroscopy

The wave number range of infrared rays used for chemical structure determination is 400–4000 cm$^{-1}$. According to research (Nzai and Proctor, 1998), a Fourier transform infrared detector (FTIR) is used to determine the phospholipid content in vegetable oil. First, dissolve a standard phospholipid mixed sample (including 39% PC, 22% PI, 27% PE) in n-hexane to prepare a standard solution with various concentrations for determining the wavelength of various chemical bond vibrations of phospholipids, and draw a standard curve. It is found that both RSD and related coefficients are good with FTIR regardless of the sample concentration. Then find the most suitable wave band for determination from different wave bands. Studies have shown that the results are the best when the wavelength is 1200–970 cm$^{-1}$. Therefore, it is decided to determine phospholipid content with this wave band. In the determination, use salad oil as blank reference to remove interference. This method has high accuracy and good reproducibility. Although the FTIR technique has good prospects in the determination of total phospholipid content, various phospholipids have the same structure and it is difficult to carry out quantitative analysis respectively.

### 5.5.1.1.7 Nuclear magnetic resonance (NMR)

The nuclear magnetic resonance (NMR) technique was first used for phospholipid analysis as early as the 1990s. The instrument used in this method is a 31P nuclear magnetic resonance spectrometer. Its principle is as follows (Wang et al., 2003): as the base groups on phosphate radicals of different components of phospholipid are different, they have different actions on P, resulting in different chemical displacements of different phospholipid components in the magnetic field and achieving the qualitative and quantitative analysis of the different components of phospholipids; and a certain integral area value also represents the quantitative relationship between different base groups.

Glonek (1998) analyzed the content of each component in phospholipids rich in PC extracted from alcohol with 202.4 MHz NMR. Phospholipids are dissolved with methanol–chloroform and high valence ions in it are precipitated with EDTA to eliminate interference for 31P; and phospholipids are analyzed by nuclear magnetic resonance at 202.4 MHz. In the analysis process, each component receives a separate peak signal, so a phospholipid mixture can be analyzed without prior separation, with the advantages of high speed, sensitivity, less interference factors, and higher accuracy. However, this method requires a large amount of the sample and has certain requirements on the concentration of other components in phospholipid mixture. 31P NMR is a new qualitative and quantitative analysis method in the phospholipid detection field and needs to be further explored and developed.

### 5.5.1.2 Preparation Process and Technique

Apart from rich phospholipids, oil residue also contains water, fat, protein, carbohydrate, ash (minerals), fatty acids, pigments, oil-soluble vitamins, and sterols, etc. Therefore, phospholipids must be isolated and purified. Currently, phospholipid isolation and purification methods include organic solvent extraction method, membrane separation method, supercritical extraction method, and organic solvent precipitation method.

### 5.5.1.2.1 Organic solvent extraction

Organic solvent extraction is an important extraction method. It refers to the method of isolating the required component according to different solubilities of different components in the same solvent by choosing a certain solvent and adding it to a liquid mixture. The phospholipid organic solvent extraction method is based on phospholipid's insolubility in methyl acetate, acetone, and other organic solvents and solubility in aliphatic hydrocarbon, aromatic hydro-carbon, and halogenated hydrocarbon organic solvents, such as ether, benzene, chloroform, and petroleum ether, and its partial solubility in aliphatic alcohols. With these solvents, fat and other impurities from the materials dissolved in the solvent can be removed.

In industrial production, solvent extraction operation generally includes the following three steps:

1. Mixing: material liquid has close contact with the extraction agent;
2. Separation: separate extraction phase from raffinate phase;
3. Solvent recovery: extraction solvent is removed from extraction phase (sometimes from raffinate phase) and collected.

The extraction process must involve mixer, separator, and collector. An agi-tator tank is generally used as a mixer. Material liquid and extraction solvent can also be mixed with pipes by turbulence or with a jet pump by eddy-mixing. Disc-plate-type centrifuge is generally used for separation. Th recovery gen-erally involves the distillation apparatus used in unit operations of chemical engineering. According to the contact modes of material liquid and extraction agent, the extraction operation process can be divided into unipolar extraction and multipolar extraction. The latter can also be divided into a multilevel cross-flow extraction process, a multistage countercurrent extraction process, and/or a combination of the two. In the three extraction processes, the second has the highest recovery and needs the least solvent. This is very economical and thus widely used in the industry.

This method is the most widely used in phospholipid production and is ap-plicable to phospholipid extraction from various natural materials, and only the solvent is slightly different for different materials. After peanut material is pro-cessed through various steps, phospholipid colloid will further coagulate and precipitate; after being collected by a bag filter, it can be used as a raw material for the production of peanut phospholipids. These crude peanut phospholip-ids contain no water, so dehydration is not needed in phospholipid refinement. Phospholipids of 90–95% can be obtained through purification with acetone; after solvent isolation, column chromatography, acetoxylation with alcohol, or hydrolysis with phospholipase A2, etc., phospholipids can be used W/O and O/W food or as an additive for cosmetics.

### 5.5.1.2.2 Membrane separation method

In the separation techniques, the membrane separation method is a very impor-tant unit operation of the material separation technique and its biggest feature is its driving force (mainly pressure). This method separates different components of homogeneous mixture with a semipermeable membrane, which has no phase change in this process. The membrane itself is a complex composed of a con-densed matter of one uniform phase or two or more phases and the mobile phase separated by the membrane is liquid or gas.

The membrane separation method is widely used in phospholipid product development. For example, first dissolve phospholipids with ethane, propyl alcohol, or another organic solvent to form phospholipid microcapsules; then carry out separation with a semipermeable membrane with a certain aperture diameter according to different molecular masses of different components in phospholipids. In recent years, this method has been used in the preparation of high purity products. For example, lecithin concentration can be improved from 20% to 51% by making an ethane–isopropanol solution of phospholipid pass a polypropylene semipermeable membrane. However, existing membrane function is still unable to separate components with similar molecular mass, so membranes with specific functions need to be developed for industrial application.

### 5.5.1.2.3 Supercritical extraction method

The supercritical extraction method is a new separation technique developed in recent years. It is used to extract and separate substances using a supercritical fluid as a solvent. Both the temperature and pressure of the supercritical fluid are higher than the critical point. The supercritical fluid has similar density to an ordinary fluid and many substances have a good solubility in it. Meanwhile, it also keeps the transfer properties and easy penetration characteristics of gas. Small changes in temperature or pressure near the critical point will cause a very significant change in the density of the supercritical fluid. With this characteristic, it is easy to separate solvent and extract. In recent years, the supercritical extraction method has been widely used in the food, chemical, pharmaceutical, and environmental fields (Lv et al., 2000). Supercritical $CO_2$ is the most commonly used extraction solvent in the fine chemical industry, especially in the extraction of fragrance, food additives, and surfactants. Many studies have shown that supercritical $CO_2$ is very effective in the removal of oil, but phospholipid classification is still in the research stage mainly because phospholipids are not soluble in supercritical $CO_2$.

Fluid (solvent) is balanced with phospholipid oil in vegetable oil at around the critical point and has transmission performance, and the solubility of phospholipids varies in a wide range with changes in pressure and temperature, thus achieving the purpose of separation of phospholipids. In the supercritical $CO_2$ extraction method, supercritical $CO_2$ is used as a solvent to remove oils and other nonpolar and weakly polar impurities in the material to obtain high purity phospholipids. The operating pressure is up to approximately 30 MP and the operating temperature is only 40–60°C, which will not cause deterioration of phospholipids by heating. Because supercritical fluid has very good mass transfer performance, the extraction time is short and $CO_2$ escapes in a gaseous state after relief, without solvent residue. The insoluble content of acetone can reach 98% by one operation. This method features easy operation, low production costs, and good product quality but requires high investment in equipment. Compared with other organic solvents, phospholipid extraction with supercritical $CO_2$ has higher purity and there is no toxic and hazardous organic solvent residue in the product. As the extraction process proceeds in a mild condition at room temperature and without oxygen, oxidation of phospholipids is greatly reduced.

#### 5.5.1.2.4 Organic solvent precipitation method

Organic solvents, such as acetone, ethyl acetate, etc., can be used to precipitate phospholipids. That is because the dielectric constant of solution will be reduced after such an organic solvent is added and electrostatic attraction between phospholipids, that is, coulomb force will be increased. Due to the nature of phospholipid solvation, the water bonding to phosphatidase is replaced by the solvent, thus reducing the solubility. Currently, the most commonly used organic solvent precipitation methods are the two as follows:

1. Ethyl acetate extraction method: dissolve crude phospholipids in ethyl acetate and cool the solution to $-10°C$; then carry out centrifugal separation and precipitation to obtain phospholipids with very high purity, with lecithin content of 50.8%. As ethyl acetate is a safe and nontoxic solvent, products obtained with this purification technique can be widely used in the food, pharmaceutical, and cosmetic industries.
2. Inorganic salt compound precipitation method: this method isolates lecithin from organic solvent by taking advantage of inorganic salt's property of reacting with lecithin to produce sediment; then the inorganic salt is extracted with a suitable solvent to obtain lecithin.

### 5.5.1.3 Application and Development Prospects

Phospholipid is a lipid with important physiological functions. It has characteristics of a surfactant and also plays an important role in the physiological metabolism of animals, so it is widely used in food, pharmaceutical, chemical, and many other fields and has good market prospects.

#### 5.5.1.3.1 Current situation of application

1. Application in food industry
   Peanut phospholipid is a promising food additive. Its application in food lies in its effects of emulsifying, moistening, stabilizing, antioxidant, and starch aging prevention. As a food additive, it can increase the size, homogeneity, and shortening property of dough and extend the shelf life of food in baking; for candy, instant food, and artificial butter, etc., it can have the effects of emulsifying, dispersing, and moistening; it can also have some special effects when used in meat, eggs, dairy products, and various convenience foods. Furthermore, studies by Xia (2005) have shown that phospholipids can be used in ice cream to increase its smoothness, prevent "sand" phenomenon, and reduce the amount of egg yolk. A small amount of phospholipids added to cheese can increase its cohesive strength and prevent breaking of

the cheese. Phospholipids can also be used to prepare soluble cocoa powder to increase its nutritional function. An appropriate amount of phospholipids added to gravy, soy sauce, ketchup, dairy products, fruit juices, sausages, and tripe can make the product well mixed, prevent precipitation of juice or drinks, and increase its special flavor.

As an emulsifier, phospholipids can enable oil phase and water phase to form a stable emulsoid. Under certain conditions, phospholipids can make margarine form W/O and O/W products. The emulsifier used in early margarine production is mainly phospholipids (or eggs). At present, apart from synthetic emulsifiers, phospholipid is also one of the important emulsifiers for margarine production. The phospholipids could be added into flour dough in shape of emulsifier in shortening; then it will give the food plasticity, water absorption, emulsification, and dispersion, and other advantages; phospholipids added to the dough of bread, biscuits, and pastries will improve the water sorption of dough, enable flour, water, and oil to mix evenly, increase the shortening property and the size, improve the nutrient content and make the food crisp and delicious; phospholipids added in various candies will help to quickly emulsify syrup and oils, improve wetting effect, and give candies a smooth and nonsticky surface, reduce the viscosity of materials for the ease of operation, increase the food's uniformity and stability, and it is also a good mold release agent; appropriate amounts of phospholipids added in powder or crystal drinks will have good emulsifying and wetting effects. Phospholipids can be used as an emulsifier and stabilizer in the production of ice cream. When soy milk is heated, phospholipids can be used as a defoamer to prevent spillover.

2. Application in medicine and health products

Phospholipids widely exist in the primary and biological membranes of plant and animal cells and play a very important role in regulating biological activity of biofilm and normal metabolism of the body; the key component of lecithin and sphingomyelin is choline, which is also a precursor compound of acetylcholine and plays an important role in regulating life activities of cells. The application of phospholipids in medical care has been developing rapidly in the past 20 years. The research on the application of phospholipids, especially lecithin in health care products and medicines, is attracting the increasing attention of nutrition and medical experts. Phospholipids can cure and protect the nervous system, cardiovascular system, immune system, and the storage and transport organs, so they play a very important role in human health and disease prevention and control.

Phospholipid can regulate blood lipids, prevent and relieve cardiovascular and cerebrovascular diseases, promote nerve conduction, invigorate the brain, promote fat metabolism, and prevent fatty liver; it also has a lubricating effect in vivo, etc. Reports (Qi et al., 2005) have shown that phospholipid is also the precursor of acetylcholine and other active substances in the brain and nervous system and prostate. It can dissolve and remove some lipid peroxides in the human body, activate brain cells, and regulate the endocrine system. Therefore, phospholipids in antiaging health products may extend life and have varying degrees of mitigation and treatment for various geriatric and Alzheimer's and cerebrovascular diseases. According to related research reports, phospholipids can also increase and improve blood clotting and can be used for the production of blood enriching products. In addition, soybean phospholipids can also protect the liver, prevent fatty liver and cirrhosis, partially dissolve gallstones and prevent gallstone formation, and also have some detoxication effect to eliminate tobacco, alcohol, aniline, and other toxic substances. Studies have shown that phospholipid products can be used for the treatment of diabetes, arthritis, low blood pressure, gastric dilatation and neurosis, and other chronic diseases.

3. Application in feed industry

Phospholipids can be used in the feed industry as a nutritional additive in feed for livestock, poultry, and aquaculture. Phospholipids can promote gastrointestinal absorption and metabolism of poultry and supplement trace elements for livestock and poultry. Phospholipids are a promoter of animal growth and development and can promote poultry to lay eggs and livestock growth. According to Lan et al. (2003), phospholipids added in feed can promote the development of animal's nerve tissues, organs, bone marrow, and brain, increase production of livestock, poultry, and freshwater fish, and reduce methionine consumption. Phospholipids added in feed can supplement energy in animal bodies and improve the nutritional value of the feed.

Phospholipids added in the feed of laying hens will promote their digestion and absorption and improve egg production. An and Yue (2004) have shown that phospholipid is rich in choline, inositol, linoleic acid, and vitamin E, and phospholipids added in chicken feed can effectively prevent the occurrence of fatty liver syndrome. The apparent metabolizable energy of soy phospholipids containing 65% of acetone insolubles and 90% of crude fat is 30 MJ/kg. Phospholipid supplementation can improve the digestion and utilization of fat and other nutritients in piglet feed, such as dry matter, protein, and energy. Phospholipids added in the feed of weaned piglets can improve feed utilization and reduce weaning stress response. The digestive function of early weaned piglets is not well developed and it is not easy to digest fat in the feed, so large amounts or poor quality of fat will easily cause diarrhea. Nutrients and the emulsifying effect of phospholipids can effectively promote nutrient absorption and energy utilization. Phospholipids added in feed can improve its palatability, improve digestion rate of feed and promote the absorption of fat-soluble vitamins. Phospholipids can also reduce dust in feed processing and reduce the loss of nutrients during processing. Phospholipids added in fish feed will play a very significant role in promoting the heathy growth, weight gain, and reproduction of fish. Therefore, phospholipids added in fish feed can not only improve the energy in feed, promote fish growth, and lower feed conversion rate but also reduce the occurrence of liver metabolic diseases. The amount of soy crude phospholipids in fish feed is preferably 2–4%. Fish in rapid growth after hatching need rich phospholipids to form cell components. When the biosynthesis of phospholipids cannot fully meet the needs of larval fish, phospholipids need to be added in the feed. In addition, phospholipids in feed can also promote crustacean's utilization of cholesterol and improve the growth and survival rate of crustaceans. Shrimps have different phospholipids demands at different growth stages. Juvenile shrimps are unable to synthesize enough phospholipids for the needs of growth and metabolism, so they have high demands for phospholipids.

4. Application in cosmetics
   Appropriate amount of phospholipids added in cosmetics will have a good effect of moisturizing the skin and preventing dry skin. The safety and nutritional function of phospholipid is the main reason for its popularity in cosmetics. On the one hand, the application of phospholipid in cosmetics is to use its surfactant property, naturalness and colloidal property; on the other hand, phospholipid is an inherent part of skin cells, plays an important role in regulating cell metabolism and cell membrane permeability, and has certain physiological effects on the skin. The roles that phospholipids play in cosmetics are summed up as follows: antioxidant, emulsifier, softener, emollient, penetrating agent, stabilizer, lubricant, superfatting agent, embedding agent of liposomes, nutritional supplement, and vitamin sources.

   According to the relevant reports (Hao and Xu, 2008), phospholipids are widely considered to be the basic ingredient of the skin's natural moisturizing factor (NMF) and a necessary ingredient to maintain normal physiological functions of the skin and play an important role in cell metabolism coordination. Phospholipid has hygroscopicity and can easily penetrate into the skin, so it can bind moisture in the skin and prevent the skin from drying, thereby keeping the skin velvety and physiologically healthy. The research results of marking phospholipids with radioiodine have shown that phospholipids in cosmetics can be partially absorbed by the body. Phospholipids on the skin can form a monomolecular film or a thin film of oligomeric molecules, which have colloidal properties and can prevent the degreasing effect of detergent. Vegetable phospholipids contain a large amount of essential fatty acids, so they can be used as an additive to treat eczema and other skin diseases.

   Phospholipids are an ideal natural quality material for cosmetics and can be widely used in the production of skin care creams, skin washing agents, cleansing agents, sunscreen agents, and other skin care agents, as well as shampoo, hair dye, perm agent, hair modification agent, hair conditioner, and other cosmetics. Meanwhile, they can also be used in lipstick, mascara, make-up powder, and other hairdressing products and bath products.

5. Application in the leather industry
   Phospholipids are obtained mainly from animal and vegetable oil and have a similar molecular structure to natural oils. Concentrated phospholipid is

generally used in the leather industry because it contains some neutral oil. Therefore, it can be used as an important raw material for manufacturing fat liquor. The surface activity of phospholipid itself enables it to be used as a surfactant in the production of fat liquor and finishing auxiliary. Fat liquor that contains phospholipids is among the best of fat liquors because of its good flexibility, water resistance, and optical properties, so great importance is attached to it in the leather industry. In particular, such fat liquor also has good biodegradability and will not cause environmental pollution after use, so it is preferred in the tanning process for reasons of environmental protection.

According to Shen et al. (1994), phospholipids used as leather greasing material are mainly natural oil (crude) phospholipids and their derivatives. At present, fat liquors used for leather are mostly composite fat liquors prepared with phospholipids or modified phospholipids and other greasing materials. Phospholipids are chemically modified and specially treated to give them finer emulsion particles, so they can evenly penetrate into the leather lining for greasing and filling when leather is greased.

6. Other applications

Apart from the above, phospholipids are also used in agriculture and agricultural products. Phospholipids added in crop protection agents can enhance the effectiveness of active ingredients and reduce their amount of use. In addition, phospholipids can also prevent diseases and pests of vegetables, rice, and other crops. Phospholipids are often used with pesticides to emulsify them and improve their adhesiveness and penetration. Moreover, a small amount of concentrated phospholipids added in the lubricating oil of automobiles can prevent the formation of gum; an appropriate amount of phospholipids in the rubber can prevent it from aging; phospholipids can improve negative ions' decontamination ability and they are widely used in detergents; phospholipids added in dyeing of plastic products and knitting can prevent prevent pigment settling, etc.

### 5.5.1.3.2  Market development prospects

Phospholipids were first studied in Germany as early as 1930. Since the 1960s, the United States, Europe, and Japan have achieved industrialization of crude and refined phospholipids. Since the late 1980s, phospholipids have become a hot research topic in the United States, Europe, and Japan. The research and production of phospholipids in China was started in the early 1950s, but currently the relevant domestic research is mainly focused on soybean lecithin and there is little on peanut phospholipids. There are still many problems in the research of peanut phospholipids. First, there is little research on the activity and preparation of peanut phospholipids. Currently, there is a lot of Chinese and foreign research on the activity and preparation of phospholipids but little on the similarities and differences of functional properties of phospholipids extracted from peanuts, soybean, and other plants. There is a lot of Chinese research on soybean phospholipids, which are mostly crude phospholipids, and little on refined products with high purity. Second, there is little research on phospholipid modification. By chemical modification of peanut phospholipids, their physical and chemical properties can be further improved, thereby greatly broadening the field of application. Therefore, China should attach great importance to research on the production process and product development of modified peanut phospholipids. Third, China's phospholipid enterprises have a small scale of production, a single brand, and there has not been leading enterprise or group production and operations. In the aspect of quality, concentrated phospholipids as a basic raw material have low quality, dark color, heavy smell, poor liquidity, and transparency.

Based on the analysis of the above problems, there are two aspects that are worthy of further research. One is theoretical research. Currently, there is a lot of research on the detection, extraction, preparation, activity, and qualitative analysis of soybean phospholipids, but little research on peanut phospholipids. Therefore, basic theoretical research on peanut phospholipids is a good direction. The other is development and application. Domestic development and utilization of plant phospholipids are mainly focused on soybean phospholipids and rapeseed phospholipids. Early primary products of soybean phospholipids are mainly used in food, cosmetics, and other industries for their surface activity. In recent years, with the strengthening of people's medical and nutrition awareness, as well as the call for green products, the development and application of high-grade soybean phospholipid products has attracted much attention. Due to their poor quality, rapeseed phospholipids are hardly used in food and they are mainly used as additives in the feed and rubber industries. However, there is still no substantial development and utilization of peanut phospholipid resources. The development of peanut phospholipids is an area worthy of research.

## 5.5.2 Sterol

Phytosterols occur widely in seeds of various oil plants and also in other plant foods, such as vegetables and fruits. The content varies in different types of plants and is high in wheat germ oil, corn germ oil, rice bran oil, and sesame oil, mainly β-sitosterol, stigmasterol, and campesterol. In cereals, phytosterols exist mainly in the form of fatty acid esters, phenolic esters, and glycosides.

### 5.5.2.1 Detection Techniques

#### 5.5.2.1.1 Gas chromatography

Gas chromatography (GC) can analyze the content and composition of phytosterols simultaneously and is suitable for the analysis of volatile substances. Phytosterol molecules contain polar hydroxyl, which is a substance with high polarity and low volatility. It needs to be converted into its corresponding volatile derivative by an appropriate chemical treatment to be well isolated. And common agents are used in the determination of phytosterol with DC, which features easy operation, full derivatization, good reproducibility, high accuracy, and low cost, and thus is worthy of further interest (Feng et al., 2006).

Wang et al. (2011) established a method of determining the phytosterol content of fat-containing food with GC: with β-cholesterol as an internal reference, saponify it for 1 h in a solution of potassium hydroxide in ethanol at 70°C; then extract it with n-hexane and determine the phytosterol content with capillary GC, which has high accuracy and precision. This method can be used to effectively determine the phytosterol content of functional food with phytosterol fatty acid ester. Bao et al. (2002) used 2.0 mol/L solution of potassium hydroxide in ethanol for 45 minutes' saponification of the sample in a water bath at 85°C; then carried out extraction with petroleum ether and analysis with capillary GC; and finally determined the optimum conditions for sample saponification. They determined the sterol content of sesame oil with this method. The relative standard deviation was 0.86–4.13%, and the recovery rate was 89.59–103.09%. This method features quick, simple, and accurate operation. Peng (2006) analyzed the phytosterol components and their contents with a gas chromatograph with a flame ionization detector (FID) and elastic capillary column. It was found in the analysis of phytosterol multielement standard solution using this method that each component, especially campesterol and stigmasterol with very similar polarity, can be completely separated. They can hardly be separated with other methods. All three kinds of edible rapeseed oil contained brassicasterol, campesterol, and β-sitosterol, and no stigmasterol was detected. In all samples, β-sitosterol content is the highest among the three sterols. Feng et al. (2006) established a phytosterol GC method and analyzed the phytosterol content of common vegetable oils in China. The GC conditions used were Shimadzu gas chromatograph, capillary column, FID, split ratio of 39:1, inlet and detector temperature of 300°C; program temperature rise from 210 to 275°C at 10°C/min and retention of 25 min. The processing conditions for sample injection pretreatment determined through studies are 30 minutes' saponification with saturated solution of potassium hydroxide in ethanol; determination of sample and sample treated with BSTFA + TMCS agent. Xu et al. (2010) determined the phytosterol content of royal jelly with GC-MS method. Lipid is obtained by freezing and extraction of royal jelly sample; after saponification and derivatization, use HP-5MS column for separation; then identify sterols with GC-MS and measure it with GC-FID internal reference method. The GC conditions are: capillary column, temperature program from 150 to 240°C at 4°C/min and then to 310°C at 12°C/min; keeping for 13 min; inlet temperature 310°C; carrier gas flow rate 1.0 mL/min; split ratio 20:1; FID temperature 210°C; pretreatment of sample: 30 minutes' saponification with solution of potassium hydroxide in ethanol at 80°C; extraction with n-hexane; fix the volume after the solvent is evaporated.

### 5.5.2.1.2   HPLC

For the determination of sterol with HPLC, sample injection can be carried out directly without derivatization. It features simple operation, high accuracy, and good reproducibility. If it is used for high performance liquid chromatogram preparation, some proof samples separated to each single sterol can also be collected. In the chromatographic analysis of the phytosterols, complete separation of stigmasterol and campesterol is the key to liquid chromatography analysis. Site C-22 of stigmasterol is a double-bond and site C-24 is connected to ethyl; site C-22 of campesterol is a single-bond and C-24 is connected to methyl. Therefore, they have small polarity differences and are difficult to separate (Lv, 2006; Wang, 2001).

Mu et al. (2007b) determined the phytosterol content of deodorized distillate of soybean oil with HPLC. The chromatographic conditions determined by the studies are C18 reverse phase column, mobile phase methanol: water = 100:4; detector wavelength of 205 nm; column temperature of 25°C, flow rate of 1.0 mL/min, injection volume of 20 μL. Feng and Liu (2005) determined phytosterol under the HPLC conditions of C18 reversed-phase column, mobile phase of pure methanol, flow rate of 0.8 mL/min; detection wavelength of 210 nm; sample concentration of 0.1 mg/mL; injection volume of 20 μL. The sample was well separated under isocratic conditions, and the degree of separation of campesterol and sitosterol is 1.0. A high pressure infusion pump can be upgraded to preparation type and each component of sterol can be prepared by further optimizing the conditions. Liu et al. (2006a) established a method of separating and detecting the content of five sterols with HPLC-evaporative light scattering detector. The method is to adopt column C18 and mobile phase of methanol; good linear relationship within the range of 75–5920 ng/L; RSD is smaller than 0.45%; average recovery rate is 98.3–101.0%. With this method, the sample can be directly dissolved and injected without pretreatment. It has solved the problem of system stability in UV detection and can determine five phytosterols simultaneously. Dai et al. (2005) separated soy sterol and its isomers with HPLC and carried out GC-MS analysis of the unknown fractions. The chromatographic conditions determined are C18 column; mobile phase acetonitrile:isopropyl alcohol = 95:5; flow rate of 1 mL/min; column temperature of 25–45°C; detection wavelength of 210 mn; injection volume of 10 μL. Under this condtion, campesterol, stigmasterol, and β-sitosterol are well separated and the components such as γ-sitosterol, 24-β-campesterol, campesterol, and brassicasterol, etc., are identified with GC-MS. Shao et al. (2007) established a method of analyzing VE and sterol content of deep processed deodorized distillates of rapeseed oil. The method is to adopt column C18 reverse phase column, mobile phase of 96:4 (V/V), flow rate of 1.0 mL/min, column temperature of 30°C; diode array detection and external standard determination. This method determines the VE and sterol content of deodorized distillates, featuring a simple and quick operation with accurate and reliable results.

Based on the experience of practical determination, this research group has provided the specific steps of determining the phytosterol content of peanut oil with RP-HPLC and determined the VE content of 45 varieties of peanut oil from different regions of China.

1. Peanut oil extraction

   After shelling and crushing of peanuts, leach it for 6 h in n-hexane at a solid–liquid ratio of 1:4 and at a temperature of 60°C; carry out extraction twice and then recover n-hexane by rotary evaporation to obtain peanut oil.

2. HPLC conditions

   Chromatographic conditions for determination of phytosterols: column: Waters C18 (4.6 × 250 mm, 5 μm); mobile phase: 100% acetonitrile; UV detection wavelength: 210 nm; injection volume 20 μL; flow rate: 1.5 mL/min; column temperature: 30°C.

3. Standard curve drawing

   Drawing of phytosterol standard curve: accurately weigh a certain quantity of stigmasterol, campesterol, β-sitosterol proof samples; dissolve them with absolute ethanol and mix them evenly; make their mass concentration to be respectively 0.2, 0.6, 0.8, 1.0, 1.2 mg/mL for campesterol, 0.05, 0.15, 0.25, 0.40, 0.50 mg/mL for stigmasterol, and 0.5, 1.0, 1.5, 2.0, 3.0 mg/mL for β-sitosterol.

   Inject samples respectively under the chromatographic conditions of (2); take peak area X as the abscissa and concentration C (mg/mL) as the ordinate; carry out a regression fit with Microsoft Excel and obtain the linear regression equations for stigmasterol, campesterol, and β-sitosterol.

4. Preparation of sample solution

   Accurately weigh 5.00 g of various peanut oils and add 30 mL absolute ethyl alcohol, 10 mL potassium hydroxide solution (1 + 1), and 5 mL 10% ascorbic acid solution and fully shake it; boil it under reflux for 60 min; extract unsaponifiable matter and dissolve the extract with 2 mL ethanol; filter it with 0.45 μm filter membrane and seal the filtrate for HPLC determination.

5. Sample analysis

   Take sample solution of 20 μL and carry out HPLC analysis under the chromatographic conditions in (2); determine the nature according to the consistency of retention time and quantify it according to the standard curve. Determine the content of each phytosterol component in peanut oil, where $c$ is the content of a certain phytosterol component found on the standard curve (mg/mL); $V$ is constant volume of concentrated sample (mL), and $m$ is the sample mass (g).

6. Establishment of the phytosterol standard curve regression equation for the chromatogram of phytosterol standard mix and peanut sample is shown in Fig. 5.35 and the linear regression equation is shown in Table 5.17.

With the method established above, this research group analyzed and determined the content of each phytosterol component of 45 peanut varieties collected from different regions of China (Fig. 5.36). Among the 45 varieties of peanuts, the total phytosterol content is 46.44–200.11 mg/100 g, with an average of 109.70 mg/100 g, the highest is in "Yuanza 9102" and the lowest is in "Longhua 243." The campesterol content is 3.89–67.39 mg/100 g, with an average of 24.10 mg/100 g, the highest is in "Yuanza 9102" and the lowest is in

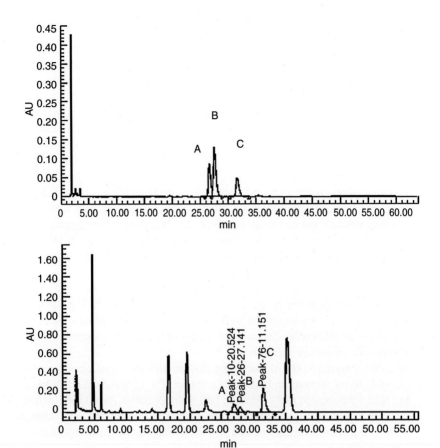

**FIGURE 5.35** Chromatogram of phytosterol standard mix and peanut sample. A: stigmasterol, B: campesterol, C: β-sitosterol.

**TABLE 5.17** VE and Phytosterols Linear Regression Equation

| Proof Sample | Linear Regression Equation | Linear Range | $R2$ |
|---|---|---|---|
| Campesterol | $C = 4 \times 10{-}7X{-}0.412$ | 0.20–1.20 | 0.986 |
| Stigmasterol | $C = 3 \times 10{-}7X{-}0.138$ | 0.05–0.50 | 0.991 |
| β-sitosterol | $C = 3 \times 10{-}7X{-}0.119$ | 0.50–3.00 | 0.993 |

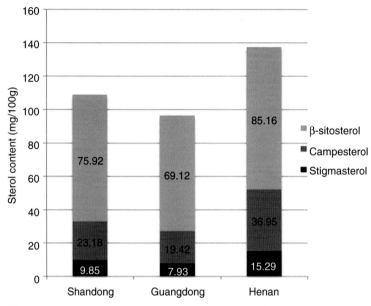

**FIGURE 5.36**    Result of statistical analysis of phytosterol content of peanut varieties from different regions.

"Longhua 243." The stigmasterol content is 0.31–22.62 mg/100 g, with an average of 10.11 mg/100 g, the highest is in "Kainong 37" and the lowest is in "Silihong." The β-sitosterol content is 38.82–115.70 mg/100 g, with an average of 75.49 mg/100 g, the highest is in "Yuanza 9102" and the lowest is in "Longhua 243." The variation coefficients of the contents of total phytosterol, campesterol, stigmasterol, and β-sitosterol in different peanut varieties are 29.56, 48.94, 47.51, and 24.99%, respectively.

In this study, the results of the determination of phytosterol content of 45 peanut varieties from Shandong, Guangdong, and Henan have shown that the total phytosterol in peanuts is 46.44–200.11 mg/100 g, including campesterol, stigmasterol, and β-sitosterol; the phytosterol in content order is β-sitosterol > campesterol > stigmasterol, wherein β-sitosterol accounts for 61.98–71.65% of the total phytosterol in peanuts. Compared with the result of Peng et al.'s (2006) detection using CGC, this experiment shows that peanuts also contain stigmasterol as well as β-sitosterol and campesterol, which is consistent with the research findings of Feng et al. (2006) and Nelson and Carlos (1995). In addition, as shown by the averages, the orders of the three regions in terms of the contents of total phytosterol, campesterol, stigmasterol, and β-sitosterol are all Henan > Shandong > Guangdong.

Among the peanut varieties tested, the varieties from Shandong, Guangdong, and Henan have difference in phytosterol content. According to previous studies, the components and contents of peanut varieties are different in different regions due to the genetic material of peanuts, the growing environment, temperature, humidity, light, water, fertilizer, and soil nutrients, etc., so the phytosterol contents also show differences (Feng et al., 2006; Liu et al., 2011c).

### 5.5.2.1.3 Other methods

1. TLC

   With TLC, free sterols, sterol esters, and other different components can be separated; unsaponifiable matters are extracted from nigre or distillates and then derivatized for spotting analysis. With this method, the approximate proportional relationship can be directly judged according to the spot size and color shades; or spot for development with standard solution, and carry out dual wavelength scanning of a sheet with a CS-910 sheet scanner under reference wavelength of 700 nm and sample wavelength of 550 nm, to obtain a standard scanning graph; carry out accurate stabilization with the standard curve, but TLC is mainly used for qualitative analysis (Feng et al., 2006).

2. UV spectrophotometry

   To find a simple, fast, accurate, and cost-effective determination method, we have conducted an experiment to study Lieberman–Burchard colorimetry for chemical determination of serum cholesterol and proposed spectrophotometry to determine total sterol content of soybeans by improvement. The principle of the S-P-Fe method is to form a stable purple compound under the effect of concentrated sulfuric acid and ferric iron; such a compound has a characteristic absorption peak at 520–550 nm wavelength, and the mass concentration and absorbance have a proportional relationship. Using phosphorus and sulfur ferron agent as color-developing agent; sterol and its ester react with color-developing agent to produce a purple compound, whose absorbance is proportional to the sterol compound content. A spectrophotometer may be used for determination. The specific measurement procedure is as follows: after adding the appropriate chromogenic reagent, keep color developing for 15 min at room temperature; measure its absorbance ($A$) at wavelength of 520 nm with a spectrophotometer; draw a standard curve with stigmasterol as the abscissa and absorbance ($A$) as the ordinate. After the sample is processed, measure its absorbance at the wavelength of 520 nm; determine the stigmasterol content on the basis of the standard curve; color stability time is 1 h (Liu et al., 2005; Xu et al., 2010).

   Total sterol content of the sample is calculated according to formula (5.8):

   $$\text{Total sterol content of the sample}(\%) = \frac{m \times V_1 \times B}{M \times V_2} \times 100\% \qquad (5.8)$$

   where $m$ is stigmasterol content corresponding to sample absorbance ($A$), in µg; $V_1$ is metered volume, in mL; $V_2$ is volume taken for absorbance determination, in mL; $B$ is dilution times; $M$ is sample weight, in µg,

3. Digitonin method

The digitonin method is also known as the gravimetric method and it is generally used to determine total sterol content. One sterol molecule and one digitonin molecule can form a molecular complex and generate a white precipitate of sterol digitonin complex for quantification by weighing. The specific determination process is as follows: accurately weigh a certain amount of sample; add an appropriate amount of absolute ethyl alcohol and shake it up for dissolving; put the solution in a 100 mL volumetric flask and dilute it to the scale and shake it up; then accurately take 10 mL of the solution and put it in a 100 mL conical flask; heat the flask in a water bath until it is nearly boiling and immediately add 9 mL of nearly boiling 1% solution of digitonin in ethanol and shake it up; slowly add 2 mL of water by dripping and leave it standing overnight; filter it with a weighed glass crucible; wash the precipitate respectively with 4, 2, and 3 mL of acetone–water–ethanol (73:18:9) mixture and dry it at 100°C; accurately weigh it and multiply it by 0.253 to get the total sterol weight in the sample. The sterol content obtained with this method is slightly high and this method requires high safety requirements and long analysis time, with many disturbance factors (Feng et al., 2006).

### 5.5.2.2 Preparation Process and Technique

Phytosterols are mainly extracted from vegetable oils and their by-products. Table 5.18 shows the sterol contents of deodorized distillates of different plants. Deodorized distillates of soybean oil, rapeseed oil, and rice bran oil are rich in sterol. Deodorized distillates of soybean oil are used as the main material for sterol extraction in Japan, Europe, and North America. Sterol extraction from deodorized distillates is generally on the basis of the difference between raw materials' physical and chemical properties and biochemical reactions, such as the difference of saponifiability, solubility in organic solvent, complexation ability, and complex solubility, difference of hydrophilic properties in the presence of surfactant, and difference of vapor pressure, molecular free paths, and

**TABLE 5.18** Sterol Contents of Seven Deodorized Distillates

| Deodorized Distillate Sources | Sterol Content (%) |
| --- | --- |
| Soybean oil | 9.0–12.0 |
| Rapeseed oil | 24.9 |
| Rice bran oil | 4.9–19.2 |
| Peanut oil | 5.9 |
| Palm oil | 2.0 |
| Sunflower seed oil | 0.8–1.2 |
| Olive oil | 0.6 |

adsorption force, etc. The selection of an extraction method is mainly based on the raw materials, sterol composition, degree of separation, purity and feasibility of industrial production, and these methods include solvent crystallization, complexometry, molecular distillation, dry saponification, adsorption method, enzymatic method, high pressure fluid absorption method, and chromatographic separation. Table 5.19 shows the comparison between several sterol extraction methods.

**TABLE 5.19 Comparison of Several Sterol Extraction Methods**

| Method | Reagent (Enzyme) | Advantage | Disadvantage |
|---|---|---|---|
| Solvent extraction method | Methanol, ethanol, ethyl ether, acetone, etc. | Direct separation, easy operation | Low product purity and high sterol yield ; too much solvent required |
| Complexation | Complexing agent: organic acids, acid halides and halide salts, etc.; Solvent: benzine | High purity and high yield | Troublesome solvent recovery in alcohol phase saponification |
| Molecular distillation | Sulfuric acid, lower carbon alcohol and acetone | Particularly suitable separation in the laboratory; it also can be used for simultaneous extraction and separation of sterols and VE. | — |
| Enzymatic extraction | Nonspecific immobilized lipase | Pretreatment of materials not required, VE and sterol | — |
| Dry saponification | Hydrated lime or quicklime, ethanol | A series of drying procedures not required; save ethanol and ensure safe and nontoxic production process | Yield is not high. |
| Supercritical $CO_2$ extraction | 99.5% $CO_2$ | Direct extraction, environmental friendliness, no pollution, high purity and high yield | Higher costs |

Note: "—" means no data.

### 5.5.2.2.1 Solvent extraction method

It is a traditional method with easy operation and can be used for direct separation. Its drawbacks are that it requires more solvent and it is difficult to recover the solvent; the sterol yield is not high and it is difficult to achieve industrial production. It is suitable for qualitative and quantitative analysis of sterols in the laboratory. The technical process is shown in Fig. 5.37.

Gao and Liu (2007) extracted phytosterols from crude rapeseed oil using the solvent extraction method and the results showed that the optimal conditions for extraction are: crude rapeseed oil of 60 mL, saponification time of 60 min, extraction solvent ether of 300 mL; extract crude phytosterols with absolute ethyl alcohol and get white sterols by reduced pressure distillation. Liu et al. (2011b) extracted phytosterols from *Jatropha curcas* seed oil using the solvent extraction method and the results showed that the factors affecting phytosterol extraction in order of importance are: volume of potassium hydroxide solution in ethanol > saponification temperature > concentration of solution of potassium hydroxide in ethanol > saponification time. The optimal extraction conditions are saponification temperature of 90°C and saponification time of 1 h; 15 mL solution of potassium hydroxide in ethanol at the concentration of 1.25 mol/mL. Under such conditions, the extraction rate of phytosterols can be up to 121.1 mg/100 g. Liu et al. (2009) carried out studies to determine the optimal process of extracting sterols from apple seed oil using the saponification method: solid–liquid ratio of 1:4, KOH-ethanol concentration of 0.5 mol/L, saponification time of 120 min, and saponification temperature of 85°C. Peng et al. (2008) studied the effect of water content of solvent on stigmasterol extraction and extracted stigmasterol using multistage countercurrent crystallization. The results showed that the stigmasterol has the best selectivity when the azeotrope system of dichloroethane and n-heptane contains 7% water. Wan and Jiang (2008) extracted β-sitosterol and stigmasterol using the recrystallization method and optimized the conditions for stigmasterol crystallization. The results showed that the β-sitosterol purity can be up to 90% after four crystallization

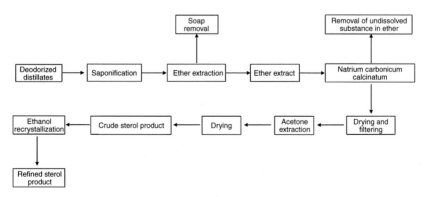

**FIGURE 5.37** Process flowchart of phytosterol extraction with solvent extraction method.

operations. During stigmasterol crystallization, the ratio of mixed phytosterols to cyclohexanone (g/mL) is 1:3.3; the stigmasterol purity can be up to 96% after five crystallizations of 5 h at 25°C. Hu et al. (2003) conducted experiments and found the optimal conditions for phytosterol extraction from deodorized distillate of soybean oil:distillate:methanol (W/V) 10:10; cooling temperature of −3°C and esterification time of 5 h. Han (2004) studied phytosterol extraction from deodorized distillate of sunflower seed oil and the results showed that the composition of solvent and cosolvent has great influence on the sterol extraction rate and purity; water-saturated hexane solution has good crystallinity; a small amount of ethanol can strengthen the effect of single water cosolvent; the sterol extraction rate can be up to 84% with primary crystallization of S/EDD.

### 5.5.2.2.2  Supercritical $CO_2$ extraction method

Phytosterol extraction from deodorized distillates includes three steps: converting phytosterol fatty acid ester to free phytosterol by saponification; esterification of free fatty acid; recovery of phytosterol or its concentrate by distillation. Three extraction methods can be used for the final extraction and separation of phytosterol or its concentrate: physical extraction–plant crystal separation; chemical extraction–phytosterol extraction with solvent; physical and chemical extraction–crystal separation of phytosterol from additive through an addition compound. The supercritical fluid extraction (SFE) method has features and advantages that molecular distillation technique does not have, such as low production cost, suitability for industrial production, high extraction rate, high product purity, no introduction of hazardous substances, a simple process, and low energy consumption (Wang et al., 2010). The technical process is shown in Fig. 5.38.

Zhao and Ding (1999) studied the optimal conditions for supercritical $CO_2$ extraction of phytosterol and the results showed that the phytosterol purity can reach 50–95% and the recovery reaches 85–41% when the extraction temperature is 40–60°C and the pressure is 15–30 MPa. Mu et al. (2007) studied supercritical $CO_2$ extraction of soybean oil deodorized distillates and found that the optimal conditions were: pressure of 20 MPa, temperature of 45°C, $CO_2$ flow

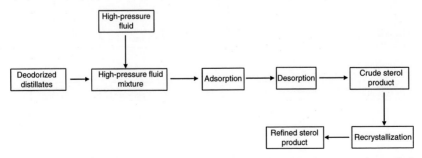

**FIGURE 5.38**  Process flowchart of supercritical CO2 extraction with solvent extraction method.

rate of 20 kg/h; under such conditions, the phytosterol recovery rate was above 65% and the purity was 80–85%. Huang et al. (1994) carried out supercritical $CO_2$ extraction of sterol from wheat-germ oil of high acid value. The results showed that the extraction temperature and pressure will affect the content of the extract and the sterol mixture mainly contains β-sitosterol and campesterol.

### 5.5.2.2.3 Ultrasonic extraction

Ultrasound is an elastic mechanical oscillation wave at a frequency of higher than 20,000 Hz and can produce and deliver a strong energy. The basic principle of ultrasonic extraction is to use ultrasound to produce a multistage effect, such as strong oscillations, emulsification diffusion and crushing, on a liquid to form a lot of holes in it; these holes will have a cavitation effect destroying plant cells, so that the solvent can penetrate into the plant cells; strong vibration, cavitation, and stirring accelerate the leaching extraction of the active ingredients of plants. Compared with heat reflux extraction, ultrasonic extraction has the features of a short production cycle, high extraction rate, and no heating required.

Zhang and Yu (2007) extracted phytosterols from rice bran using the ultrasonic extraction method and optimized the process conditions to be as follows: ethyl acetate as the solvent, solid–liquid ratio (m / v) of 1:15; extraction temperature of 70°C; extraction time of 90 min; ultrasonic power of 160 W. The phytosterol yield reached 46.12 mg/10 g, showing ultrasonic extraction to be superior to heat reflux extraction. Chen (2007) extracted ergosterol from red rice under the optimal conditions as follows: chloroform as the solvent; solid–liquid ratio of 1:250 (g/mL); reaction temperature of 50°C; extraction time of 20 min. The phytosterol yield was 9 mg/g. Ultrasonic extraction over a certain time period can extract most water-soluble components; extraction of a longer time period will damage the biomolecules of cells and the crushing efficiency will not be increased accordingly. An appropriate solid–liquid ratio will improve the sterol yield and reduce unnecessary waste. Wan et al. (2011) studied the process of total sterol extraction from Cape jasmine under the following conditions: ultrasound power of 100 W; benzine as the solvent; solid–liquid ratio of 1:8; extraction times: 1; ultrasonic treatment time of 25 min. The sterol yield reached 3.12 mg/g. Total sterol extraction from Cape jasmine using the ultrasonic technique has the features of high yield and comprehensive sterol components.

### 5.5.2.2.4 Microwave extraction

Zhao et al. (2010) established the optimal process model for corn silk β-sitosterol extraction using microwave extraction as follows: microwave extraction time 5 min; ethanol concentration 95%; extraction temperature 20°C; solid–liquid ratio 1:30 (g/mL) and β-sitosterol content 0.24%. Ethanol shows low toxicity, good safety, low cost, and easy industrialization and recovery, so it is used as the extraction solvent. The amount of β-sitosterol extracted decreases with the increase of extraction duration and also with the increase in temperature. This is mainly because the electromagnetic waves transfer energy to the inside of the material and heat it instantly under microwave radiation; heat is generated when the extraction duration is long, and the active ingredients of β-sitosterol are thermally degraded or isomerized, resulting in a decrease in the active ingredient content of extract. Pang and Xu (2010) took husked pumpkin seeds as the sample and extracted phytosterols from them using the microwave-assisted extraction method. The optimal process conditions for phytosterol extraction using the microwave-assisted extraction method are as follows: factors in the order of their influence on phytosterol extraction rate: microwave time > microwave temperature > liquid-solid ratio > microwave power; optimized process parameters for extraction: ethyl acetate as extraction solvent; solid–liquid ratio of 1:15; microwave time of 6 min and power of 600 W; temperature of 50°C. The phytosterol yield under these conditions reached 0.892 mg/g.

### 5.5.2.2.5 Complexation

Industrial production of phytosterols is carried out with this method in Japan and North America and this method is also used in the industrial production of a small amount of rice bran sterol in China. Experiments have proved that this method has the features of high purity and high yield of products. The complexing agents used include organic acids, halogen acids, urea, and halogen alkaline-earth metal salts. Halide salts are generally used for complexing, including calcium chloride, zinc chloride, magnesium chloride, ferrous chloride, etc. Complexation reaction solvents are benzine and isooctane. And the reaction temperature varies. Alcohol phase saponification is replaced by water phase saponification in this method to make it easy for solvent recovery and to reduce the cost. The technical process is shown in Fig. 5.39.

Fu (2004) extracted phytosterols using the calcium chloride complexation method and the optimal conditions are as follows: using ethyl acetate as the solvent; mass ratio of ethyl acetate to transesterification product (g/g) is 1/1; mass ratio of calcium chloride to transesterification product (g/g) is 2/16; using methanol as protonic solvent and the mass ratio of methanol to calcium chloride (g/g) is 1/2–2/2; reaction temperature of 40°C, reaction time of 1h. Liu et al. (2010a) established the method of extracting phytosterol from jatropha seed oil with complexometry, with the optimal process as follows: using ethyl acetate as aprotic solvent and methanol as protonic solvent; mass ratio of saponifiables to calcium chloride is 8:1; solid–liquid ratio of saponifiables to ethyl acetate

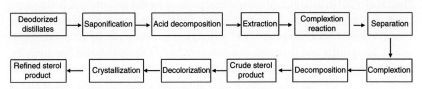

**FIGURE 5.39** Process flowchart of phytosterol extraction with complexation.

(g:mL) is 16:27; ratio of saponifiables to methanol (g:mL) is 16:1; reaction temperature of 60°C, reaction time of 1 h. The extraction rate under such conditions was up to 2.72%.

### 5.5.2.2.6    Molecular distillation

Molecular distillation is a new and efficient liquid–liquid separation technology and a new process of nonequilibrium separation in high vacuum. It has the features of low solvent consumption, low temperature of distillation, high vacuum degree, short heating time of materials, and high separation efficiency. It has been widely used in the food, pharmaceutical, essence, and fragrances industries. Molecular distillation can reduce the cost of materials and protect quality of the thermal materials, so it is particularly suitable for the separation of substances with high boiling point, and heat-sensitive and readily oxidizable substances. Currently, this technology is widely used in the oil industry (Fizet, 1996). The technical process is shown in Fig. 5.40.

Struve et al. (1983) carried out transesterification of deodorized distillates with 5–15% of sterol with methanol and then carried out molecular distillation of the mixture; mixed fraction containing more than 50% of sterols can be obtained at temperatures of 220–280°C and vacuum degree of 0.01–0.5 mbar; if the temperature is 0–200°C and the vacuum degree is 0.5–1 mbar, the sterol content of fraction is 5–40%. The mixed sterols contain β-sitosterol, stigmasterol, campesterol, cholesterol, or their mixtures. After degasification of deodorized distillates, Fizet made the sterol and fatty acid within it undergo esterification; removed nonsterol components by two distillations and then decomposed the sterol fatty acid esters to get sterols. Huang et al. (2010a) carried out base catalysis and ester exchange of rice bran oil and extracted phytosterol from rice bran oil using the molecular distillation technology. The optimal process condtions for phytosterol preparation with a wiped film molecular distillation

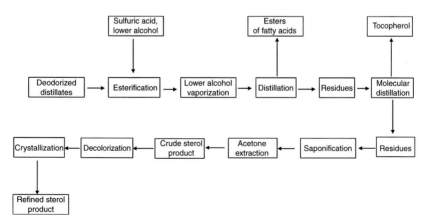

**FIGURE 5.40**    Process flowchart of phytosterol extraction with molecular distillation.

device are as follows: preheating temperature of 60°C and condensing surface temperature of 10°C; feeding rate of 2 mL/min and wiping rate of 250 r/min; the optimal temperature and pressure for primary and secondary distillation are respectively 80°C, 50 Pa, and 160°C, 1 Pa. Under such conditions, the phytosterol content was increased from 3.15% in the crude oil to 61.37%. Ding et al. (2007) concentrated phytosterol in deodorized distillates of soybean oil using the molecular distillation method. The phytosterol content of heavy phase after molecular distillation reached 26.12% and the yield reached 60.09%. On the premise of a system pressure of 0.1 Pa and a preheating temperature of 60°C, the optimal process conditions for a small scale test were determined as follows: system pressure of 0.1 Pa; distillation temperature of 170°C; feeding speed of 0.4 L/h, and wiping speed of 300 r/min.

### 5.5.2.2.7   Other extraction methods

1. Dry saponification

   After saponification of deodorized distillates with hydrated lime or quick-lime at 60–90°C, directly crush them with a machine into cream; carry out low temperature leaching with ethanol as an extraction agent. This can save a lot of ethanol and the production process is safe and nontoxic (Fig. 5.41).

2. Adsorption

   In the adsorption method, different adsorbents are used to separate sterols from the distillate to increase the sterol purity. There are the following three methods: ① use a synthetic inactive porous carbon molecular sieve, with average pore diameter of greater than 20 Å, generally 30–90 Å, such as coking carbonaceous polymers or carbonized resin, to selectively adsorb sterols and use aromatic hydrocarbon as the eluant; ② use an active carbon molecular sieve as adsorbent, such as with an aperture of 15–20 Å and surface area of 300–1500 cm$^2$/g; selectively adsorb all sterols without any other substances; use aromatic hydrocarbon as the eluant; ③ use a magnesium silicate molecular sieve (Florisil vector) as the adsorbent; use MTBE as the eluant; remove acidic substances from the materials before adsorption, which can be achieved by liquid–liquid extraction; the extraction solvents DMSO and n-hexane enter from the top and bottom of column, respectively, and the feed

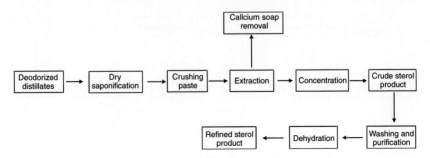

**FIGURE 5.41**   Process flowchart of dry saponification method.

enters from the middle part of column; DMSO is rich in acids and n-hexane is rich in sterols (Wei, 2002). The technical process is shown in Fig. 5.42.

3. Enzymatic method

   Catalyze the esterification of deodorized distillates with a nonspecific immobilized lipase to increase the conversion rate of FAME to 96.5%, thereby improving the proportion of VE and sterol and overcoming the shortcoming of the low recovery rate of VE and sterol extraction by solvent leaching, chemical treatment, and molecular distillation. This can recover more than 90% of sterols and VE from the materials; the enzymatic treatment conditions were mild and pretreatment of the material is not needed. The treatment method is shown in Fig. 5.43 (Li and Wang, 2006).

4. Sterol ester synthesis

   Free phytosterol is insoluble in water and has low solubility in oil. The absorption rate is low when it directly acts on the human body, limiting its practical application in food. The C-3 hydroxyl group of sterol is an important active group and can combine with fatty acid to form sterol fatty acid esters. Phytosterol esters are derivatives of phytosterols and have the same or even better biological activity than phytosterols. Studies have shown that the absorption and utilization rate of phytosterol esters is about six times that of free phytosterols and they have better liposolubility and better cholesteryl lowering effect than free phytosterols; they also have good oil solubility, higher safety and bioavailability, and can improve product appearance and taste, thus resolving the limitations of phytosterol application in food. Therefore, phytosterols can be used as an additive in functional food (Wang et al., 2010).

   The methods of fatty acid phytosterol esters synthesis are mainly the chemical method and the enzymatic method. The chemical method can be used in the following ways: ① direct esterification of phytosterols and fatty acid; ③ ester exchange with FAME; ③ phytosterols react with fatty acid halide or fatty acid anhydride. The chemical method has simple process

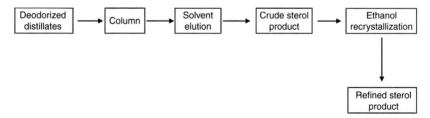

**FIGURE 5.42**  Flow of adsorption method.

**FIGURE 5.43**  Process flowchart of enzymatic method.

conditions and it is easy to control and achieve industrial production. With the extension of the fatty acid carbon chain, the difficulty in reaction and the reaction temperature gradually grow; side reactions may occur in long-time reactions at a high temperature, resulting in complex product components, a great loss of materials, and difficult separation and purification, so it is not conducive to getting safe and high activity sterol ester products. The reaction condition for the synthesis of phytosterol fatty acid esters by enzymatic catalysis is mild, with few side reactions, forming a high quality product with easy purification and separation. But the enzyme is expensive and organic solvent is added during the reaction, which requires demanding reaction conditions and high production cost. The research is still at the laboratory stage (Liu et al., 2011a).

Liu (2002) synthesized phytosterol esters of conjugated linoleic acid with phytosterols and conjugated linoleic acids as the raw materials, sulfonic acid and stannic chloride as the esterification catalyst or with sodium methylate and sodium hydroxide as the ester exchange catalyst. The process was as follows: carry out esterification at 110°C and 1.33 Pa; keep ester exchange reaction at 40–250°C for 2–20 h. Vu et al. (2004) used as lipase AYS as a catalyst; the mole ratio of phytosterol to conjugated linoleic acids is 1:3; solvent amount is 0.3 mL/g; speed is 175 r/min; keep reaction for 48 h at 55°C. When n-hexane is used as the solvent, the esterification rate is low, only 28.3%. Therefore, lipase of *Candida rugosa* is not suitable for the synthesis of linoleic acid phytosterol esters. Chen (2005a, 2005b) optimized the process conditions for the synthesis of oleic acid phytosterol esters through orthogonal experiments to be as follows: using 2% sodium bisulfate as a catalyst; mole ratio of oleic acid to phytosterol of 1.3:1; reaction temperature of 135°C; reaction time of 8 h; the esterification rate is 84.13%. Chen (2007) studied microwave synthesis of phytosterol esters by synthesizing sterol acetic esters through direct esterification of phytosterol and acetic anhydride with pyridine as a catalyst. The results showed a reaction temperature of 85°C, raw material ratio n (sterol):n (acetic anhydride): n (pyridine) = 1:12:10; reaction time of 15 min; and sterol esterification rate of 96.5%. Chen also studied the reaction of phytosterol with excess acetic acid and the optimal synthesis conditions are: reaction time of 5 min, microwave heating temperature of 95°C; sterol–acetic acid ratio of 3:30; and esterification rate of 93.2%.

Numerous studies on enzymatic synthesis of phytosterol esters have attracted much attention in recent years. Yu et al. (2011) studied the technical parameters and conditions for the ester exchange reaction of sunflower seed oil and phytosterols. Novozym 435 lipase was used and the optimal process parameters were determined as follows: sterol concentration of 5%, enzyme amount of 1%; reaction time of 3 h; reaction temperature of 100°C; stirring speed of 600 r/min; esterification rate of ester exchange reation under optimal conditions reached 89.7%. With this method, no harmful or toxic chemicals are used in the reaction, avoiding the impact of reagent residual on human health. Added to foods and oil, it can not only meet the nutritional needs of people but also has health care function. Liu et al. (2011a) studied the catalytic synthesis of phytosterol esters in nonaqueous phase with lipase. The optimal reaction conditions include: molar ratio of acid to alcohol 3:1; solvent amount 6 mL/g (substrate); catalyst dosage 11% (substrate percentage); reaction temperature of 45°C; reaction time of 48 h; esterification rate reaches 94.62%. Therefore, lipase can efficiently catalyze the synthesis of phytosterol esters in nonaqueous phase reaction medium. According to the studies of Li and Wang (2006), catalyzed by Novozym 435, the sterol esterification rate in distillates reached 85.3%; after purification, transparent and odorless pale yellow sterol esters of 95.3%, with high quality are obtained. Wang et al. (2010) carried out the esterification reaction with Novozym 435 lipase as the catalyst. The optimal process conditions include reaction temperature of 85°C; phytosterol mass fraction of 5%; reaction pressure of 8 MPa; reaction time of 1 h; stirring speed of 600 r/min; and esterification rate of 92.1%. Compared with conventional methods, this process reduces the reaction temperature and shortens the reaction time.

China is a large oil producer. The rational use of deodorization by-product deodorized distillates has become one of the major issues to be addressed. The catalytic preparation of high quality phytosterol esters from deodorized distillates will greatly improve the economic value of deodorized distillates and the economic benefits of oil production. Sterol ester synthesis technology has developed rapidly in recent years, but there are still many problems that need to be resolved in the esterification process. For example, some esterification reactions require vacuum conditions, high temperatures, and long reaction times; it is difficult to separate and purify by-products of esterification of some acids and sterol esters, which has significantly reduced the purity of the sterol esters. Although enzymatic synthesis of phytosterol esters has improved the efficiency of esterification, lipase is expensive, and the production cost is high, causing difficulties to industrial production. There have been research reports on the application of microwave in chemical reaction and organic synthetic chemistry. The advantages include energy saving, simple operation, less organic solvent consumption, high reaction rate, and less waste produced, and therefore this has attracted extensive attention (Zhang et al., 2009e).

### 5.5.2.3 Application and Development Prospects
#### 5.5.2.3.1 Current situation of application

Phytosterol applications involve the pharmaceutical, cosmetics, food, optical products, animal feed, paint, pigment, resins, paper, textile, pesticide and herbicide industries. Currently, phytosterols are applied mainly in the pharmaceutical industry as the raw material of steroid hormone drugs, cholesterol lowering drugs, anti-inflammatory, antipyretic, and anticancer drugs, etc.

In the food industry, phytosterols are mainly used as a functional active ingredient for the prevention of cardiovascular diseases. In September 2000, FDA approved the health statement that foods containing phytosterols alkyl and phytosterol esters can reduce the risk of cardiovascular diseases. The "Wholesome" label allowed sterol esters' application in spreads and salad seasoning. Since then a phytosterol-containing product margarine Benecol was the first to be developed and put on the market in Finland. Later, Unilever Lipton, USA launched the relevant functional sterol product "take control." Subsequently, functional sterol products were rapidly developed. Currently, SDM and Cargill of America, Cognis of Germany, Forbes Medi-Tech of Canada, Pharmaconsult and Raisio of Finland, etc., have stepped into the functional sterol food industry and developed and launched many kinds of functional sterol products. Their products include spreads, milk drinks, and cheese, etc. In 2005 and 2006, FDA approved sterols' application in puddings, noodles, biscuits, and egg products, followed later by other food ranges. There are more and more diversified functional sterol products, such as sauces, desserts, beverages, ice cream, snack bars, whole wheat bread, cereal, candy, and cooking oil, etc. At present, the link between phytosterol dosage, composition, and food ingredients (such as fatty acid composition and content) has been established. Compared with other countries, the development of functional phytosterol foods in China is still in its infancy. In 2007, phytosterols were approved as a new resource food in China; early in 2010, a new resource food certification under the new national regulations was approved. In recent years, as people have recognized the cholesterol-lowering effect of phytosterols, many enterprises in China have been in competition and relevant products have come onto the market. The development of functional phytosterol foods in China is mainly focused on oil. In addition, scientific colocation of phytosterols and milk can prevent cardiovascular and cerebrovascular diseases and help with treatment. The principle is that it promotes plasmin source activators as a trigger of fibrin dissolution to prevent thrombosis and help with treatment, to prevent cholesterol absorption and lipid accumulation, to discharge excess cholesterol in blood out of the body, and to prevent and treat high blood pressure, hyperlipidemia, and other diseases. Table 5.20 shows some of the existing phytosterol products on the market.

Phytosterols are also applied in the feed and chemical industries. In the feed industry, phytosterols can stimulate an animal's appetite, form plant hormones in water, that is, the phytosterol–ribonucleoprotein complex (which can improve the stability of the decomposition of the original plant hormone in vivo), promote the synthesis of animal protein, and be favorable for animal health and growth. In the chemical industry, phytosterols are mainly used as the critical components of high-end cosmetics, cleaning, and beauty products (Yang et al., 2011; Pang and Jiang, 2010).

**TABLE 5.20** Part of Phytosterol Products on the Market

| Product Name | Product Type | Sources | Purity | Producer |
|---|---|---|---|---|
| New-generation phytosterol corn oil | Vegetable oil | Corn | 13,000 ppm | COFCO |
| Arowana phytosterol corn oil | Vegetable oil | Corn | 10,000 ppm | Yihai Karry (Vegapure) |
| Jonquil phytosterol corn oil | Vegetable oil | Corn | 15,000 ppm | |
| Duoli phytosterol corn oil | Vegetable oil | Corn | 8000 ppm | Shanghai Standard Foods Co., Ltd. |
| Haishi phytosterol corn oil | Vegetable oil | Corn | — | Shanghai Cereals & Oils Sea Lions Industrial Co., Ltd. |
| Mengniu Deluxe Chunxian Milk | Milk | — | 1500 ppm | Inner Mongolia Mengniu Dairy Group |
| Autran phytosterol milk | Milk | — | — | Qingdao New Hope Qin Brand Dairy Co., Ltd. |
| Soy protein drinks | Beverage | Soybean | — | Hangzhou Wahaha Group Co., Ltd. |
| Qianshutai β-sitosterol | Capsule | — | 42,000 ppm | Shanghai Tianlong Biotechnology Co., Ltd. |
| Baiai stigmasterol capsules | Capsule | Soybean | 650,000 ppm | Harbin Baiai Technology Co., Ltd. |
| Phytosterol capsules | Capsule | Corn and soybean | — | Jiangxi Wenir Nutrition High-Tech Co., Ltd. |
| Zhonghong natural phytosterol tablets | Tablet | Soybean | 230,000 ppm | Zhonggu Tianke Bioengineering Co., Ltd. (under COFCO) |
| Bronson phytosterol | Tablet | Soybean, corn and rice bran | β-sitosterol 494,000 ppm Campesterol 287,000 ppm Stigmasterol 217,000 ppm | Shenyang Zhongkang Technology Co., Ltd. (BRONSON) Nowfood |
| NOW phytosterol formula of America | Capsule | — | β-sitosterol 392 mg, campesterol 200 mg, stigmasterol 800 mg | Nowfood |

*(Continued)*

**TABLE 5.20 Part of Phytosterol Products on the Market (*cont.*)**

| Product Name | Product Type | Sources | Purity | Producer |
|---|---|---|---|---|
| Nutryfarm natto phytosterol tablet | Tablet | — | 100 mg/tablet | Nutryfarm, Canada |
| GNC Xuezhiqing Capsule | Capsule | Corn | 400 mg/capsule | GNC |
| Soya Plus | Beverage | Soybean | 0.75 g/250 mL | Alpro, Belgium |

Note: "—" means no data.

### 5.5.2.3.2 Market development prospects

The existing problems are in the following areas.

1. Detection methods
   The current detection methods are mainly developing in the direction of sophistication and mainly include gas or liquid phase methods, which requires cumbersome pretreatment and large measuring equipment.
2. Raw materials extraction
   Compared with the phytosterol content of the above materials, sterol content of peanuts is similar to soya sterol content, but there is little research on the extraction and application of peanut phytosterols.
3. Product form
   A variety of new phytosterol products continue to come out in foreign countries while the research and development in China are still lagging behind. Currently, sterol products in the domestic Chinese market are mostly sold in the form of a capsule or tablet; or a plant rich in sterol is processed into finished products and its health care effect is emphasized, such as phytosterol corn oil. In recent years, sterols have been added as a functional food additive in dairy products, beverages, etc., to strengthen their physiological efficacy, but there are few such products and it is still at the preliminary stage with insufficient market development. China has abundant oil and other phytosterol resources available for development, and currently functional foods are quite popular among consumers. Therefore, the development of functional sterol products has good prospects.

### 5.5.2.3.3 Future trends

According to the current research status and problems of phytosterols, the development trends are proposed to be as follows.

1. Simplification of detection methods
   Aimed at the tedious and time-consuming pretreatment using a gas or liquid phase method, the direction of future research is to develop quick and easy sterol detection methods.

2. Rational and efficient use of phytosterols in peanuts

   Soybeans have been used as one of the major materials for phytosterol extraction. The phytosterol content (252 mg/100 g) of peanuts is similar to that of soybeans (260 mg/100 g), but there is less exploitation of sterols in peanuts. Therefore, it is necessary to carry out extraction, development, and utilization of functional phytosterols in peanuts.

3. In-depth study of phytosterol ester synthesis

   The presence of esterified phytosterols helps broaden the scope of its utilization. Research on phytosterol ester synthesis is still in its infancy, although the amount of research literature has begun to gradually increase since 2009. Therefore, further studies in this direction are needed, such as selection of fatty acid types and phytosterol types for esterification and changes in the physiological activity of specific phytosterol esters after esterification, etc.

4. Diversification of phytosterol product forms

   The application of phytosterols in foods in China is still in the exploratory stage. There are few types of products and only a few large-scale enterprises have been attempted. The development of functional phytosterol foods has good market prospects. In short, people should fully tap into the functional ingredients of the by-products of plants, turn waste into treasure, and further broaden the range of materials and applications and increase the added value of products.

# Chapter 6

# Peanut Allergy

**W. Xue\*, Y. Cong\*, A. Shi\*\*, L. Deng\*\*, X. Sheng\*\*, Yunhua Liu\*\***

*\*College of Food Science and Nutrition Engineering, China Agricultural University, Beijing, China; \*\*Institute of Food Science and Technology, Chinese Academy of Agricultural Sciences, Beijing, China*

## Chapter Outline

## 6.1 INTRODUCTION

Food allergy, also known as food hypersensitivity, means that the body gives an abnormal immune response to food entering human body, causing physiological disorder or tissue damage and further triggering a series of clinical symptoms, such as dermatitis, asthma, abdominal pain, diarrhea, and even severe shock. More than 90% of food allergies are caused by eight types of highly allergenic foods, including eggs, fish, shellfish, milk, peanuts, soy, nuts, and wheat. Major food allergens are from allergenic proteins, food additives in food, and allergen-containing food.

Like other food allergies, peanut allergy is an immediate type allergy. But the difference is that peanut allergy is lifelong, that is, the allergy to peanut will not disappear with age. The most common target organ involved in peanut allergy is gastrointestinal tract. Nearly 100% of allergic patients show allergic reactions in the perioral skin and oropharyngeal mucosa. Other major allergic target organs include the skin and the respiratory system. Sometimes peanut allergy can cause anaphylactic shock and even endanger life.

## 6.2    PEANUT ALLERGEN SPECIES

According to statistics, 8% of children and 2% of adults worldwide have food allergies. More than 20% of people in developed countries are affected by food allergy problems. Epidemiological studies have shown that approximately 0.6% (proportion of the total population) of the people in the United States and 0.5% in the Great Britain are allergic to peanuts. In the survey of food allergies among 2-year-old children in Australia, David et al. (1998) found that 1.9% of allergies were caused by peanut products. The same survey among people in some Asian countries and regions (Mainland China, Thailand, Japan, the Philippines, Singapore, Malaysia, Indonesia, Hong Kong, and Taiwan, China) showed that peanut protein products are among the causes of children's food allergies. Epidemiological studies have shown that approximately 3% of anaphylaxis is induced by food and one-fifth of life-threatening anaphylaxis is caused by peanuts. The survey of patients in the Allergy Division of Xiehe Hospital carried out by Hong Li et al. (2001) showed that approximately 4% of food allergic patients in the Beijing area were allergic to peanuts. Currently, food allergies have become a major health concern of the World Health Organization (WHO) and the Food and Agriculture Organization (FAO). With the process of globalization, food production, circulation, and consumption styles have led to internationalization and the peanut allergy will become a common concern of all countries. Therefore, it has important practical significance to undertake research into the allergy-causing mechanism of peanut protein, explore desensitization methods, and ensure the safety of peanut products.

Food allergens are the proteins or glycoproteins with acidic isoelectric point (pI) with a molecular weight of 10–70 kDa (Lehrer et al., 1996). To explore the key to peanut allergens, a total of seven allergenic proteins (including glycoproteins) have been identified worldwide since Sachs et al. found the first one and named it Peanut-I in 1981. According to the international unified allergen naming standard, they are Ara h1 (Kleber-Janke et al., 1999), Ara h2 (Stanley et al., 1997), Ara h3 (Rabjohn et al., 1999), Ara h4 (Kleber-Janke et al., 1999), Ara h5 (Kleber-Janke et al., 1999), Ara h6 (Kleber-Janke et al., 1999), and Ara h7 (Kleber-Janke et al., 1999). They were identified by IgE assay. The sera of nearly 90% of patients identify Ara h1 and Ara H2, so they are major peanut allergens (King et al., 1995). In North America, the probability of Ara h1 identification by the sera of peanut allergic patients is 95% (Burks et al., 1991). But the survey of three European populations showed that the probability of Ara h1 identification was 35% (De Jong et al., 1998), 65% (Kleber-Janke et al., 1999), and 70% (Clarke et al., 1998) respectively. The difference between the major allergenic peanut proteins of different countries is related to both their different peanut varieties and the serum types used. Therefore, further research is still needed in this field. Peanut products may also be allergenic. Allergenic proteins were detected with the Westernblot approach in cooked, stored, and processed peanut products, and new antigens were found (Schäppi et al., 2001).

## 6.2.1 Ara h1

Ara h1 accounts for 12–16% of the total amount of peanut protein and is a kind of glycoprotein with molecular weight of 63.5 kDa (Chun-Wook Park et al., 2000) and isoelectric point of 4.55 (Burks et al., 1991). It has good thermal stability, has resistance to hydrolysis, and is nondigestible (Wijk et al., 2004); its sequence similarity with pea protein is 40% and it probably exists in the form of macromolecular proteins (150–200 kDa or greater) (Li, 2002).

The studies of Ara h1's primary, secondary, tertiary, and quaternary structures (Burks and Helm, 1997) showed that Ara h1 occurs in the form of trimeric complexes; it has a clear β fold at the secondary structure level, of which 31% are α helix, 36% are β fold, and 33% are uncurled structure; at the quaternary structure level are complexes containing three monomers. When purified Ara h1 is heated to 80–90°C, the secondary structure folding is increased and solubility is decreased. Thermal treatment experiments show that Ara h1 is heat resistant. Although the allergen protein conformation has changed, its allergenicity is not decreased or increased (Shefcheck and Musser, 2004).

Shin et al. (1998) drew out the 23 linear IgE binding epitopes of Ara h1 with an IgE serum pool of individuals allergic to peanuts. His studies showed that the substitution of a single amino acid in the epitope will result in the loss or enhancement of IgE binding capacity; the hydrophilic residues in the middle of epitope have a major impact on IgE binding capacity. If the amino acids at the sites 144, 145, and 147–150 are substituted by alanine or methionine, the IgE binding capacity of the serum pool will drop; if the alanine at site 152 substitutes arginine, the IgE binding capacity will increase. Studies (Burks et al., 1997) showed that Ara h1 has four dominant epitopes and its amino acid sequences are at the sites 25–34, 65–74, 89–98, and 498–507.

Currently, it is known that Ara h1 has two forms of cDNA homolog; a transcriptional promoter fragment that plays a major role in gene recombination has been determined and Ara h1's gene structure has been made clear (Viquez et al., 2003). In addition, in-depth research has been carried out on Ara h1's antigenic epitope, structure, and its prokaryotic expression of gene, and the GC/MS and polyclonal antibody immunoassay for Ara h1 detection are established.

## 6.2.2 Ara h2

Ara h2 is identified by the serum IgE of more than 90% of peanut allergy sufferers (Stanley et al., 1997; Burks et al., 1992) and is a major allergenic protein, with high homology with Ara h6 and Ara h7 (Kleber-Janke et al., 1999). Compared with other allergenic 2S albumins, Ara h2 occurs in the form of continuous

**TABLE 6.1** Sequence Homology of Protein and Ara h1 From Different Sources

| Protein | Sources | Homology |
|---|---|---|
| δ-Lupine globulin | Lupine | 39 |
| Mabinlin I(chain B) | Caper | 32–35 |
| 2S albumin | Sunflower | 34 |
| α-Amylase inhibitor | Wheat | 29 |
| CM3 protein | Wheat | 27 |

polypeptide chains (Stanley et al., 1997) and is a major allergenic ingredient. It has PI of 5.2 and homology of 39% with lup-globulin (Koppelman et al., 2003). GenBank, Swiss-prot, and EMBL data show that Ara h2 has great homology with many seed storage proteins: 40% homology with δ-lup-globulin of lupine, 30–35% homology with albumins and Mabinlins, as well as a great homology with alpha-amylase inhibitor in wheat (as shown in Table 6.1).

Epitope mapping analysis of linear IgE binding epitope of Ara h2 can be performed with a peptide scanning technique. Studies (Lehman et al., 2003) showed that the 10 epitopes of Ara h2 can bind to IgE and 63% of the amino acids in the epitopes are polar uncharged or nonpolar residues; among them, three epitopes, aa27-36, aa57-66, and aa65-74, can be identified by serum IgE of patients, with their lengths ranging from 6 to 10 amino acids; these three epitopes can bind to more IgE than others and are considered to be the major immunological binding sites of Ara h2.

Ara h2 has a large amount of internal arginine codons and its gene expression level varies with different strains and expression vectors (Lehman et al., 2003). Although Ara h2 is recognized as an important component of the peanut allergen, there is little information of the spatial structure, thermal stability, and immunological detection information of the proteins in the family.

## 6.2.3 Ara h3

Ara h3 is identified by the serum IgE of more than 45% of peanut allergy sufferers and is composed of a series of proteins with molecular weight of 14–45 kDa, which can be divided into acidic subunits and basic subunits. 62–72% of sequences in Ara h3 are the same as the protein glycinin and legume proteins. In particular, the residues at the site 24/26 are considered to be an important part of the tertiary structure of the seed storage protein (Shin et al., 1998). The cDNA sequences of Ara h3 show that its 1530 nucleotides and ORF correspond to 510 amino acids; the ORF starts with CGG and ends with TAA at the nucleotide site 1533 (Lehman et al., 2003). Ara h3 has four exons (289, 293, 645, and 387bp) and three introns (339, 87, and 168bp) (Koppelman et al., 2003). From

**TABLE 6.2** Major IgE Binding Sequences of Ara h3

| Epitope | Amino Acid Sequence | Ara h3 Binding Site |
|---------|---------------------|---------------------|
| 1 | TETWNPNNQEFECAG | 33–47 |
| 2 | GNIFSGFTPEFLEQA | 240–254 |
| 3 | VTVRGGLRILSPDRK | 279–293 |
| 4 | DEDEYEYDEEDRG | 303–317 |

the cDNA reduced amino acid sequences and transcription process of different polypeptides, We found that basic subunits binding to IgE and some acidic subunits are allergenic (Koppelman et al., 2003).

The IgE binding area demonstrated by amino acid residues of recombinant Ara h3 is sites 21–25, 134–154, 231–269, and 271–328. The major binding epitopes of IgE binding amino acid residues in Ara h3 are shown in Table 6.2 (Rabjohn et al., 1999). Clinical trials showed that the serum IgE of 44% of peanut-allergic patients (8/18) could identify this recombinant protein (Rabjohn et al., 1999).

### 6.2.4 Components of Other Peanut Allergens

The content of other peanut allergenic components is low and there are fewer relevant studies. It is known that Ara h4 has a molecular weight of 61 kDa and is identified by 53% of peanut allergy sufferers. Recombinant Ara h5 has a molecular weight of 14 kDa and 80% homology with peanut suppressor protein on the amino acid sequence. In the serum IgE binding experiments of peanut allergy sufferers, 70% of binding bands (13 binding bands/19 bands) have the molecular weight of 17–63 kDa and the sera of allergy sufferers also have a high ratio at bands of 15, 10, 30, and 18 kDa (Wijk et al., 2004). The results showed that the protein components with other molecular weights were also important peanut allergens. Studies (Li et al., 2001) showed that some patients were allergic to seven allergenic components and most patients were allergic to two or more components; only a small number of patients were allergenic to only one of them, and different peanut allergy sufferers were allergic to different allergenic components. Therefore, it is also of great significance to carry out studies on other peanut allergen components.

### 6.3 ALLERGIC MECHANISM AND CLINICAL MANIFESTATIONS OF PEANUT ALLERGENS

#### 6.3.1 Allergic Mechanism

Peanut allergies are IgE-mediated hypersensitivity and the mechanisms are mainly type-I hypersensitivity mechanisms. Susceptible individuals' exposure to allergenic peanut proteins will result in the generation of IgE with special proteins and make them bind to high-expression IgE acceptors on the surfaces of mastocytes and granulocytes, thereby making the body allergic. In case of a second exposure to the same or similar allergens, the specificity of allergen molecules will identify IgE on the surface of allergic cell membrane, induce degranulation, and release inflammatory mediators, such as histamine, prostaglandins, and leukotrienes, etc. In addition, these cells also produce interleukins IL-4 and IL-13, cytokines, and chemokines to attract other inflammatory cells. These chemical reactions will lead to a series of clinical allergy symptoms.

Through IgE-related reaction mechanisms, cross-reaction can occur between peanuts and other vegetable proteins. Studies show that 38–79% of patients have clinical manifestations of progress from IgE reaction with a single bean (skin test positive/RAST) to reaction with all beans. Cross-reaction also occurs between peanuts and nuts. One study showed that they have a common allergic reaction rate of 2.5% (Sicherer and Sampson, 2000). Therefore, peanut allergy sufferers must maintain high vigilance when taking in allergens that may cause cross-reactions.

## 6.3.2 Clinical Manifestations

The most common exposure to allergens is food intake (91%), followed by skin contact (8%) and inhalation (1%). Allergies caused by oral ingestion of peanuts are usually serious and can be life threatening, while allergic symptoms caused by skin or air contact are relatively milder. The critical allergic reactions usually occur in adolescents or adults with asthma and atopic dermatitis history. Clinical symptoms of peanut allergy develop rapidly. They are shown within a few seconds after exposure to peanut protein and more than 95% of the symptoms occur within 20 minutes. Allergic reaction of one-third of patients shows diphase, that is, allergy symptoms reappear 1–8 hours after the original symptoms subside. 75% of the children have symptoms after eating peanuts for the first time and the average age at diagnosis is 14 months. The first eating rarely causes fatal reactions.

Clinical symptoms often involve multiple systems, including the skin, cardiovascular, digestive, and respiratory systems. 31% of the initial reactions involve two systems and 21% involve three systems. The main clinical symptoms include atopic dermatitis (46%), articarial/angioedema (32%), asthma (15%), systemic anaphylaxis (5%), and gastrointestinal symptoms (3%) (Burks, 2008). Almost all patients have skin manifestations (89%), including rashes, red rash, and acute urticaria (or angioedema). Some patients have only skin manifestations and the level of peanut specific IgE in their sera is lower than that of patients with respiratory and (or) gastrointestinal tract symptoms. 52% of patients had respiratory symptoms, including upper and lower respiratory tract symptoms, laryngeal edema, repeated coughing, voice changes, and asthma; 34% of patients have gastrointestinal symptoms, such as acute vomiting, abdominal pain, or diarrhea. In addition, the most serious manifestation of allergic reaction is cardiovascular abnormalities, such as hypotension and arrhythmias.

## 6.4 ALLERGEN DETECTION METHODS

There are many methods to judge whether the proteins in a food are allergens, and they are divided into in vivo and in vitro tests. In vivo tests include the skin prick test (SPT) and double-blinded placebo-controlled food challenges (DBPCFC); in vitro tests include the histamine releasing test (HRT) and the commonly used IgE assay.

### 6.4.1 In Vivo Tests

#### 6.4.1.1 Skin Prick Test (SPT)

SPT for Type-I hypersensitivity reaction is to let trace amounts of harmless suspected allergens enter the skin; if the corresponding anti-IgE antibody binds to the surface of subcutaneous mast cells, the allergens will bind to it; after a series of changes, mast cell degranulation will be caused and histamine and other chemical mediators will be released, thereby causing local vasodilation, increased permeability, wheals, and flush reaction. According to the clinical response, the presence of specific anti-IgE antibodies is confirmed, thereby determining the allergens. The rationality of SPT is that a small amount of allergens induce typical skin reactions, such as blister and erythema, and the biological activity of allergens is quantified by measuring the area of blister or erythema within certain concentration ranges (Poulsen et al., 1993, 1994). The results can be expressed by endpoint titers (ie, maximum concentrations for negative reactions) or 10 mg/mL histamine equivalent prick (10HEP).

Although SPT is simple and easy, it usually needs up to several tens of pricks and causes great pain and suffering to subjects. In addition, the patient must be consistent with and under strictly controlled experimental conditions, such as skin integrity, no use of drugs with great impact on skin; and it should be ensured that the test substance has no infectiousness or toxicity except for allergen characteristics. Skin test results are sometimes not entirely consistent with the medical history. Symptoms may not appear even when SPT proves positive; or conversely, allergy symptoms occur when SPT proves negative. Therefore, SPT results should be analyzed in combination with a serological test, to avoid unnecessary food restrictions.

#### 6.4.1.2 Food Challenge Test

An open food challenge (OFC), single-blind placebo-controlled food challenge (SBPCFC) or DBPCFC can be performed. In 1976, May took the lead in using DBPCFC for the diagnosis of food allergies and found that the SPT of only 29% of the children with positive DBPCFC proved positive. Bock et al. (1990) conducted DBPCFC on 500 pediatric cases and believed that this test was highly safe. DBPCFC was rated by the Academy of Allergy and Clinical Immunology (EAAC) as the only conclusive test for food allergen judgement and is considered the standard method for the diagnosis of food allergy.

DBPCFC should be conducted 7–14 days after an exclusion diet and after one night's fasting. Ten milligrams of food antigens is generally given first. If no related symptoms occur, the dose is doubled every 15–60 min until it reaches 8–10 g. If still no symptom occurs, such food allergies can be excluded.

## 6.4.2    In Vitro Tests

It is more safe and ethical to test and evaluate biological activity of food allergens with in vitro tests than in vivo tests. Currently, HRT and IgE assay are mainly applied in research and practice.

### 6.4.2.1    Histamine Releasing Test (HRT)

The rationale behind HRT is that allergens stimulate allergic target cells and cause the cells to release histamine; histamine levels can be determined with fluorescence assay or radioimmunoassay; the results are compared with a suitable control to determine the standard of positive histamine release rate, thereby determining and identifying the allergen activity. Typically, anti-IgE antibody is used as a positive control, and sample medium (eg, BBS) as a negative control, to exclude nonspecific release of histamine in the test. In addition, some allergens or extracts to be tested contain histamine in themselves and allergens or their coexisting substances may have target cell toxicity and induce nonspecific release of histamine. Therefore, sample controls and nonsensitized target cells controls must be set.

### 6.4.2.2    IgE Assay

The binding of allergens to IgE is the central link of allergen's role in biological activity. Therefore, the determination of specific IgE in allergens plays an important role in the diagnosis and evaluation of allergies.

#### 6.4.2.2.1    RAST inhibition and EAST inhibition

Since the successful purification of IgE and the preparation of anti-IgE antibody, the radioallergosorbent test (RAST) has been designed for application. Enzyme linked allergosorbent assays (EAST) improved later and fluorescent and chemiluminescent labeling assays also have been established (Ali et al., 2005). These methods are all immunoassays for detecting allergens and specific IgE interaction. The binding of specific anti-IgE antibodies to food allergens occurs on the surface of a solid phase carrier and is detected. It is a standardized in vitro diagnosis of food allergies and has been widely used.

On this basis, radioallergosorbent test inhibition (RAST inhibition) and enzyme linked allergosorbent assays inhibition (EAST inhibition) were further established. In in vitro allergen testing, these methods are the most widely used and most sensitive methods used clinically and by researchers on allergy worldwide. They are also the key techniques for evaluating total allergenic activity of allergens. A major deficiency of RAST and EAST inhibitions is their dependence on human serum. It is difficult to ensure the consistency of sera, so it is also difficult to standardize these two methods. Although they can be used to detect food allergens, the uncertainty of the specificity of human anti-IgE antibodies has greatly limited the application of these methods to a wider area.

### 6.4.2.2.2 IEF-PAGE and SDS-PAGE immunoblotting tests

The isolation and identification of a single allergen can be performed through isoelectric focusing (IEF) and SDS-PAGE electrophoresis (including dielectro-phoresis) in conjunction with an immunoblotting assay. Currently, SDS-PAGE immunoblotting is available for the assay of almost all food allergens. Dielec-trophoresis is often used for detecting homologous protein molecules with simi-lar molecular weights but different isoelectric points and amino acid sequences. Similar to RAST and EAST, immunoblotting can also eliminate cross-reaction between different food allergens from different sources by IgE inhibition test.

Immunoblotting test, also known as enzyme linked immune electro transfer blot, is also called Western blot because it is similar to the blotting method Southern blot for nucleic acid detection established by Southern. This method has combined high definition of SDS-PAGE, high specificity and sensitivity of ELISA and is an effective assay. It is widely used in the analysis of antigen components and immune competence and can also be used in disease diagnosis (see Fig. 6.1)

### 6.4.2.2.3 Double immunodiffusion assay (Ouchterlony)

This is a qualitative assay to compare two or more food allergens. Put antibod-ies and various protein samples simultaneously into the hole and make them

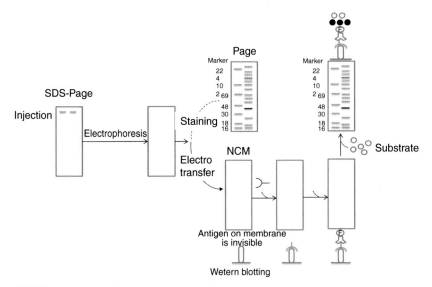

**FIGURE 6.1** Immunoblotting test schematic diagram.

diffuse to each other and form a precipitation line when they reach equilibrium. A double immunodiffusion test can be used to distinguish consistent, inconsistent, and partially consistent protein samples. Malmheden et al. (1994) analyzed the wheat gluten in pasta and buckwheat foods with double immunodiffusion assay. The deficiency of this method is that quantitative analysis cannot be carried out, and it has low sensitivity and is time-consuming (24–48 h).

### 6.4.2.2.4 Dot immunoblotting assay

According to recent reports, dot immunoblotting assay is widely used in detecting peanut allergens in a variety of foods. Samples are dropped onto a polyester substrate embedded in peanut protein antibodies and an immunoassay is conducted with secondary antibodies, and then the bound peanut proteins can be detected. This method has very high sensitivity and low cost.

### 6.4.2.2.5 Rocket immunoelectrophoresis (RIE)

This assay is performed with antibody-containing gel. Proteins in the sample move according to their respective electrophoretic mobility rates until "rocket-like" precipitate forms on the gel according to the antibody/antigen ratio constant. The strength of color of the rocket-like precipitation zone is directly related to the abundance of the sample protein. The precipitate formed can be stained with Coomassie Brilliant Blue or by immunostaining (the latter has higher sensitivity). There are rare reports on RIE's application in food allergen testing. Malmheden et al. (1994) tested the peanut protein presence in a variety of foods, such as eggs, milk, and hazelnuts, but did not give any data recovery rate and reproducibility. The deficiency of RIE lies in the complicated gel production and staining process.

### 6.4.2.2.6 Enzyme-linked immunosorbent assay (ELISA)

ELISA is an enzyme immunoassay method. In 1971, Engvall and Perlmann published a paper on IgG quantitative determination with enzyme-linked immunosorbent assay (ELISA), which developed the enzyme-labeled antibody technique used for antigen localization in 1966 into a method for determining trace substances in liquid samples. ELISA plays an important role in the determination of antigen residues. ELISA is a suppression model and was first introduced by Knights to describe the characteristics of hydrolysates. Sampson described in detail the determination of residue allergens in hydrolysates with this method.

In the late 1990s, ELISA developed rapidly. With the application of the automated ELISA analyzer, the specificity and sensitivity of ELISA have been greatly improved and its application in the field of food testing has been greatly expanded. The characteristic of ELISA is the solid phase and enzyme labeling of antigens or antibodies. Currently in the most commonly used ELISA, antibodies will bind to the surface of a solid phase carrier and have free reaction with the protein sample (antigens). After the reaction between antibodies and antigens, antibodies labeled by another enzyme (secondary antibody) are used to test sample proteins (antigens) bound to the surface of solid phase carrier. If the antibodies bind to the carrier of the fixed surface of solid phase, the more sample proteins (antigens) also bind and the stronger the color reaction of enzyme-labeled antibodies appear. Meanwhile, the higher the concentration of antibodies, the stronger ELISA color development will be. To test trace food allergens, biotin-labeled secondary antibodies and peroxidase-containing streptavidin are used in ELISA.

There is no uniform standard for the classification of ELISA. The commonly used assays include indirect ELISA double-antibody sandwich ELISA and competitive ELISA. Furthermore, there are also IgM antibody determination with capture-ELISA, ABS (avidinbiobin system)-ELISA, PCR-ELISA, and dot-immunobinding assay, etc. (Zhang, 2003).

1. Indirect ELISA

   Indirect ELISA is mainly used to determine antibodies. First, a solid phase carrier is coated with antigens, which must be soluble or at least extremely small particles; after washing, samples containing the antibodies to be assayed are added; then after incubation and washing, enzyme-labeled specific antibodies (eg, antihuman globulin IgG and IgM) are added; after incubation and washing, substrates for color development are added. The amount of degraded substrates is the amount of antibodies determined. The results can be quantitatively determined with a spectrophotometer.

2. Double-antibody sandwich ELISA

   In this technique, a solid phase carrier is also coated with specific antibodies; after washing, the antigen-containing samples to be assayed are added; if the corresponding antigens are present in the sample to be assayed, they will bind to the specific antibodies on the solid phase carrier; after incubation and washing, enzyme-labeled specific antibodies are added; then after incubation and washing, substrates are added for color development. The amount of degraded substrates is the amount of antigens determined. In this technique, the antigens to be assayed must have two sites that can bind to antibodies. That is because one end will interact with antibodies on the solid phase carrier and the other will interact with enzyme-labeled specific antibodies. Therefore, this technique cannot be used for the assay of haptens or other antigens with a molecular weight less than 5000 Da.

3. Competitive ELISA

   In this technique, first let specific antibodies adsorb on the surface of a solid phase carrier and then divide them into two groups after washing. A mixture of enzyme-labeled antigens and antigens to be assayed is added to one group and enzyme-labeled antigens are added to the other; after incubation and washing, substrates are added for color development. The difference between decomposed substrate amounts in the two groups is the amount of unknown antigens assayed.

In the field of food safety, ELISA has been widely used in the rapid assay of veterinary drug residues, pesticide residues, epidemic situations, and toxins of harmful microorganisms, plants, and animals, etc., and good results have been achieved. With people's in-depth awareness of an increased emphasis on food allergies, many ELISA techniques have been developed for the detection and quantitative analysis of food allergen total protein extracts and special food allergens. ELISA has become a series, trace, and commercial rapid assay of food safety and is currently one of the most widely used biological detection techniques.

## 6.4.3 PCR Analysis

PCR amplification technique is generally used for the detection and quantitative analysis of genetically modified components in a variety of foods (such as genetically modified soy and maize). And now, there are two PCR methods specifically for food allergens (hazelnut and wheat) testing (see Table 6.3). Both methods have a high sensitivity, as high as 0.001%.

ELISA can be used to detect proteins from known sources (including allergens) in food samples and PCR can detect DNA in the absence of related proteins. However, PCR has increased the probability of false positive assay results of allergen presence in food.

The advantage of the PCR technique is that it is time-saving. It takes only 7–10 d to study the DNA of a known sequence. The assay of purified allergens usually takes a month or even longer because specific antibodies need to be prepared first. The PCR technique is based on a simple and determined DNA sequence, while ELISA depends on animal antibodies with stable quality. The stability of DNA and proteins in food processing is to be considered. In summary,

**TABLE 6.3** Food Allergens cDNA Detection With PCR

| Method | Allergens Detected | Cross-Reactivity | Sample | Detection Limit |
|---|---|---|---|---|
| PCR amplification | Hazelnut (182 bp fragment in Cora 1) | No cross-reactivity with 30 kinds of fruits, nuts, beans, and other components studied | Chocolate bars, muesli, and peanut paste, etc. | Hazelnut 0.001 g/100 g |
| PCR amplification | Wheat gliadin ($\omega$-gliadin) | No cross-reactivity with oat | Oats, rice, oats products, etc. | Wheat flour: 0.001–0.01 g/100 g Gliadin: 0.04–4 mg/100 g |

the PCR technique is suitable for quantitative analysis of allergens in complex foods and should be vigorously promoted in this regard.

## 6.5   ALLERGEN FINGERPRINTS RAPID DETECTION

The earliest concept of fingerprints was from criminology and forensic science of the late 19th and early 20th centuries. It is the result of combining the latest scientific and technological achievements with the oldest traditions of medical science. In view of the practicality and scientificity of fingerprints and the similarity of food and medicine in many aspects, Wu et al. (2006) first proposed the application of fingerprint principles and techniques to related research of food safety, especially to the rapid assay of food allergens. With common allergenic foods as the object of study, their allergen total protein was extracted to study the feasibility of the application of food allergen total protein 2-DE fingerprints to the rapid detection of food allergens, and the good reproducibility and stability of this technique with different testing instruments and supplies. On this basis, the feasibility of the study of specific allergen protein dot MALDI-TOF/MS fingerprints on the basis of 2-DE fingerprints was further studied. The study shows a good analysis result and provides a solid theoretical and experimental basis for the application of fingerprints technique to the rapid assay of food allergens.

Although allergens in peanuts have a stable nature, some processing methods can have a significant impact on them. The allergenicity of peanut allergens can be changed by different physical methods, such as heat treatment and irradiation treatment, as well as enzymatic methods and genetic engineering biological methods. It is of great importance to explore the appropriate ways to reduce the allergenicity of peanuts to develop hypoallergenic peanut products and ensure the safety of peanut products.

### 6.5.1   Heat Treatment

As the major peanut allergens are proteins, heat treatment can affect the allergenicity of peanuts by changing the structure of their proteins. Studies (Chung, 2003) have shown that the allergenicity of peanut allergens is not changed at a temperature of 35–60°C, but it is increased when the temperature is above 77°C. The temperature of dry peanuts is generally not higher than 60°C, so drying and dehydration of peanuts does not increase their allergenicity. Beyer et al. (2001) have found in their studies that the peanuts' ability to bind with IgE is reduced by half after cooking. The Ara h1 monomers and trimers contents of peanuts are relatively reduced after frying. Although the Ara h2 and Ara h3 allergen content is not reduced, their ability to bind with IgE is significantly reduced. Studies (Hu et al., 2010a) have shown that the antigenicity of Ara h2 is significantly reduced after treatment at 85, 100, or 115°C and it is continuously reduced with the increase of temperature and time. Reddy et al.

(1995) tested ACE modified proteins with different diagnostic methods and found that glycosylation products from thermal processing were also important food allergens. Although allergens can be destroyed by heat treatment, there is also a loss of nutrients, organoleptic quality declines, processing characteristics decline, new allergens are generated, and other defects may occur. Therefore, the development of hypoallergenic peanut products with heat treatment desensitization method is subject to certain restrictions.

## 6.5.2 Enzymatic Technique

The enzymatic technique is to reduce the peanut allergenicity by modifying peanut epitopes and destroying the antigen determinants of peanut allergens. Studies (Cong et al., 2007) have shown that alcalase and alkaline protease are the optimal enzymes for reducing the allergenicity of roasted peanuts. Through hydrolysis under optimal conditions of 55°C temperature, pH 8.0, 3% substrate concentration, 3000 U/g protein of enzyme dosage (E/S), and 6 hours time, the allergenicity of peanut proteins is reduced by 34.5%. Studies (Chung et al., 2004) have shown that peroxidase (POD) can reduce the allergenicity of peanut proteins by causing polymerization of Ara h1 and Ara h2. You (2006) has found in her studies that hydrolysis with trypsin can effectively remove allergenic proteins from peanut milk. Under the conditions of 0:6000 enzyme dosage (E/S), sample pH 7.0, enzymolysis temperature of 55°C, and reaction time of 4 hours, the antibody titers of peanut milk proteins were reduced by 61.6%. Compared with heat treatment desensitization, the enzymatic technique has the advantages of high efficiency, mild reaction, and good controllability, etc. Due to the rapid development of the enzyme industry and the wide application of enzyme technology in food industry, efficient and specific enzymes can be screened for directional enzymolysis of allergens, or complex enzymes can be used for restrictive enzymolysis to control the hydrolysis process and optimize parameters. This research has pointed the way for the development of hypoallergenic peanut products. The enzyme technique is marked by being green and environmentally friendly and shows good safety. It is the optimal method for antigen protein degradation and will be widely applied to the development of hypoallergenic peanut products.

## 6.5.3 Irradiation Treatment

After ionizing irradiation, allergen molecules in food will produce free radicals, ions, and other activated particles, which will cause a series of chemical reactions and destroy their molecular conformation and structure, thereby achieving the effect of desensitization. There are two types of effects of irradiation on protein allergen immunogenicity in foods: (1) it destroys the primary structure of cell B or cell T antigenic epitope, (2) antigenic epitope is masked due to a cross-linking effect even if the primary spatial conformation or structure of immune proteins is not significantly changed. Currently, there are few studies on the peanut desensitization with irradiation technique. Wang et al. (2009d) have shown that small-dose irradiation of below 1.0 kGy has little effect on allergenic proteins in peanut products and solution. When the irradiation dose is above 1.0 kGy, the content of protein with molecular weight larger than 96.9 kDa in protein solutions is increased and that of other proteins is generally decreased. The advantages of irradiation as a high technology in food desensitization have become visible. It has a huge potential for development and is expected to become one of the major peanut desensitization techniques.

## 6.5.4 Genetic Engineering

Genetic engineering aims to eradicate protein allergenicity by eliminating endogenous genes of peanuts. cDNA of several major internationally recognized peanut allergens have been isolated, sequenced, cloned, and recombined. Currently, scholars are exploring gene "silence" to inhibit the expression of major allergens, thereby reducing the allergenicity of peanut products. At present, this technique has made certain progress. Hu et al. (2010b) showed that cloned and recombinant Ara h2 has good immunogenicity. Yi et al. (2011) successfully cloned, expressed, and purified peanut allergen Ara h8. The recombinant protein of this gene expression had good immunogenicity. Zhong et al. (2009) have shown that recombinant iso-Ara h3 protein has a serum IgE recognition rate of 12.5% and is an allergen protein with low allergenicity. Yi et al. (2011) cloned and expressed the peanut allergen Ara h2.02, and the recombinant protein of this gene expression had good immunogenicity. The ability of genetically engineered Ara h2 protein (S-Ara h2) to combine with IgE in the mixed sera of peanut allergic patients was significantly reduced compared with that of recombinant Ara h2 protein (R-Ara h2). Yi et al. (2011) also successfully constructed sequence-recombined Ara h2 expression vectors, and the recombinant protein of this gene expression had good immunogenicity. The allergenicity of allergens is reduced or eliminated by genetic engineering while the original integral immune competence is maintained, therefore becoming more secure in clinical trials. Genetic recombination is an important research direction for allergic reactions today and in the future and it will open up a new way for the development of hypoallergenic peanuts.

# References

Adamson, G., et al., 1999. HPLC methods for the quantification of procyanidins in cocoa and chocolate samples and correlation to total antioxidant capacity. Agric. Food Chemistry 47 (10), 4184–4188.

Ahmed, E., et al., 1998. Extraction and purification of lectin from florunner peanut seeds. Peanut Sci. (15), 44–47.

Alfred, S., et al., 1983. Sterol concentrates, the preparation thereof, and their use in the transformation of sterols by fermentation, US Patent 4374776.

Ali, R., et al., 2005. Latest approaches to the diagnosis and management of food allergies in children. J. Pak Med. Assoc. 55 (10), 458–462.

Aliou, M., 1995. Oligomeric proanthocyanidins possessing a doubly linked structure from Pavetta owariensis. Phytochemistry 38 (3), 719–723.

Aminlari, M., et al., 1977. Protein dispersibility of spray-dried whole soybean milk base: effect of processing variable. J. Food Sci. (42), 985–988.

An, J., Yue, Z., 2004. The application of soybean lecithin in feed in the modernization of agriculture. Modernizing Agric. (7), 8.

An, H., et al., 2010. Exraction and preliminary study on antioxidant activity of polyphenol from peanut. Anhui Agric. Sci. Bull. 16 (22), 45–47.

Andersen, P., et al., 1998. Fatty acid and amino acid profiles of selected peanut cultivars and breeding lines. J. Food Compos. Anal. (11), 100–111.

Angsupanich, K., et al., 1999. Effects of high pressure on the myofibrillar proteins of cod and turkey muscle. J. Agric. Food Chem. 47 (1), 92–99.

Bamdad, F., et al., 2006. Preparation and characterization of proteinous film from lentil (Lens culinaris): edible film from lentil (Lens culinaris). Food Res. Int. 39 (1), 106–111.

Bao, et al., 2002. Determination of phytosterol and cholesterol in oil by capillary gas chromatography. Chin. J. Anal. Chem. 30 (12), 1490–1493.

Basha, S., Pancholy, S., 1982. Isolation and characterization of two cryoproteins from Florunner peanut (Arachis hypogaea L.) seed. J. Agric. Food Chem. 30 (1), 36–41.

Basha, S., 1992. Effect of location and season on peanut seed protein and polypetide composition. J. Agric. Food Chem. (40), 1784–1788.

Beyer, K., et al., 2001. Effects of cooking methods on peanut allergenicity. J. Allergy Clin. Immunol. 107 (6), 1077–1081.

Bi, H., 2006. Studies on extraction, purification and antioxidant activity of trans-resveratrol in brewing grape wine dregs, Ha'er'bing. Master degree Thesis of Northeast Agricultural University.

Bock, S.A., Atkins, F.M., 1990. Patterns of food hypersensitivity during sixteen years of double-blind, placebo-controlled food challenges. J. Pediatr. 117 (4), 561–567.

Bockisch, M., 1983. Vegetable fats and oils. Fats and Oils Handbook, AOCS Press, Champaign.

Bovi, M., 1983. Genotypic and environmental effects on fatty acid composition, iodine value, and oil content of peanut (Arachis hypogaea L.) NT: University of Florida, no. 44, pp. 406.

Brantl, V., 1979. Novel opioid peptides derived from caseins (β-casomorphins. Isolation from bovine casein peptone. Hoppe-Seylers Z. Physiol. Chem. 360, 211–1216.

Brantl, V., Teschemacher, H., 1979. A material with opioid activity in bovine milk and milk products. Naunyn-Schmiedeberg's Arch. Pharmacol. 306 (3), 301–304.

Briscoe, B., et al., 2002. Pressure induced changes in the gelation of milk protein concentrates. Prog. Biotechnol. 19, 445–452.

Broadhurst, R., Jones, W., 1978. Analysis of condensed tannins using acidified vanillin. J. Sci. Food Agric. 29 (9), 788–794.

Bryant, R., et al., 2004. α- and β-galactosidase activities and oligosaccharide content in peanuts. Plant Foods Hum. Nutr. 58, 213–223.

Burks, A.W., Helm, 1997. Major peanut allergen Ara h. Biotechnol. Adv. 15 (2), 445.

Burks, A.W., et al., 1991. Identification of a major peanut allergen, Ara h I, in patients with atopic dermatitis and positive peanut challenges. J. Allergy Clin. Immunol. 88 (2), 172–179.

Burks, A.W., et al., 1992. Identification and characterization of a second major peanut allergen, Ara h II, with use of the sera of patients with atopic dermatitis and positive peanut challenge. J. Allergy Clin. Immunol. 90 (6), 962–969.

Burks, W., et al., 1998. Peanut allergens. Allergy (53), 725–730.

Byun, H., Kim, S., 2001. Purification and characterization of angiotensin converting enzyme (ACE) inhibitory peptides from Alaska Pollack (Theragra chalcogramma) skin. Process Biochem. 36, 1155–1162.

Cai, Y., 2010. Studies on extracting of resveratrol and its biological activities, Bejing. Master degree thesis of Beijing University of Chemical Technology.

Cao, Y, 2001. Studies on the extraction and purification of resveratrol from Polygonum Cuspidatum, Changsha. Master degree thesis of Hunan Agricultural University.

Chai, W., 2003. The main methods to analysis of dietary fiber and its application in China. Food Nutr. China 9 (8), 36–38.

Chen, Y., Duan, N., 1994. Study on character of peanut germplasm in Fujian. J. Peanut Sci. 4, 9–10.

Chen, M., Huang, Q., 2005. Study on anti-tumor activity of phytosterols acetate and phytosterols oletate. Cereals Oils (5), 16–18.

Chen, F., Xu, Y., 2008. Research progress on extraction and physiological function of dietary fiber. J. Fujian Fisheries (2), 51–54.

Chen, H., et al., 1995. Structural analysis of anti-oxidative peptides from soybean β-conglycinin. J. Agric. Food Chem. 43, 574–578.

Chen, L., et al., 2000. Separation and analysis of resveratrol and piceid in polygonum Cuspidatum Sieb. Et Zicc. J. Instrum. Anal. 19 (4), 60–62.

Chen, M., et al., 2002. The research on the scavenging hydroxyl free radical isolated from soy protein hydrolysate. Food Sci. 23 (1), 43–46.

Chen, X., et al., 2003. Cardiovascular Pharmacology, third ed. People's Medical Publishing House Co. Ltd, Beijing, China, pp. 352–353.

Chen, M., et al., 2005. Synthesis of phytosterols oletate. China Oils Fats 30 (6), 63–65.

Chen, S., et al., 2007a. Principal component analysis and cluster analysis on biological characters of the peanut cultivars. J. Peanut Sci. 2, 28–34.

Chen, T., et al., 2007b. The analysis on nutritional qualities of peanut cultivars. Chin. Agric. Sci. Bull. 23 (11), 141–145.

Chen, Z., et al., 2007c. Determination of procyanid in B2 in grape seed extract by RP-HPLC. Chin. Traditional Med. Patent 29 (11), 1645–1647.

Chen, J., et al., 2009. Sustainable development in peanut production of HUluodao. Agric. Econ. (5), 38.

Chen, Q., et al., 2013. High pressure-microwave assisted extraction of resveratrol from peanut roots. Food Sci. 20, 43–48.

Chen, S., et al., 2006. Studies on the extraction of dietary fiber from peanut shell and its properties determination. Food Ind. 1, 45–47.

Chen, X., 2004. The content of resveratrol, piceid in polygonum cuspidatum and resveratrol in arachis hypogaea, Fujian. Master degree thesis of Fujian Normal University.

Chen, H., 2007a. Optimization conditions for the extraction of ergosterol from fermented rice of monascus by using ultrasonic. Anhui Agric. Sci. Bulletin Bull. 13 (7), 84–85.

Chen, S. 2007. The study on synthesis of phytosterols ester by microwave heating, Xi'an. Master degree thesis of Xi'an University of Technology.

Chen, C., 2011. Research progress of purification technologies for resveratrol. Fine Specialty Chem. 19 (3), 39–42.

Cheung, H., et al., 1980. Binding oligopeptide substrate and inhibition of angiotensin-converting enzyme importance of the COOH-terminal dipeptides sequence. J. Biol. Chem. 255, 401–407.

Christian, F., 1996. Process for tocopherols and sterols from natural sources, US Patent, 5487817.

Chu, Y., Kung, Y., 1998. Study on vegetable oil blends. Food Chem. 62 (2), 191–195.

Chun, W.P., et al., 2000. A comparison study on allergen components between Korean (Arachis fasgigiata Shinpung) and American peanut (Arachis hypogaea Runner). Korean Med. Sci. (15), 387–392.

Chung, S.Y., 2003. Linking peanut allergenicity to the processes of maturation, curing, and roasting. J. Agric. Food Chem. (51), 4273–4277.

Chung, S.Y., et al., 2004. Allergenic properties of roasted peanut allergens may be reduced by peroxidase. J. Agric. Food Chem. 52, 4541–4545.

Cifuentes, A., et al., 2001. Fast determination of procyanidins and other phenolic compounds in food samples by micellar electrokimetic chromatography using acidic buffers. Electrophoresis 22, 1561–1567.

Clarke, M., et al., 1998. Serological characteristics of peanut allergy. Clin. Exp. Allergy (28), 1251–1257.

Cong, Y., et al., 2007. The effect of cooking methods on the allergenicity of peanut. Food Agric. Immunol. 18 (1), 53–65.

Cushman, D., et al., 1973. Inhibition of angiotensin-converting enzyme by analogs of peptides from Bothrops jararaca venom. Cell. Mol. Life Sci. 29, 1032–1035.

Dai, J., et al., 2005. Determination of soybean phytosterols by reversed phase chromatography and gas chromatography-mass spectrometry. Chin. J. Anal. Chem. 33 (12), 1725–1729.

David, T., et al., 1998a. Food allergy in children. Postgrad. Med. J. 81 (961), 693–701.

David, T., et al., 1998b. Resolution of peanut allergy. Br. Med. J. (317), 1317–1319.

De Jong, E.C., et al., 1998. Identification and partial characterization of multiple major allergens in peanut proteins. Clin. Exp. Allergy (28), 743–751.

Dello, S., et al., 2004. Influence of dietary fiber addition on sensory and rheological properties of yoghurt. Int. Dairy J. 14, 263–268.

Deng, J., 1981. Effect of temperature on fish alkaline protease, protein interaction and texture quality. J. Food Sci. 46 (1), 62–65.

Deng, Q., et al., 2005. Enzymic hydrolysis of ginkgo albumin protein and its antioxidant activity. Trans. CSAE 21 (11), 155–159.

Dev, Sagarika, et al., 2006. 2,2,2-Trifluoroethanol-Induced structural change of peanut agglutinin at different pH: A comparative account. IUBMB Life 58 (8), 473–479.

Ding, A., et al., 2005. Technology of extracting luteolin from peanut hull. J. Chin. Cereals Oils Assoc. 20 (4), 92–95.

Dong, T., 2008. Chemical synthesis and antioxidant activity of phytosteryl esters, Wuxi, Master degree thesis of JiangNan University.

Dong, C., Wu, J., 2011. The research on the dandelion terol extraction and its anti-inflammatory activity. HeiLongJiang XuMu ShouYi (3), 133–134.

Douglas, R., Sturrock, E., 2012. Structure-based design of domain-selective angiotensin-converting enzyme inhibitors. Drug Discovery Africa, 355–377.

Du, Y., 2011. Principal component analysis and comprehensive evaluation on quality traits of peanut parents. J. Plant Genet. Resources (4), 5.

Du, Y., 2012. Study on the preparation and gel properties of peanut protein fraction, Beijing, Master degree thesis of Chinese Academy of Agricultural Sciences.

Du, Y., et al., 2013. Major protein fractions and subunit contents in peanut from different cultivars. Food Sci. 34 (9), 42–46.

Dun, X., Chen, Z., 2004. Enzyme hydrolysis preparation of rice small polypeptides from rice residue. Food Sci. 25 (6), 113–116.

Duncan, K., Gilmour, I., 1999. Process for extraction of proanthocyanidins from botanical material, USA Patent 5968517.

Feng, Z., Liu, Q., 2004. Study on preservative properties of peanut–protein-isolate film. Food Sci. Technol. (9), 16–19.

Feng, T., Liu, H., 2005. Analysis of nature VE and plant sterols by HPLC. Anal. Instrum./Fenxi yiqi 3, 63–64.

Feng, Z., Zhong, S., 2009. Study on the application of soluble dietary fibre extracted from peanut. Food Fermentation Technol. 45 (5), 58–60.

Feng, S., et al., 2006. The establishment of phytosterols determination in plant oils and the analyze of phytosterols content in edible oil. Chin. J. Food Hygiene 18 (3), 197–201.

Feng, Z., Huang, D., 2007. The effect of reducing agent on preservative properties of peanut–protein-isolate film. Food Sci. Technol. (3), 242–246.

Feng, Y., et al., 2011. Optimization of extracting insoluble dietary fiber from peanut shell by response surface methodology. China Oils Fats 36 (5), 71–73.

Fizet, C., 1996 Process for tocopherols and sterols from natural sources, U.S. Patent 5,487,817.

Friedman, M., et al., 1982. Inactivation of soya bean trysin inhibitors by thiols. J. Sci. Food Agric. (33), 165–172.

Fu, W., et al., 2002. Determination of the procyanidins in the grape seed extract by ulteoviolet-colorimetry. Food Res. Dev. 23 (6), 90–92.

Fu, X., 2004. Plant sterols extraction and solubility determination, Hangzhou, Master degree thesis of Zhejiang University.

Galus, S., et al., 2012. Effect of modified starch or maltodextrin incorporation on the barrier and mechanical properties, moisture sensitivity and appearance of soy protein isolate-based edible films. Innovative Food Sci. Emerging Technol. 16, 148–154.

Gao, H., et al., 1995. Peanut protein stability and processed products process conditions of research. J. Chinese Foreign Technol. 12, 39–40.

Gao, Z., et al., 2007. Study on phytosterol extraction from crude rapeseed oil. Acad. Period. Farm Prod. Process. (11), 31–33.

Gao, L., et al., 2010. Production optimization of β-sitosterol microcapsules and their protective effect against lipid oxidation. Food Sci. 3 (21), 28–32.

Garini, M., et al., 2000. UVB-induced of rat erythrocytes: protective effect of Procyanidins from grape seeds. Life Sci. 67 (15), 1799–1814.

Ghanbarzadeh, B., et al., 2006. Effect of plasticizing sugars on rheological and thermal propertiesof zein resins and mechanical properties of zein films. Food Res. Int. 39, 882–890.

Ghanbarzadeh, B., et al., 2007. Effect of plasticizing sugars on water vapor permeability, surface energy and microstructure properties of zein films. LWT-Food Sci. Technol. 40, 1191–1197.

Gill, I., et al., 1996. Biologically active peptides and enzymatic approaches to their production. Enzyme Microbial Technol. 18, 162–183.

Glonek, T., 1998. 31P nuclear magnetic resonance phospholipid analysis of anionic-enriched lecithin. J. Am. Oil Chem. Soc. 75 (5), 569–575.

Gosal, W., Ross-Murphy, S., 2005. Gelation of whey protein induced by high pressure. Curr. Opin. Colloid Interface Sci. (34), 188–194.

Govindaraj, G., Jain, V., 2011. Economics of non-oil value chains in peanut: a case of peanut-candy and salted-peanut small-scale units in India. J. Agric. Sci. (1), 37–54.

Grimble, G., Silk, D., 1990. Intravenous protein hydrolysates-time to turn the clock back. Clin. Nutr. 9 (1), 39–45.

Gu, L., et al., 2002. Fractionation of polymeric procyanidins from lowbush blueberry and quantification of procyanidins in selected foods with an optimized normal-phase HPLC-MS fluorescent detection method. Agric. Food Chem. 50 (17), 4852–4860.

Guan, M., et al., 2008. Determination of phosphatidylcholine in lecithin capsule by UV. China Brewing (23), 95–96.

Guandlupe, P., Rosario, G., 2005. High pressure and the enzymitic hydrolysis of soybean whey proteins. Food Chem. 85 (4), 641–648.

Guang, C., Phillips, R., 2009. Purification, activity and sequence of angiotensin I converting enzyme inhibitory peptide from alcalase hydrolysate of peanut flour. J. Agric. Food Chem. 57 (21), 10102–10106.

Guo, H., 2009. Enzymatic Preparation and Antioxidant Activity of Wheat Germ Protein Hydrolysates, Wuxi. Doctor degree thesis of JiangNan University.

Guo, H., et al., 2010. Molecular mechanism study of targeting of two angiotensin-converting enzyme inhibitory peptides. Food Sci. 31 (23), 1–5.

Han, J., 2004. Extraction of phytosterols from sunflower seed oil deodorizer distillates. China Oils Fats 12, 26.

Han, B., 2010. A preliminary study on the preparation of peanut meal oligosaccharides and polysaccharides, Beijing. Master degree thesis of China Agricultural University.

Han, X., et al., 2006. Determination of resveratrol in peanut stem by thin layer chromatography fluorescence scanning. Chem. Res. 18, 628–630.

Han, X., et al., 2010. Determination of resveratrol in peanut root by thin layer chromatography fluorescence scanning. Chem. Res. 21 (5), 88–89.

Han, B., et al., 2011. Optimization of primary extraction of oligosaccharides from defatted peanut cake. Sci. Technol. Food Ind. 32 (07), 285–288.

Han, S., Luan, W., et al., 1998. The initial analysis of the resources of fat and protein amino acids of peanut varieties introduced from abroad. J. Shandong Agric. Sci. 6, 30–34.

Hao, L., Xu, Z., 2008. Application of natural phospholipids in cosmetics. Cereals Oils Process. (10), 87–89.

Hao, H., et al., 2005. Changes in phenols of cider during storage analyzed by HPLC. Food Sci. Technol. 5, 74-76+93.

Hao, P., 2006. Analysis of the plant sterol component and content of edible rape oil with gas cheromatography. J. Anhui Agric. Sci. 34 (19), 4830.

Hollecker, L., et al., 2009. Simultaneous determination of polyphenolic compounds in red and white grapes grown in Sardinia by high performance liquid chromatography-electron spray ionisation-mass spectrometry. J. Chromatogr. A 1216, 3402–3408.

Holley, K., Hammons, R. 1968. Strain and seasonal effects on peanut characteristics. Res. Bull. 32 Ga Agric. Exp. Stn.

Hong, X., et al., 2009a. Study on determination conditions of proanthocyanidins from grape seeds by colorimetry using ferric ions as catalyst. J. Henan Univ. Technol. Natural Sci. Ed. 30 (6), 41–45.

Hong, Y., et al., 2009b. Study on extraction of fiber in peanut. China Food Additives 2, 102–104.

Hu, Y., Zhang, S., 2002. New progress on the effect of resveratrol anti-tumor. Foreign Med. Sci. (Cancer Section) 29 (3), 174–177.

Hu, X., 2006. Study on extraction of resveratrol from polygonum cuspidatum siobet zucc and contents investigation, Ya'an. Master degree thesis of Sichuan Agricultural University.

Hu, X., et al., 2001. Prospects of phytocholesterols. China Western Cereals Oils Technol. (5), 34–36.

Hu, X., et al., 2003. Phytosterol extraction from soybean oil deodorizer distillate. China Oils Fats 28 (12), 47–49.

Hu, C., et al., 2010a. Effect of heat treatment on the antigenicity and conformation of peanut allergen Ara h 2. Spectrosc. Spectral Anal. 30 (9), 2550–2554.

Hu, C., et al., 2010b. Preliminary investigation on the biosynthesis of recombinant peanut allergen Ara h 2. Food Sci. 31 (15), 203–207.

Hu, Y., 2011. Grain and Oil Processing Technology. Chemical Industry Press, Beijing.

Huang, G., 2004. Research on zein about extraction by ultrasound, modification and medicinal release performance, Guangzhou. Doctor degree thesis of South China University of Technology.

Huang, D., 2007. Research on determination methods of resveratrol from grape seed ultrafine-power, Xi'an. Master degree thesis of Northwest A & F University.

Huang, S., Fu, J., 1992. Storage protein in peanut seeds. Peanut Sci. Technol. (1), 1–6.

Huang, S., Jiarui, F., 1992. The relation of storage protein to vigor and its mobilization pattern in germinating peanut seeds. Acta Bot. Sin.

Huang, C., et al., 1994. Extrantion of sterols by supercritical carbon dioxide. GuangZhou Chem. (1), 32–35.

Huang, M., et al., 1996. Color, Taste and Odor Chemistry of Food, second ed. China Light Industry Press, Beijing, China, p. 45.

Huang, J., et al., 2006. Status and trends in peanut resveratrol research and development. Food Nutr. China 2, 20–23.

Huang, S., et al., 2008. Technology of extracting luteolin from peanut hull by orthogonality experiments. Chem. Ind. Times 22 (3), 47–49.

Huang, M., et al., 2010a. Study on the preparation technique of phytosterol in rice bran oil by molecular distillation. China Food Additives (5), 216–219.

Huang, W., et al., 2010b. Advances on extraction, determination and biological action of resveratrol. Food Mach. 26 (6), 148–152.

Hui, D., et al., 2007. A study on effect of molecular distillation on concentration of phytosterols. Cereals Oils Process. 6, 92–94.

Hui, A., et al., 2010. Extraction and preliminary study on antioxidant activity of polyphenol from peanut. Anhui Agric. Sci. Bull. 16 (22), 45–47.

Hui, C., et al., 2011. Study on enzyme extraction of insoluble dietary fiber from peanut meal. Food Ind. 1.

Jamdar, S., et al., 2010. Influence of degree of hydrolysis on functional properties, antioxidant activity and ACE inhibitory activity of peanut protein hydrolysate. Food Chem. 121 (1), 178–184.

Jangchud, A., Chinnan, M.S., 1999. Properties of peanut protein film: sorption isotherm and plasticizer effect. LWT-Food Sci. Technol. 32 (2), 89–94.

Jeanne, E., et al., 2005. Influence of traditional processing methods on the nutritional composition and antinutritional factors of red peanuts (Arachis hypogea) and small red kidney beans (Phaseolus vulgaris). J. Biol. Sci. 5 (5), 597–605.

Jian, J., et al., 2010. Study progress on the modification of peanut protein. Acad. Period. Farm Prod. Process. 1, 12.

Jiang, H., et al., 1998. Evaluation of groundnut germplasm. Chin. J Oil Crop Sci. 20 (3), 31–35.

Jiang, H., et al., 2006. Genetic diversity of peanut genotypes with resistance to bacterial wilt based on seed characters. Chinese J. Oil Crop Sci. 28 (2), 144–150.

Jiang, Y., et al., 2007. Effect of processing parameters on the properties of transglutaminase-treated soy protein isolate films. Innovative Food Sci. Emerging Technol. (8), 218–225.

Jiang, Y., et al., 2010. The research trend of natural antioxidant proanthocyanidin. Guangxi J. Light Ind. 10, 14–15.

Jianmei, Y., et al., 2005. Effects of processing methods and extraction solvents on concentration and antioxidant activity of peanut skin phenolics. Food Chem. 90, 199–206.

Jiao, J., et al., 2010. Study progress on the modification of peanut protein. Acad. Period. Farm Prod. Process. (1), 40–42.

Jie, S., et al., 2011. Study on properties of lectin in seed of peanut (Arachis hypogaea L.). Sci. Technol. Food Ind. 32 (8), 134–137.

Jin, Q., et al., 2007. The research on rapeseed sterol ester to reduce lipid. Oil Eng. (10), 2–4.

Jong, D., Zijverden, V., 1998. Identification and partial characterization of multiple major allergens in peanut proteins. Clin. Exp. Allergy 28 (6), 743–751.

Jung, W., et al., 2006. AngiotensinI-converting enzyme inhibitory peptide from yellowfin sole (Limanda aspera) frame protein and its antihypertensive effect in spontaneously hypertensive rats. Food Chem. 94, 26–32.

Kakgawa, H., et al., 1985. Inhibitory effects of tannins on hyaluronidase activation and on the deregulation from rat mesentery mast cell. Chem. Pharm. Bull. 33 (1), 5079–5082.

Kannel, W., 1996. Blood pressure as a cardiovascular risk factor. Jam Med. Ssoc. (275), 1571–1576.

Karchesy, J., Hemingway, R., 1986. Condensed Tannins: (4β-8; 2β-O-7)-Linked Procyanidins in Arachis hypogeal L. J. Agric. Chem. 34, 966–970.

Kella, N., Poola, I., 1985. Sugars decrease the thermal denaturation and aggregation of arachin. Int. J. Peptide Protein Res. 26 (4), 390–399.

Kester, J.J., Fennema, O., 1989. Resistance of lipid films to water vapor transmission. J. Am. Oil Chem. Soc. 66 (8), 1139–1146.

King, T.P., et al., 1995. Allergen nomenclature. J. Allergy Clin. Immunol. (96), 5–14.

Kleber-Janke, T., et al., 1999. Selective cloning of peanut allergens, including profilin and 2S albumins, by phage display technology. Int. Arch. Allergy Immunol. 119 (4), 265–274.

Koppelman, S., et al., 2003. Peanut allergen Ara h 3: isolation from peanuts and biochemical characterization. Allergy 58 (11), 1144–1151.

Kreimever, J., et al., 1998. Separations of flavan-3-ols and dimeric proanthocyanidins by capillary electrophoresis. Planta Med. 64 (1), 63.

Krishna, T., et al., 1986. Variation and inheritance of the arachin polypeptides of groundnut (Arachis hypogaea L.). Theor. Appl. Genet. 73, 83.

La, Y., et al., 2003. The properties and applications of phospholipids. China Leather 32 (21), 34–36.

Lan, T., 2007. Study on extraction procession of resveratrol, Tianjin. Master degree thesis of Tianjin University.

Lee, J., et al., 2008. Characterization of protein-coated polypropylene films as a novel composite structure for active food packaging application. J. Food Eng. 86, 484–493.

Lehman, K., et al., 2003. High-yield expression in Escherichia coli, purification, and character-relation of properly folded major peanut allergen Ara h2. Protein Expression Purification (31), 250–259.

Lehrer, S.B., et al., 1996. Whey are proteins allergic? Implications for biotechnology. Crit. Rev. Food Sci. Nutr. 36 (6), 553–564.

Li, Y., Li, C., 2007. Research progress on determination method of Proanthocyanidins. Sci. Technol. Assoc. Forum 8, 28–29.

Li, Y., Li, C., 2008. Recent studies on procyanidins extraction, separation and purification methods. Food Eng. 1, 9–11.

Li, M., Wang, J., 2006. The producing of phytosterol-ester from the deodorizer distillate catalyzed by enzyme. Mod. Food Sci. Technol. 22 (3), 96–99.

Li, H., Zhang, H., 2001a. Analysis of sensitizing components for peanut allergen. Chin. J. Microbiol. Immunol. 21, 12–15.

Li, M., et al., 2004. Conditions optimization of peanut protein isolate production by alkaline extraction and acid precipitation. Chin. Fats 29 (11), 21–23.

Li, G., et al., 2005a. Alcalase hydrolysates of peanut protein isolates inhibit angiotensin I-converting enzyme activity. Food Sci. 26 (6), 55–60.

Li, M., et al., 2005b. Determination of Flavonoid and Luteolin in the extract from peanut shell. Chem. Res. 16 (2), 75–77.

Li, T., et al., 2006. Present research situation and expectation of the ultrasound extraction. J. Anhui Agric. Sci. 34 (13), 3188–3190.

Li, Y., et al., 2008. Current situation and prospect of resveratrol research. China Brewing 7, 10–12.

Li, J., et al., 2009. Ultrasonic extraction of resveratrol from peanut. J. Chin. Cereals Oils Assoc. 24 (6), 118–122.

Li, H., et al., 2010a. Optimization of extraction of soluble dietary fibre from peanut hull by aspergillus niger solid-state fermentation. Food Sci. 31 (19), 277–282.

Li, H., et al., 2010b. Optimization of extraction processing of soluble dietary fiber from peanut stems by microwave technology. Food Sci. 31 (2), 221–225.

Li, Q., et al., 2011. Principal component and genetic distance determine in 27 peanut germplasm resources. J. Plant Genet. Resources 12 (4), 519–524.

Li, X., et al., 2011a. Advances in resveratrol research of grape. Acta Hortic. Sin. 38 (1), 171–184.

Li, Y., et al., 2011b. Study on hydrolysis of potato starch by α-amylase and glucoamylase. J. Northwest A&F University (Nat. Sci. Ed.) 39 (7), 147–152.

Li, W., et al., 2012a. Evaluation methods of gel properties of peanut protein isolate with different varieties. China Oils Fats 37 (7), 20–23.

Li, W., et al., 2012b. Research on processing characteristics and quality evaluation of protein suitability of peanuts, Beijing. Doctor degree thesis of Chinese Academy of Agricultural Sciences.

Li, H., Zhang, H., 2001. Analysis of sensitizing components for peanut allergen. Chinese J. Microbiol. Immunol. 2 (Suppl 21), 12–15.

Li, F., 2002. Research progress on the treatment of food allergen. Foreign Med. Sci. (Section Pediatr.) 29 (2), 103–105.

Li, L., 2005. Study on extraction, toxicology and functional properties of dietary fibers from seaweeds, Qingdao. Doctoral Degree Thesis, Ocean University of China.

Li, G. 2005. Studies on angiotensin-converting enzyme inhibitory peptides derived from food proteins, Wuxi. Master degree thesis of JiangNan University.

Li, P., 2008, Influence of high pressure microfluidization on the functional properties of araehin and its mechanism study, Nanchang. Doctor Degree Thesis, Nanchang University.

Li, H., 2010. Study on the extraction technology of luteolin from peanut shell by ultrasonic wave. Hubei Agric. Sci. 49 (5), 1183–1185.

Li, Y. 2010. Thiolysis-HPLC analysis of proanthocyanidins in fruit and their products, Shijiazhuang, Master degree thesis of Agricultural University of Hebei.

Liang, X., et al., 2006. Differential induction of resveratrol in susceptible and resistant peanut seeds infected by Aspergillus flavus. Chin. J. Oil Crop Sci. 28 (1), 59–62.

Liang, J., 1999. Sterilization of pepper powder. Sci. Technol. Food Ind. 1 (1), 63–64.

Liao, C., Liao, Z., 2004. Application of peanut hull: I. Preparation of heavy metal ion adsorbent. J. Guangxi Teachers Educ. Univ. (Nat. Sci. Ed.) 21, 68–70.

Liao, D., et al., 2006. Enzymatic hydrolysis and screening of prodrug-type angiotensin I-converting enzyme inhibitor from chicken egg yolk. Fine Chem. 23 (8), 757–759.

Liao, G., et al., 2010. Optimization of ultrahigh pressure extraction for polydatin and resveratrol from Polygonum cuspidatum by using uniform design. China J. Chin. Mater. Med. 35 (24), 3282–3285.

Liao, C., 2004. Industrial application of peanut hull. Technol. Dev. Chem. Ind. 33 (2), 22–35.

Liao, W., 2008. Study on extraction and purification of Resveratrol from grape skin, Urumqi. Master degree thesis of Xinjiang Agricultural University.

Licher, M., 2005. Strategic options for the Virginia Peanut Industry after the 2002 Farm Bill: a Linear Programming Model. Master of Science in Agriculture and Applied Economics Thesis, Faculty of the Virginia Polytechnic Institute and State University.

Lin, L., et al., 2002. Immunomodulatory proanthocyanidins from *Ecdysanthera utilis*. J. Nat. Prod. 65 (4), 505–509.

Lin, W., et al., 2006. Research and application of twin. Trans. Chin. Soc. Agric. Eng. 6.

Liompart, M.P., et al., 2007. Evaluation of supercritical fluid extraction, microwave-assisted extraction and sonication in the determination of some phenolic compounds fron various soil matrices. J. Chromatogr. A 74 (1-2), 243–251.

Liu, Y., Cao, Y., 2002. Research prospect of development of biological protein peptides. Food Sci. 23 (8), 319–320.

Liu, R., Li, H., 2007. Physiological function of plant phytosterols and its application in food. China Healthc. Innovation 2 (18), 109–110.

Liu, X., Ma, Y., 2005. Quality determination and adulteration identification of peanut cake. Henan Animal Husbandry Vet. Med. 26 (6), 33.

Liu, J., Xi, Y., 1994a. Characteristics and Changes of Food Chemistry and Food Ingredients, first ed. Science Press, Beijing.

Liu, J., Xi, Y., 1994b. Food Chemistry-Food Composition Features and Changes. Science Press, Beijing, pp. 114–140.

Liu, C., et al., 2004. Physical and mechanical properties of peanut protein films. LWT-Food Sci. Technol. (37), 731–738.

Liu, D., et al., 2005. Research on extraction technology of resveratrol and procyanidin from red skin of peanut. J. Food Sci. 26 (7), 144–148.

Liu, D., et al., 2008. Simultaneous preparation of peanut oil and defatted protein powder by low-temperature prepressing, extraction and low-temperature desolventizing. Chinese Oils Fats 33 (12), 13–15.

Liu, L., et al., 2005. Determination of total sterol in soybean by spectrophotometry. China Oils Fats 30 (4), 46–47.

Liu, F., et al., 2006a. Determination of sterol by HPLC with evaporative light-scatter detector. J. Anal. Sci. 18 (3), 230–232.

Liu, H., et al., 2006b. Determination of resveratrol by RP-HPLC from peanut cake, peanut oil and raisins. Chin. Traditional Herbal Drugs 37 (8), 1188–1189.

Liu, C., et al., 2009a. Ultrasonic extraction and HPLC detection of resveratrol extracted from peanut shell. J. Anhui Agric. Sci. 37 (28), 13797–13798.

Liu, H., et al., 2009b. Extraction and spectrophotometry determination of sterol in apple seed oil. Food Sci. 30 (6), 146–150.

Liu, S., et al., 2010a. Extraction of plant sterols from jatropha seed oil by adduct formation with calcium chloride. Chin. J. Spectrosc. Lab. 27 (6), 2477–2480.

Liu, Z., et al., 2010c. Membrane separation technique for purifying proanthocyanidins from peanut skin. Food Sci. 31 (20), 183–186.

Liu, Z., et al., 2010d. Optimization on the technique of proanthocyaid in ethanol extraction from peanut skin. Food Fermentation Ind. 36 (6), 166–170.

Liu, H., et al., 2011a. Bio-synthesis of phytosterol laurate in non aqueous phase reaction. Food Fermentation Ind. 37 (1), 37–41.

Liu, S., et al., 2011b. Phytosterol extraction from Jatropha seed oil. Food Sci. Technol. 36 (3), 224–226.

Liu, Y., et al., 2010b. Preparation of peanut protein film. China Oils Fats 35 (12), 28–32.

Liu, Y., et al., 2011c. Identification and evaluation on the agronomic characteristics of peanut varieties in different areas. J. Jinling Instit. Technol. 27 (1), 34–38.

Liu, L., 2001. Hypertension, first ed. People's Health Publishing House, Beijing, pp. 30–34.

Liu, L., 2002. Sterol esters of conjugated linoleic acid and process for their production, US Patent 6413571.

Liu, F., 2006. Study on production textured soy protein by medium-sized twin-screw extruder, Har'er'bin. Master degree thesis of Northeast Agricultural University.

Liu, Y., 2009. Oil Producing and Processing Technology. Science Press, Beijing.

Liu, Y., 2011. Study on the preparation and modification of peanut protein films, Beijing. Master degree thesis of Chinese Academy of Agricultural Sciences.

Long, Z., 2010. Peanut color sortor design based on TCS230 color sensor, Qingdao. Master degree thesis of Qingdao University of Science and Technology.

Lu, C., et al., 2000a. Composition of amino acid in storage proteins and changes in 17.5 kDa polypeptide synthesis during peanut seed development. J. Trop. Subtrop. Bot. 8 (4), 339–345.

Lu, W., et al., 2000b. The technological study on the extraction of soybean phospholipid with supercritical $CO_2$. Food Sci. 21 (3), 28–29.

Lu, L., 2002. Research progress on natural antioxidants-oligomeric proanthocyanidins. Food Sci. 23 (2), 147–149.

Lu, Z., 2006. The separation of sterols derived from water caltrop by high performance preparative liquid chromatography, Changchun. Master degree thesis of Northeast Normal University. Tianjin.

Luan, W., Han, S., 1990. Studies on the Nutritive Value of Peanut Germplasm Resources in Shandong Province China. Crop Genetic Resources 2, 22–25.

Luan, W., et al., 1986. Peanut varieties study, character and main characters of the differences between different types. Crop Variety Res. 2, pp. 8–11.

Lusas, E.W., 1979. Food uses of peanut protein. J. Am. Soc. (56), 425–430.

Lv, W., 2000. Extraction of soybean phospholipid by supercritial $CO_2$. Food Sci. 21 (3), 28–29.

Lv, L., 2002. A review on the recent studies on oligomeric proanthocyanidins. Food Sci. 23 (2), 147–150.

Lv, Z., 2006. The seperation of sterols derived from water caltrop by high performance preparative liquid chromatography, Changchun. Master Degree Thesis, Northeast Normal University.

Ma, C., Duan, H., 1999. Determination of phospholipids in soybean lecithin. China J. Chin. Mater. Med. 24 (11), 671–672.

Ma, D., 2006. Studies on the Extraction and Separation technology of Emodin and Resveratrol from Polygonum Cuspidatum. Tianjin. Master degree thesis of Tianjin University.

Ma, X., 2007. Study of isolation and purification of resveratrol from giant knotweed, Tianjin. Master degree thesis of Tianjin University.

Ma, T., 2009. Study on preparation and solubility of peanut protein concentrate, Beijing. Master Degree Thesis, Chinese Academy of Agricultural Sciences.

Ma, Z., 2010. A discussion about several problems in the use of Disc-type Centrifuge. Cereal Food Ind. 17 (1), 11–14.

Malmheden, et al., 1994. Analysis of food proteins for verification of contamination or mislabelling. Food Agric. Immunol. (6), 167–172.

Marquez, M., Fernandez, V., 1993. Enzymic hydrolysis of vegetable proteins: mechanism and kinetics. Process Biochem. (28), 481–490.

Mason, T., Paniwnyk, L., 1996. The uses of ultrasound in food processing. Adv. Sonochem. (4), 177–203.

Mason, M.E., et al., 1969. Non volatile flavour components of groundnut. J. Agric. Food Chem. (17), 732–782.

Matsui, T., 1999. Preparation and characterization of novel bioactive peptides responsible for angiotensin-I-converting enzyme inhibition from wheat germ. J. Peptide Sci. 5, 289–297.

Meng, L., et al., 2005. The effect of environment temperature on properties of protein films. Sci. Technol. Cereals Oils Foods 13 (1), 1–4.

Miao, W., 2008. Study on extraction and purification of resveratrol from grape skin, Wulumuqi. Master Degree Thesis, Xinjiang Agricultural University.

Mo, Q., 1996. A study on the papain hydrolysis of peanut meal. Cereal Feed Ind. 12, 37–38.

Mo, Y., et al., 2008. HPLC determination of polyphenols in Tea' PTCA (Part B: Chem. Anal.). Cereal Feed Ind. 44 (7), 593–596.

Molina, E., et al., 2001. Emulsifying properties of high Pressure treated soy protein isolate and 7S and 11S globulins. Food Hydrocoll. (15), 263–269.

Monteiro, P., Prakash, V., 1994. Effect of proteases on Arachin, conarachin I and conarachin II from peanut (Arachis hypogaea L.). J. Agric. Food Chem. (42), 268–273.

Moore, K., Knauft, D., 1989. The inheritance of high oleic acid in peanut. J. Heredity 80, 252–253.

Mu, D., et al., 2007a. Determination of phytosterol by HPLC. Food Eng. 1, 62–64.

Mu, D., et al., 2007b. Study on supercritical- $CO_2$ extraction of phytosterol. Food Fermentation Ind. 33 (1), 118–120.

Munilla-Moran, R., Saborido-Rey, F., 1996. Digestive enzymes in marine species. 1: Protease activities ingut from redfish (Sebastes mentella), sea bream (Sparus aurata) and turbot (Scophthalmus maximus. Comparative Biochem. Physiol. B: Biochem. Mol. Biol. 113 (2), 395–402.

Nakagomi, K., et al., 1998. A novel angiotensin-l-converting enzyme inhibitory peptide isolated from tryptic hydrolysate of human plasma. FEBS Lett. 438, 255–257.

Nelson, R., Carlos, A., 1995. Lipid, protein, and ash contents, and fatty acid and sterol composition of peanut (Arachis hypogaea L.) seeds from Ecuador. Peanut Sci. 22, 84–89.

Neucere, N.J., et al., 1969. Effect of roasting on the stability of peanut proteins. J. Agric. Food Chem. (17), 25–28.

Newell, et al., 1967. Precursors of typical and atypical roasted peanut flavor. J. Agric. Food Chem. 15, 767–772.

Nzai, J., Proctor, A., 1998. Determination of phospholipid in vegetable oil by Fourier transform infrared spectroscopy. J. Am. Oil Chem. Soc. 75 (10), 1283–1289.

Ókeefe, S., et al., 1993. Comparison of oxidative stability of high-oleic and normal-oleic peanut oils. J. Am. Oil Chem. Soc. 70, 489–492.

Ondetti, M., et al., 1971. Angrotensin-converting enzyme inhibitors from the venom of Bothrops jararaca: isolation, elucidation of structure and synthesis. Biochemistry 10, 4033–4039.

Osés, J., et al., 2009. Development and characterization of composite edible films based on whey protein isolate and mesquite gum. J. Food Eng. (92), 56–62.

Oshima, Y., Ueno, Y., 1993. Ampelopsins D, E and cis-ampelosine, oligosibenes from Ampelopsis brevipedungulata var. hanceii roots. Phytochemistry 33 (1), 179.

Ou, S., et al., 2003. Effect of ferulic acid on the properties of edible films prepared from soy protein isolate. J. Chin. Cereals Oils Assoc. 18 (3), 47–50.

Pan, Y., et al., 2002. Improved method of the determination of dietary fiber. Food Sci. 23 (11), 106–108.

Pan, Y., et al., 2005. Study on the extraction insoluble dietary fibre from peanut residue and its adsorption to NO2-1. Mod. Food Sci. Technol. 21 (2), 30–34.

Pang, M., Jiang, S., 2010. Research progresses on the oxidative stability of phytosterol and its applications in food. Food Sci. 31 (23), 434–438.

Pang, L., Xu, Y., 2010. Optimization of microwave-assisted extraction parameters for phytosterol from pumpkin seeds. J. Chin. Cereals Oils Assoc. 25 (8), 47–50.

Pang, G., Wang, Q., 2001. Research progress on theoretical basis of bioactive peptide and its prospect. Food Sci. 22 (2), 80–84.

Park, W., et al., 2000a. Somatic mutations of the trefoil factor family 1 gene in gastric cancer. Gastroenterology 119, 691–698.

Park, Y., et al., 2000b. Activity of monomeric, dimeric, and trimeric flavonoids on NO production TNF-alpha secretion, and NF-kappaB-dependent gene expression in RAW264.7 macrophages. FEBS Lett. 465 (2–3), 93–97.

Patchett, A., et al., 1980. A new class of angiotensin-converting enzyme inhibitors. Nature 298, 280–283.

Peng, H., 2006. The plant sterols components and content of several common nuts analyzed by gas cheromatography. Cereals Oils 11, 28–29.

Peng, H., et al., 2006. Analysis of the plant sterol component and content of edible rape oil with gas cheromatography. J. Anhui Agric. Sci. 34 (19), 4830–4831.

Peng, L., 2008. Influence of high pressure microfluidization on the functional properties of arachin and its mechanism study, Nanchang. Master degree thesis of Nanchang University.

Peng, Y., et al., 2008. The influence of water in the solvent on the purification of stigmasterol. Guangzhou Chem. Ind. 36 (5), 47–48.

Pezet, R., Perret, C., 2003. δ-viniferin, a resveratrol dehydrodimer: one of the major stilbenes synthesized by stressed grape vine leaves. J. Agric. Food Chem. 51 (18), 5488–5492.

Poulsen, L.K., et al., 1993. Precise area determination of skin-prick tests: validation of ascanningdevice and software for a personal computer. Clin. Exp. Allergy 23 (1), 61–68.

Poulsen, L.K., et al., 1994. Quantitative determination of skin reactivity by two semiautomaticde-vices for skin prick test area measurements. Agents Actions, no. 41 (Spec No) C, pp. 134–135.

Pruss, G., et al., 1997. Plant viral synergism: the potyviral genome encodes a broad-range pathogenicity enhancer that transactivates replication of heterologous viruses. Plant Cell 9, 859–868.

Pruss, H., 1997. Chromium, zinc, and grape seed extract (Flavonoids) can overcome age-related increases in SPB normotensive rate. J. Am. Coll. Nutr. 16 (5), 43.

Qi, W., et al., 2005. Physicochemical properties, development and application of soybean phospholipid. China Oils Fats 30 (8), 35–38.

Qi, K., 2012. Development and present situation of subcritical solvent organisms extraction technology. Cereals Food Ind. 19 (5), 5–8.

Qu, B., et al., 2008. The development of peanut industry and security and safeguard strategies for edible oil. Food Nutr. China 11, 13–15.

Quist, E., et al., 2009. Angiotensin converting enzyme inhibitory activity of proteolytic digests of peanut (Arachis hypogaea L.) flour. LWT-Food Sci. Technol. 42 (3), 694–699.

Rabjohn, P., et al., 1999. Molecular cloning and epitope analysis of the peanut allergen Ara h 3. J. Clin. Investig. 103 (4), 535–542.

Ramesh, M., Lonsane, B., 1990. Characteristic and novel feather of thermostable α-amylase produced by Bacilluslicheniformis M27 under Solid-state fermentation. Starch 42, 233–238.

Rao, Y., 2008. Studies on the Purification Process of Resveratrol and Quality Standards of Feng-tongpian.Wuhan: Master degree thesis of Huazhong University of Science and Technology.

Reddy, B., et al., 1995. Nutritive value of ragi (Eleusine coracana) grain in broilers. 17th Annual Poultry Conference and Symposium of Indian Poultry Science Association, Bangalo.

Ricardo, J., et al., 1991. Procyanidin dimers and trimers from grape seeds. Phytochemistry 30 (4), 1259–1264.

Rohrbach, M., et al., 1981. Purification and substrate specificity of bovine angiotensin-converting enzyme. J. Biol. Chem. 256, 225–230.

Ruiz, J., et al., 2004. Angiotensin converting enzyme-inhibitory activity of peptides isolated from manchego cheese. Stability under simulated gastrointestinal digestion. Int. Dairy J. 14, 1075–1080.

Savage, G., Keenan, J. 1994. The composition and nutritive value of groundnut kernels. In: Smart, J. (Ed.), The Groundnut Crop: Scientific Basis for Improvement, pp. 173–213.

Schäppi, G., et al., 2001. Hidden peanut allergens detected in various foods: findings and legal measures. Allergy 56, 1216–1220.

Selvendran, R., Robertson, J., 1994. Dietary fiber in foods: amount and type in: Amadv, Barry J. Cost-92 Metabolic and physiological aspects of dietary fiber in food. Luxembourg: Commission of the Europe communities, 11–20.

Shan, H., et al., 2008. Research on extraction technology of resveratrol from roots of peanut. Trans. Chin. Soc. Agric. Mach. 39 (2), 93–97.

Shao, P., et al., 2007. Determination of Vitamin E and sterols in refined products from deodorizer. Food Sci. (1), 54.

Shefcheck, K., Musser, S., 2004. Confirmation of the allergenic peanut protein, Ara h 1, in a model food matrix using liquid chromatography/tandem mass spectrometry (LC/MS/MS). J. Agric. Food Chem. 52 (10), 2785–2790.

Shen, Y., et al., 1994. Preparation and application of the multi-functional Xlmplex Lecithin Fatliquor. China Leather 23 (10), 42–43.

Shen, B., 1996. Study on soybean anti-oxidation polypeptides. China Oils Fats 21 (6), 21–24.

Sheng, G., 1993. Function and application of soybean polypeptides. Sci. Technol. Food Ind. 6, 21–26.

Shi, B., Du, X., 2006. The progress on research and utilization of plant proanthocyanidins. J. Sichuan Univ. Eng. Sci. Ed. 38 (5), 16–22.

Shi, L., 1999. A natural oil made to last. Detergent Cosmetics 104, 23–26.

Shin, D.S., et al., 1998. Biochemical and structural analysis of the IgE binding sites on Ara h1, an abundant and highly allergenic peanut protein. J. Biol. Chem. 273, 13753–13759.

Shokarii, E., et al., 1991. Immunological characterization of a 36 kD polypeptide in peanuts (Arachis hypogaea L.). Peanut Sci. (18), 11–15.

Shu, Y., Chen, M., 2003. The progress on research of resveratrol and polydatin. J. Zhengzhou Coll. Animal Husbandry Eng. 23 (4), 22–25.

Sicherer, S., Sampson, H., 2000. Peanut and tree nut allergy. Curr. Opin. Pediatr. 12 (6), 567–573.

Singh, H., Mac, R., 2001. Use of sonication to probe wheat gluten structure. Cereal Chem. 78 (5), 526–529.

Singleton, J., Stikeleather, L., 1995. High-performance liquid chromatography analysis of peanut phospholipids Injection system for simultaneous concentration and separation of phospholipids. J. Am. Oil Chem. Soc. 72 (4), 481–483.

Stanley, J., et al., 1997. Identification and mutational analysis of the immunodominant IgE binding epitopes of the major peanut allergen Ara h 2. Arch. Biochem. Biophys. 342 (2), 244–253.

Stevens, R., et al., 1972. AngiotensinI-converting enzyme of calf lung: method of assay and partial purification. Biochemistry 1, 2999–3007.

Stikeleather, 1995. High-performance liquid chromatography analysis of peanut phospholipids I. Injection system for simultaneous concentration and separation of phospholipids. Jaocs 72 (4), 481–483.

Struve et al., 1983. Sterol concentrates, the preparation thereof, and their use in the transformation of sterols by fermentation. US 4,374,776, pp. 2–22.

Su, W., et al., 2004. Studies on extraction of resveratrol from polygonum cuspidatum by medium-pressure silicagel column chromatography. Chem. Ind. Forest Prod. 24 (1), 39–42.

Su, J., et al., 2010. Structure and properties of carboxymethyl cellulose/soy protein isolate blend edible films crosslinked by Maillard reactions. Carbohydrate Polym. 79, 145–153.

Sun, C., Zhao, X., 2011. Optimization of the extraction procedure of resveratrol from vine branches by response surface. Sino-Overseas Grapevine Wine 1, 8–11.

Sun, L., et al., 2004. Determination of Proanthocyanidins with iron catalyzed colorimetric method. Shandong Sci. Technol. Food 4, 11–12.

Sun, T., et al., 2009. Design and development of a large -scale CO2 SCFF plant. World Sci. Technol./Modernization Traditional Chin. Med. Materia Med. 11 (6), 892–896.

Sun, J., et al., 2011. Study on properties of lectin in seed of peanut (Arachis hypogaea L.). Sci. Technol. Food Ind. 32 (8), 134–137.

Sun, C., 2010. Research progress on procyanidins. Food Mach. 26 (4), 146–148.

Takahashi, T., et al., 1999. Toxicological studies on procyanidin B-2 for external application as a hair growing agent. Food Chem. Toxicol. 37, 545–552.

Tang, Y., et al., 2002. Study on the determination of resveratrol in polygonum cuspidatum by electrochemiluminescence. Northwest Pharmac. J. 17 (1), 6–8.

Tang, L., et al., 2005. HPLC determination of luteolin of peanut hull from different regions. J. Peanut Sci. 34 (2), 1–4.

Tello, D., et al., 1994. Three-dimensional structure and thermodynamics of antigen binding by anti-lysozyme antibodies. Biochem. Soc. Trans. 2 (4), 943–946.

Tello, P., 1994. Enzymatic hydrolysis of whey protein: kinetic models. Biotechnol. Bioeng. 44, 523–528.

Teng, Z., et al., 2003. The content statistics and classify analyse for major nutrition matters of peanut. Food Res. Dev. 24 (4), 84–85.

Thompson, R., et al., 1972. Plant proanthocyanidins, Part I introduction; the isolation, structure, and distribution in nature of plant procyanidins. J. Chem. Soc. Perkin Trans., 1387–1399.

Tian, S., Liang, H., 2005a. Effect of microbial transglutaminase on gel property of soybean protein isolates. China Oils Fats 30 (8), 42–45.

Tian, S., Liang, H., 2005b. Effect of microbial transglutaminase on modification of soybean protein isolates. Mach. Cereals Oil Food Process (6), 54–59.

Tian, J., et al., 2009. Preliminary study on the trypsin inhibitors in the sweet potato and peanut. Guangxi Zhiwu/Guihaia 29 (1), 70–73.

Tian, S., et al., 2005c. The effect of relative humidity on properties of protein-based films. Sci. Technol. Food Ind. 4 (26), 68–71.

Tian, Y., 2008. Extraction, purifacation and properties of proanthocyanidins from Sea buckthorn. Ha'er'bin. Master degree thesis of Hei Long Jiang University.

Tombs, M.P., 1965. An electrophoretic investigation of groundnut proteins: the structure of arachin A and B. J. Food Biochem. (96), 119–133.

Tong, Y., He, D., 2011. Study on extracting technologies of haematochrome and proanthocyanidin in peanut skin. Farm Mach. 5, 116–118.

Tong, Y., et al., 2009. Investigation on oxidation stability of different edible vegetable oil. China Oils Fats 34 (2), 31–34.

Tracey, Zang, Z., 2003. Technical characteristics and application of Z series of photoelectric color separator. Cereals Feed Ind. (5), 8–9.

Tu, Z., et al., 2012. Purification technology research of peanut non-starch polysaccharides. Sci. Technol. Food Ind. 33 (10), 243–245.

Tu, Z., 2012. Effect of dynamic high-pressure microfluidization on the morphology characteristics and physicochemical properties of maize amylose. Starch Biosytnthesis Nutr. Biomed. J. 65 (5–6), 390–397.

Viquez, O.M., Konan, K.N., Dodo, H.W., 2003. Structure and organization of genomic clone of a major peanut Arah1. Mol. Immunol. 40, 565–571.

Vu, P., et al., 2004. Lipase-catalyzcd production of phytosteryl esters and their crystallization behavior in corn oil. Food Res. Int. 37 (2), 175–180.

Wan, J.C., Jiang, B., 2008. Separation and purification of β-sitosterol and stigmasterol from soybean phytosterol using crystallization. Food Sci. Technol. (8), 50.

Wan, B., et al., 2011a. Procyanidine and its applications. Food Nutr. China 6, 15–16.

Wan, Y., et al., 2011b. Extraction of sterol from Gardenia jasminoides seed by the ultrasound technology. Sci. Technol. Food Ind. 32 (3), 327–333.

Wan, S., et al., 2004. Synthetically evaluation on peanut nutrients and study on its industrialization development strategy. J. Peanut Sci. 33 (2), 1–6.

Wan, S., 2007. Peanut Quality. China Agriculture Sciences Press, Beijing.

Wang, M., Gu, W., 1999. Preparation of soluble peptide by enzyme corngluten meal. Sci. Technol. Cereals Oils Foods 7 (1), 1–3.

Wang, H., et al., 1997. Isolation and identification of luteolin from methanolic extracts of peanut hull. J. Chin. Cereals Oils Assoc. 12 (3), 49–52.

Wang, S., et al., 2001. Gas chromatographic determination of plant sterols in fatty food. J. Instrum. Anal. 20 (4), 43–45.

Wang, Y., et al., 2003. Separation, purification and determination of soy phoshatidylcholine. J. Cereals Oils 2, 6–8.

Wang, R., et al., 2006. Flow injection chemiluminescence determination of luteolin. J. Southwest China Normal Univ. Nat. Sci. Ed. 2, 73–76.

Wang, Y., et al., 2007. Research progress in determination methods of dietary fiber. J. Henan Univ. Technol. Nat. Sci. Ed. 28 (6), 79–82.

Wang, C., et al., 2009a. Extraction technology of resveratrol from peanut roots stems and leaves. Food Res. Dev. 30 (9), 77–80.

Wang, F., et al., 2009b. Study on enzymatic extraction of the resveratrol from peanut red skin. J. Chin. Instit. Food Sci. Technol. 9 (5), 76–80.

Wang, Q., et al., 2009c. Research progress on resveratrol and derivatives in peanut. J. Chin. Cereals Oils Assoc. 24 (10), 145–151.

Wang, S., et al., 2009d. Effect of radiation process on allergic proteins of peanut. J. Peanut Sci. (2), 2.

Wang, L.Z., et al., 2010. Physical assessment of composite biodegradable films manufactured using whey protein isolate, gelatin and sodium alginate. J. Food Eng. 96, 199–207.

Wang, T., et al., 2010. Lipase-catalyzed synthesis of phytosterol ester in supercritical CO2. Food Sci. 31 (22), 293–296.

Wang, Q., et al., 2011. The review about procyanidin from peanut red skin. Food Res. Dev. 32 (4), 184–186.

Wang, L., et al., 2012a. Effect of peanut quality on protein gel property. Trans. Chinese Soc. Agric. Eng. 28 (17), 260–267.

Wang, L., et al., 2012b. Research on processing characteristics and quality evaluation of protein suitability of peanuts, Beijing. Doctor degree thesis of Chinese Academy of Agricultural Sciences.

Wang, Q., et al., 2013a. Isolation, purification and molecular mechanism of a peanut protein-derived ACE-inhibitory peptide. PLoS One 9 (10), 2014.

Wang, Q., et al., 2013b. Structural characterization and structure-activity relationship of ACE inhibitory peptides from peanut. Food Sci. 34 (9), 170–174.

Wang, L.Z., et al., 2010. Physical assessment of composite biodegradable films manufactured using whey protein isolate, gelatin and sodium alginate. J. Food Eng. 96, 199–207.

Wang, H., 2001. Pleureryn, a novel protease from fresh fruiting bodies of edible mushroom Pleurotus eryngii. Biochem. Biophys. Res. Commun. 289, 750–755.

Wang, C., 2010. Study on the components of proanthocyanidin by HPLC-MS methods. Agric. Technol. Equip. 188, 8–9.

Wang, Q., 2012. A suitable type dissolved protein processing of peanut quality measurement and evaluation methods. China, 102809635.

Wei, F., et al., 2003. Study on the extractive technique of oligomeric proanthocyanidins. Sci. Technol. Food Ind. 9, 54–56.

Wei, S., 1997. Relationships between elastin degradation and skin aging and the development of skin care product. China Surfact. Detergent Cosmetics 5, 245–256.

Wei, J., 2002. Study on extraction method of phytosterol. Hangzhou. Master degree thesis of Zhe Jiang University.

Wei, F., 2003. Study on extraction of oligometric proanthocyanidins. Sci. Technol. Food Ind. 9, 32.

Wen, Z., Yang, L., 2011. Study on microwave-assisted extraction technology and properties of water-soluble dietary fiber from peanut hulls. J. Chin. Cereals Oils Assoc. 26 (4), 99–103.

Wihodo, M., Moraru, C., 2013. Physical and chemical methods used to enhance the structure and mechanical properties of protein films: a review. J. Food Eng. (114), 292–302.

Wijk, F., et al., 2004. The effect of the food matrix on in vivo immune responses to purified peanut allergens. J. Toxicol. Sci. 86 (2), 333–341.

Worthington, et al., 1972. Varietal differences and seasonal effects of fatty acid composition and stability of oil form 282 peanut genotype. J. Agric. Food Chem. 20, 727–730.

Wu, L.P., Chen, X.F., 2008. Study on enzymatic extraction of dietary fiber from peanut shell and its properties. Sci. Technol. Food Ind. 29 (1), 194–196.

Wu, J., Ding, X., 2002. Characterization of inhibition and stability of soy-protein-derived angiotensin: converting enzyme inhibitory peptides. Food Res. Int. 35, 367–375.

Wu, H., et al., 2006. A rapid method of detection of food allergen by fingerprinting. J. Trop. Med. 6 (1), 10–12.

Wu, H., et al., 2009. Effect of peanut protein concentrate modification with transglutaminase on gel hardness and elasticity. China Oils Fats 34 (6), 31–35.

Wu, J., 2003. Study on the Enzymatic Hydrolysis of Soy Protein and Antioxidative Activity of Its Hydrolysate. South China University of Technology, Guangzhou.

Wu, L., 2008. Extraction and modification of dietary fiber in peanut hull, Xian, Shaanxi University of Science and Technology.

Wu, H., 2009. Study on the preparation and gel-forming mechanism of peanut protein concentration and its application, Beijing. Master degree thesis of Chinese Academy of Agriculture Sciences.

Xia, J., 2005. The application of protein and phospholipids of soybean in food. Food Nutr. China 1, 31–32.

Xiang, H., et al., 2005. Synthesis of resveratrol imprinted polymer and its application in separation of active ingredient in polygonum cuspidatum Sieb. et Zucc. extracts. Chin. J. Appl. Chem. 22 (7), 739.

Xiao, S., et al., 2008. Ultrasonic assisted extraction of luteolin from peanut hull. J. Hebei Polytech. Univ. Nat. Sci. Ed. 30 (2), 96–99.

Xiao, S., et al., 2010. Solid-phase extraction of luteolin from peanut hull using molecular imprinted polymers. Chem. Ind. Eng. Prog. 29 (2), 293–296.

Xin, L., et al., 2009. Research progress of application of HSCCC in the separation and purification of natural products. Sci. Technol. Food Ind. 11, 335–338.

Xu, G., Chang, L., 1998. Analysis of soybean phospholipids by thin-layer chromatography. Chin. J. Anal. Chem. 26 (1), 81–84.

Xu, S., et al., 2001. Oil Chemistry. Beijing Light Industry Press, Beijing, pp. 316–389.

Xu, X., et al., 2003. Gelling properties of proteins modified by transglutaminase. Food Sci. 24 (10), 38–43.

Xu, W., et al., 2007. Quantification of luteolin in peanut shell by indirect atomic absorption methods. Chin. J. Anal. Chem. 35, 1685.

Xu, J., et al., 2008a. Experimental research and mechanism analysis on the extraction process for flavonoids in ginkgo leaves. Food Sci. Technol. 11 (33), 221–223.

Xu, W., et al., 2008b. Determination of luteolin in peanut hull by molybdate spectrophotometry method. Phys. Testing Chem. Anal. B: Chem. Anal. 12, 1229–1230.

Xu, Y., et al., 2009. Extraction method of resveratrol in peanut stem. Cereals Oils Process. 4, 120–122.

Xu, X., et al., 2010. Sterol composition analysis of royal jelly by gas chromatography coupled with mass spectrometry. Food Sci. 31 (18), 317–320.

Yamamoto, D., et al., 2007. Prediction of interaction mode between a typical ACE inhibitor and MMP-9 active site. Biochem. Biophys. Res. Commun. 354 (4), 981–984.

Yan, H.C., 1998. A study on vegetable oil blends. Food Chem. 62 (2), 191–195.

Yan, J., 2010. Extracting resveratrol from polygonum cuspidatum by enzymolysis with complex enzymes. Guangzhou Med. J. 41 (3), 66–69.

Yang, W., Wang, C., 2010. Study on anti-hyperglycemic activity of polysaccharides from peanut meal with hot water(WP). Sci. Technol. Food Ind. 3 (12), 330–332.

Yang, X., et al., 1998. Crystal structure of the catalytic domain of protein-tyrosine phosphatase SHP-1. J. Biol. Chem. 273 (43), 199–207.

Yang, J., et al., 2009a. Separation and purification of resveratrol from polygonum cuspidatum by macroporous adsorption resin and enzyme treatment. Ion Exchange Adsorption 25 (3), 274–281.

Yang, Y., et al., 2009b. Purification and antibacterial properties of luteolin extracted from peanut shell. Food Sci. Technol. 34 (12), 211–216.

Yang, X., et al., 2011. Functions and application prospect of phytosterol. China Dairy 111, 54–56.

Yang, J., 2009. Development status and suggestions of peanut industry in China. Food Nutr. China 1, 17–19.

Yao, Y., Yu, H., 2005. Study on comparison of antioxidative action of tea polyphenol and TBHQ squeezing peanut oil and peanut and soybean reconciled oil. Cereal Food Ind. 11, 19–21.

Yao, X., et al., 2011. Hepatoprotective effect of crude peanut polysaccharide on carbon tetrachloride or alcohol-induced acute liver injury in mice. Food Sci. 32 (9), 261–265.

Yao, Y., 2004. Study on comparison of antioxidative action of tea polyphenol and TBHQ squeezing peanut oil and peanut and soybean reconciled oil. J. Cereal Food Ind. (11), 4.

Yao, Y., 2005. Comparison of peanut oil and olilve oil in nutritional value. Chinese Oils Fats 30 (4), 66–68.

Ye, Z., et al., 2011. Optimization of ultrasonic extraction process of resveratrol from peanut roots stems and leaves. J. Anhui Agric. Sci. 39 (4), 2077–2079.

Yi, H., et al., 2011. Cloning expression, purification and characterization of peanut allergen Ara h 8. Cell. Mol. Immunol. 27, 352–355.

Yin, Y., et al., 1998. Separation, purification and characterization of 2s protein from peanut (Arachis Hypogea L) seeds. J. South China Univ. Technl. (Natural Sci.) 4, 2–6.

Yin, D., et al., 2011. Principal component analysis and comprehensive evaluation on quality traits of peanut parents. J. Plant Genet. Resources 12 (4), 507–512.

Yin, W., 2005. The standards of analysis for dietary fiber in China Bejing, Session of the Third Term of Nutrition and Health Care Meeting, pp. 60–63.

Yin, D., 2012. Study on the preparation and gel properties of peanut protein fraction, Beijing. Master degree thesis of Chinese Academy of Agricultural Sciences.

Ying, Y., Ming, J.C., 2000. Characterization of grape procyanidins using HPLC-MS and MALDI-TOFMS. Agric. Food Chem. 48, 3990–3996.

Yman, I., et al., 1994. Analysis of food proteins for verification of contamination or mislabelling. Food Agric. Immunol. 6 (2), 167–172.

You, L., 2006. Separation and removal of allergen from peanut milk, Beijing. Master degree thesis of China Agricultural University.

Yu, D., et al., 2001. Study on synthesis of phytosterol esters using enzymatic method. Sci. Technol. Food Ind. 6, 318–320.

Yu, L., et al., 2009. Extraction and antioxidant activity evaluation of water soluble dietary fiber from peanut hull. Food Sci. 30 (22), 27–32.

Yu, G., et al., 2010a. Effect of transglutaminase on characteristic of soybean protein isolate gel. J. Northeast Agric. Univ. 41 (10), 100–107.

Yu, L., et al., 2008. Study on extraction of water soluble dietary fiber from peanut hull. Mod. Chem. Ind. 28 (2), 324–327.

Yu, L., et al., 2010c. Study on the extraction of water soluble dietary fiber and antioxidant activity from peanut hull by enzyme. Food Res. Dev. 31 (10), 158–163.

Yu, L, et al., 2008. Study on extraction of water soluble dietary fiber from peanut hull. Mod. Chem. Ind. 28, 2, 324–327.

Yu, S., 2008. Chinese Peanut Varieties and Their Genealogy, first ed. Shanghai Scientific & Technical Publishers, Shanghai.

Yuan, Y., Cao, Y., 2011. Developments and situation of ultrasonic technology in food industry. Sci. Technol. Food Ind. 3, 442–445.

Yuan, Y., et al., 2002. Key Technology of Modern Chinese Medicine Modernization, first ed. Chemical Industry Press, Beijing.

Yuan, J., et al., 2010. Gel property of transglutaminase-modified rapeseed protein. Food Sci. 31 (18), 10–13.

Yuan, D., et al., 2005. Application of modification vegetable protein in food protein. Sci. Res. Dev. Food 26 (6), 13–15.

Zhang, X., Feng, J., 2004. Study on the production of functional protein short peptide by enzymatic method. Animal Sci. Animal Med. 21 (3), 48–51.

Zhang, Y., Wang, Q., 2007a. Peanut protein hydrolyzing by Alcalase to prepare peanut oligopeptides. J. Agric. Eng. 23 (4), 258–263.

Zhang, Y., Wang, Q., 2007b. Research progress of functional oligopeptides. China Oils Fats 32 (2), 69–73.

Zhang, Y., Wang, Q., 2007c. Separation of antihypertensive peptides derived from peanut protein with ultrafiltration technology. China Oils Fats 32 (7), 28–31.

Zhang, Z., Yu, W., 2007. Studies on technology of ultrasonic extraction of phytosterol fromrice-bran. Food Res. Dev. 28 (1), 43–45.

Zhang, Z.S., Yu, W.T., 2007. Apple polyphenols inhibit plasma CETP activity and reduce the ratio of non-HDL to HDL cholesterol. Mol. Nutr. Food Res. 52 (8), 950–958.

Zhang, Z., Zhu, S., 2007. Nutrition of peanut and diet formula. Food Nutr. China (11), 57–58.

Zhang, L., et al., 2003. Study on determination of conventional chemical constituents in soybean meal by near infrared spectroscopy. Inspection Quarantine Sci. 13 (3), 25–27.

Zhang, G., et al., 2006. Proanthocyanidin and its development and application. Sichuan Food Fermentation 42 (129), 8–12.

Zhang, Y., et al., 2008a. Study on antihypertensive activity of peanut oligopeptide. Food Sci. 6, 299–403.

Zhang, Y., et al., 2008b. Preparation of peanut functional oligopeptides by two-step-hydrolysis. J. Agric. Eng. 24 (5), 275–279.

Zhang, J., et al., 2008d. Research progress in function component in peanut. Cereals Oils Process. 12, 73–76.

Zhang, C., et al., 2009b. Study on optimization of extraction technology of resveratrol from peanut root by response surface methodology. Food Sci. 30 (6), 34–38.

Zhang, C., et al., 2009d. Optimization of extraction for resveratrol from peanut root fermentation by aspergillus nige. Natural Prod. Res. Dev. 21, 875–880.

Zhang, C., et al., 2009d. Technology optimization for extraction of resveratrol from peanut plant. Trans. CSAE Chin. Soc. Agric. Eng. 25 (1), 148–151.

Zhang, C., et al., 2009e. Microwave-assisted extraction of resveratrol from peanut root. Food Sci. 30, (24), 30–33.

Zhang, P., et al., 2009e. Advance in synthesis of phytosterol esters of unsaturated fatty acids. China Oils Fats 34 (7), 37–41.

Zhang, Y., et al., 2010. Desalinization process of peanut oligopeptide. J. Chin. Cereals Oils Assoc. 25 (2), 117–120.

Zhang, X., et al., 2011. Film-forming properties of maillard reaction products formed from whey protein and xylose and their inhibitory effect on lipid oxidation in walnut kernel. Food Sci. 32 (5), 58–64.

Zhang, J., et al., 2012. Comparative analysis of tocopherol and phytosterol composition of peanut cultivars from different regions. Food Sci. 22, 191–195.

Zhang, H., et al., 2001. Determination of resveratrol in grape wine by second order differential simple oscillographic voltammetry. J. Instrum. Anal. 20 (2), 21–23.

Zhang, C., et al., 2006a. Proanthocyanidin and its development & application. Sichuan Food Ferment. 42 (129), 8–12.

Zhang, C., et al., 2008a. The effect of sodium dodecyl sulfate on properties of soybean isolate protein. Cereals Oils Ind. (3), 112–114.

Zhang, C., et al., 2009b. Study on optimization of extraction technology of resveratrol from peanut root by response surface methodology. Food Sci. 30 (6), 34–38.

Zhang, C., et al., 2009c. Optimization of extraction for resveratrol from peanut root fermentation by *Aspergillus nige*. Nat. Product. Res. Dev. 21, 875–880.

Zhang, C., et al., 2009d. Technology optimization for extraction of resveratrol from peanut plant. Chinese Soc. Agric. Eng. 25 (1), 148–151.

Zhang, X., 2002. Preparation, purification and classification of soybean peptides, Guangzhou. Master degree thesis of South China University of Technology.

Zhang, W., 2003. Current Protocols for Molecular Biology. Science Press, Beijing, pp. 359–361.

Zhang, H., 2006. Determination of resveratrol in peanut skin extracts by fluorescence spectrophotometer. China Oils Fats 31 (11), 48–49.

Zhang, J., 2009. Factors influencing resveratrol in peanut root and stem. Trans. CSAE 25 (Supp. 1), 251–253.

Zhang, K., 2010. Application of poston slagging disc centrifuge in centrifuging of solid-liquid -oil in degreasing liquid, Shanghai. Post Doctor Degree, Shanghai Jiaotong University.

Zhao, Y., Ding, X., 1999. Separation of phytosterol with supercritical CO2. China Oils Fats 24 (4), 41–42.

Zhao, C., Yao, X., 2000. Nutrition and healthy function of procyanidins from grape seeds. Chin. J. Food Hygiene 12 (6), 38–41.

Zhao, Y., Zheng, B., 2006. Progress in study on plant polyphenols and its function. Light Textile Ind. Fujian (11), 108.

Zhao, Y., et al., 2006. Research progress on determination method of grape anthocyanin content. J. Gansu Coll. TCM 23 (6), 43–46.

Zhao, H., et al., 2010a. Extraction process for resveratrol from Polygonum cuspidatum. Pharmac. J West China 25 (1), 85–86.

Zhao, W., et al., 2010b. Extracting the β-sitosterol from maydis stigma by mix-factor uniformity design. Food Res. Dev. 31 (9), 19–22.

Zhao, X., et al., 2011. Research progress and application of peanut protein. J. Chin. Cereals Oils Assoc. 26 (12), 118–122.

Zhao, E., et al., 2012. Comparative total phenol content and antioxidant activity of different solvent extracts from peanut shell. Food Sci. 33 (11), 79.

Zhao, L., 2010. Peanut color sortor design based on TCS230 color sensor, Qingdao. Master degree thesis of Qingdao University of Science and Technology.

Zheng, J., Geng, L., 1997. Simplification of the analytical method for dietary fibre. J. Nutr. 19 (2), 207–211.

Zheng, H., Zhong, G., 2009. Research on peanut protein modification and its development prospect. Cereals Oils 1, 8.

Zhong, H., et al., 2009. Cloning, expression and immunological characterization of peanut (Arachis hypogaea L.) Allergen Gene iso-Ara h 3. Plant Physiol. Commun. 45 (10), 958–962.

Zhou, H., Liu, C., 2006. Microwave-assisted extraction of solanesol from tobacco leaves. J. Chromatogr. A 1129 (3), 135–139.

Zhou, H., et al., 2010. Research on the extraction eechnology with microwave of resveratrol from red skin of peanut. Acad. Period. Farm Prod. Process. 199, 26–29.

Zhou, R., 2003. Peanut Processing Technology, first ed. Chemical Industry Press, Beijing.

Zhou, X., 2005b. Studies Modification Peanut Protein 3, 42–45.

Zhou, R., 2005a. Present situation and the development suggestions of peanut production and processing industry in China. China Oils Fats 30 (2), 5–9.

Zhu, F., et al., 2009. Microwave-assisted extraction of proanthocyanidins from peanut skin. Food Sci. 30 (20), 89–92.

Zhuang, X. 2006. Application of near infrared spectroscopy in determination of dietary fiber. 2006 National Symposium on Biochemistry and Biotechnology, Huangshan, pp. 391–394.

Zou, Y., et al., 2011. Optimization of ultrasonic extraction process of resveratrol from peanut roots, stems and leaves. J. Anhui AsS. Sci. 39 (4), 2077–2079.

# Index

Printed in the United States
By Bookmasters